Generation Digital

Generation Digital

Politics, Commerce, and Childhood in the Age of the Internet

Kathryn C. Montgomery

The MIT Press
Cambridge, Massachusetts
London, England

First MIT Press paperback edition, 2009

MIT Press books may be purchased at special quantity discounts for business or sales promotional use. For information, please email special_sales@mitpress.mit.edu or write to Special Sales Department, The MIT Press, 55 Hayward Street, Cambridge, MA 02142.

This book was set in Stone Sans and Stone Serif by SNP Best-set Typesetter Ltd., Hong Kong. Printed and bound in the United States of America.

Library of Congress Cataloging-in-Publication Data

Montgomery, Kathryn C.
Generation digital : politics, commerce, and childhood in the age of the internet / Kathryn C. Montgomery.
p. cm.
Includes bibliographical references and index.
ISBN 978-0-262-13478-1 (hc. : alk. paper)—978-0-262-51256-5 (pb. : alk. paper)
1. Mass media and children. 2. Mass media and teenagers. 3. Digital media—Social aspects. 4. Internet and children. 5. Internet and teenagers. 6. Internet—Social aspects. 7. Technology and children. 8. Technology and youth. I. Title.
HQ784.M3M66 2007
303.48′330830973—dc22

2006030106

10 9 8 7 6 5 4 3

Contents

Preface

This book draws on my multiple experiences as a media scholar, an advocate, and a parent. In many ways it chronicles my own dual journeys—through the labyrinth of media politics and policy, and through the new culture of digital media. In 1990, after twelve years as a professor of film and media in California, I moved to Washington, DC, where I founded the nonprofit Center for Media Education (CME) with my husband, Jeffrey Chester. One of our goals was to participate in critical policy decisions that would determine the future of electronic media. At that time, the media system already was undergoing tremendous technological change. In addition, major revisions were about to take place in U.S. telecommunications laws. But while the communications companies had a phalanx of high-paid lobbyists heavily engaged in pushing their agendas in Washington, only a tiny handful of groups were representing the public interest.

Throughout the 1990s, CME worked with dozens of nonprofit organizations, professional associations, academic institutions, and foundations to help build a new public-interest movement. During the earliest days of the debate over the "Information Superhighway," CME cofounded the Telecommunications Policy Roundtable, a coalition of civil liberty, education, and computer groups that called for "public interest principles" to guide the development of this new medium. After Action for Children's Television closed its doors in 1992, CME also became the leading advocate for children in national media-policy debates.

My move to Washington took place one year after the publication of my first book, *Target: Prime Time*. Written while I was a professor at UCLA, that book chronicles the thirty-year history of advocacy groups' efforts to influence the content of prime-time television programming. Several years later, we were in the beginning stages of our four-year campaign over the Children's Television Act (CTA), which ultimately would lead to new government rules for children's educational programming. While attending a meeting, I found myself in a conversation with a television network executive I had interviewed for my book. She asked me whether I was still studying pressure groups. I thought for a moment then responded: "Well, I guess you'd have to say I've become one." I had made a conscious choice to cross the line between "objective and neutral scholar" and "politically involved activist." I had done so because I wanted to make a difference in the media world and saw myself following in the footsteps of many people about whom I had written and long had admired.

My daughter, Lucy Chester, was born on April 29, 1993, within a day or two of the official release of the World Wide Web, making her a bonafide member of the Digital Generation. Becoming a parent both complemented and complicated my role as a children's advocate. Watching Lucy grow up has enabled me to remain closely connected to her generation's media culture—and to gain an appreciation for its ability to engage her attention, captivate her imagination, and occupy her time—as it follows her from the toddler through the teen years. Motherhood also has made me appreciate the challenges that all parents face in trying to raise their families in a media-saturated world.

The Center for Media Education's policy efforts in the children's media arena were part of our constituency-building strategy to mobilize support for broader public policies in the new digital media system. We also pushed for specific policy goals aimed at ensuring that the new media system would fulfill its potential as a positive force in children's lives. As Congress was debating the new Telecommunications Act in the early 1990s, we encouraged education and library groups to become involved, which led to passage of the "e-rate" provisions, providing schools and libraries with affordable access to the Internet.

In the earliest days of the Web, we also began to follow the emerging online marketplace, finding that advertisers already had set their sights on children in the new world of cyberspace. With my staff, I pored over the dozens of trade publications and new-media research reports that were hailing the importance of "digital kids" in the e-commerce strategies of the digital economy. I attended conferences and trade shows, interviewed dot-com executives in Silicon Valley and Silicon Alley, and monitored children's Web sites. Alarmed by some of the new online-marketing practices, I worked with my colleagues at CME on a study to alert the press, the public, and policymakers. After the 1996 publication of our report *Web of Deception,* CME in partnership with Consumer Federation of America led a national campaign that brought about passage of the 1998 Children's Online Privacy Protection Act.

During my twelve-year tenure at the helm of the Center for Media Education, I participated in some of the major media-policy developments of the era. From time to time, I found myself in the middle of highly controversial debates that were played out under the glare of the media spotlight. For example, in 1996 and 1997 I led a coalition of health, education, and child-advocacy groups in a campaign over the V-chip television ratings as we faced off with the Motion Picture Association of America's Jack Valenti in a series of pitched battles. On other occasions, I negotiated with advocates, policymakers, and industry leaders in behind-the-scenes, closed-door sessions. I was an indirect participant in numerous policy campaigns led by my colleagues—over such issues as media ownership, universal service, and spectrum allocation. But much of the time I was immersed so intensely in my own pressing work—tracking new media trends, filing petitions and complaints with regulatory agencies, meeting with White House staff, testifying before Congress, taking press calls, and raising money—that I could not keep tabs on all of the events taking place around me.

However, throughout the period I remained a curious scholar, sometimes seeing my own involvement from a variety of different perspectives, both inside and outside the process. From time to time I identified familiar patterns in the contemporary events that echoed those of earlier periods in the history of American media. As I was in the midst of some of my most fascinating experiences inside the Beltway, I knew I had to write about them someday. I kept notes, collected documents, and jotted down my observations and insights in small snippets. But I understood that these were just some of the pieces of the puzzle. To tell the larger story, I had to return to my role as a media scholar and follow the more traditional path of research.

In the late 1990s, CME launched a research initiative on new media and children. In partnership with the College of Communication at the University of Texas, we held a national conference in Washington, DC, inviting academics, health professionals, nonprofit leaders, and new-media creators to focus on "Ensuring a Quality Media Culture for Children in the Digital Age." The meeting also highlighted the dearth of academic research on children and new media, even in the midst of a booming online marketplace, and outlined a set of questions for guiding the development of a research agenda for the future. Following the conference, CME conducted its own study of teen use of the Internet, releasing our report *TeenSites.com* in 2001. In addition to our examination of commercial Web sites, the study offered one of the first glimpses of noncommercial and civic online efforts by and for young people. With additional funding, we were able to expand that study, launching a multiyear project to track the growth of hundreds of youth civic Web sites.

In the fall of 2003, I returned to teaching full time, taking a professorship in the School of Communication at American University. We closed the Center for Media Education, and I brought our Youth, Media, and Democracy project to the university. In 2004, with additional funding from the Ford Foundation and the Surdna Foundation, we published our study of the Internet and youth civic engagement, *Youth as E-Citizens*. A year later we convened a group of youth activists and policy leaders in the *FreeCulture2* conference. That project introduced me to the work of a new generation of advocates, including groups such as Downhill Battle and FreeCulture.org.

My reentry into academic life gave me the time, space, and support to complete the research and writing for this book. Out of the spotlight and the pressure cooker of intense Washington politics, I have been able to take a fresh look at this very recent history—in which I was involved intimately—and to reexamine, reflect, and in some cases reconsider the events of the decade. The process of writing the book has enabled me to see some of the broader national and international events that are influencing the policies and practices of the new media. It also has deepened my understanding of the digital marketplace and where it is headed. Some of my research has taken me into new areas, exploring, for example, what other scholars have learned about how young people are engaging with digital media, and how public health and civic organizations are using new media to reach and influence youth.

The research for this book came from a variety of different sources. In the sections based partly on my own experiences, I used personal notes, internal memos, and other private documents. However, because personal recollection is not the most reliable source of historical information, I also worked to confirm and augment my narrative by seeking out independent sources, including newspaper coverage, government documents, research articles, and interviews. I have found that carefully reconstructing the sequence of events not only has ensured greater accuracy, but also has enabled me to place these events in a broader historical, political, and social context. Where the source of my information about an event is solely from my own personal experience, I have cited so in the endnotes. In a handful of places, I have included passages that are based on off-the-record conversations with individuals involved in these events. For obvious reasons, I have not cited these specific sources, but the information is a crucial part of the story. Except for the preface, the entire book is in the third person, even in chapters 3 and 4, where some of the narrative covers my own work and that of my organization.

In the interest of complete disclosure, I must mention that two of the organizations I have written about in this book are past supporters of the Center for Media Education's work. The Henry J. Kaiser Family Foundation funded CME's 1999 public-education campaign to promote the V-chip and the Parental Television Guidelines. Chapter 6 profiles Kaiser's youth campaigns on sex education and HIV/AIDS, efforts with which neither I nor my organization ever has been involved directly. In 2002 the Center for Information & Research on Civic Learning & Engagement (CIRCLE), whose work I mention in chapters 6 and 7, provided funds for CME's study of youth civic engagement and the Internet.

This book is not intended to be a compendium of new-media content and practices, and I cover certain topics in more detail than others. Nor have I written a "how-to" book for parents. Instead, I have tried to provide background and insight about this new digital-media culture, the forces that are shaping it, and the ways in which young people are involved with it.

The dynamic nature of the new media makes it impossible to be up to date. New developments crop up every day, followed avidly by both the trade and the mainstream press. By the time the book is published, there will be further innovation, additional studies, and new controversies. But the deeper structural, economic, and social patterns this book explores likely will not change significantly in the years to come.

Acknowledgments

My work on this book began while I was at the Center for Media Education and continued after my return to academic life at American University. Because of this rather long gestation period, I need to thank many people. I apologize at the outset for any unintentional omissions. At CME, our research on children, youth, and new media involved numerous staff, interns, and consultants, including the team of people who worked on our 2001 study, *TeenSites.com,* and our youth civic-engagement research that later was published at American University. I want to thank Amy Aldman, Cathryn Borum, Cathy DeLuca, Christine Feldman, Erin Fitzgerald, Sharon Flynn, Kristina Hagemeister, Katharina Kopp, Jennifer Kotler, Emily Littleton, Melynda Majors, Yalda Nikoomanesh, Ellen O'Brien, Shelley Pasnik, Gabriela Schneider, Friederike Siller, Charlene Simmons, and Susannah Stern. I am especially appreciative of the contributions made by my two coauthors of *Youth as E-Citizens:* Barbara Gottlieb-Robles and Gary O. Larson.

The Center for Media Education's board of directors provided direction and insight for many of our research and policy efforts: Michael Brody, Patricia Koskinen, Monroe Price, Wilhelmina Reuben-Cooke, Sally Steenland, Lawrence Wallack, and Ellen Wartella. In my early explorations at CME of the culture of new media, I relied on a group of prestigious academic and professional advisory-board members: Jane Brown, Sandra Calvert, Justine Cassell, Michael Cohen, Dale Kunkel, Susan Linn, Jeff McIntyre, Rosemarie Truglio, Sherry Turkle, Brian Wilcox, and Nancy Willard. A number of these individuals have remained friends and trusted advisors for my continuing work.

I am extremely grateful to the foundations that have supported my non-profit and academic efforts over the years, and to the many individuals who believed in what we were doing: the Angelina Fund, Atlantic Philanthropies, the Benton Foundation, the Carnegie Corporation of New York, the Center for Information & Research on Civic Learning & Engagement, the Ford Foundation, the Phoebe Haas Trust, the Henry J. Kaiser Family Foundation, the Albert A. List Foundation, the Open Society Institute, the Pacific Life Foundation, the David and Lucille Packard Foundation, the Pew Charitable Trusts, the J. Roderick MacArthur Foundation, the John D. and Catherine T. MacArthur Foundation, the Charles Revson Foundation, Rockefeller Family Fund, and the Surdna Foundation.

During the research process, I interviewed and consulted with a number of people, including Neal Baer, Jodie Bernstein, Ben Brandzel, Tiffiniy Cheng, Julia Cohen, Arnold Fege, Twilight Greenaway, Jehmu Sedora Greene, Idit Harel, Bennett Haselton, Cheryl Healton, Tina Hoff, Maya Kosok, Mark Lopez, Lynne Lyman, Jane Mount, Colin Mutchler, Nelson Pavlosky, Tony Podesta, Nicholas Reville, Sarah Rosenberg, Vicky Rideout, Robin Smalley, Chandler Spaulding, Ginger Thompson, Amber Thornton, Diane Ty, Jaime Uzeta, and Holmes Wilson. David Sobel and Phil Wibur read drafts of several chapters and provided valuable feedback on some of the critical policy research.

Many friends and colleagues gave me support along the way, including Pat Aufderheide, Peter Broderick, Bobbie Eisenstock, Alan Eisenstock, Leita Luchetti, Janine Martin, and Susan Ridgeway. While I was in the early stages of conceptualizing this book, Janet Walker helped me see that I needed to tell "the whole story." I am grateful to American University for the University Research Award that supported my work during a crucial stage of writing. Special thanks to Larry Kirkman, Dean of the School of Communication, for giving me the time and resources I needed to complete the work, as well as to my other colleagues at American University who offered me their encouragement. My graduate assistants at AU—Jessica Awig, Ken Cornwell, and Jeni Hodge—gave me valuable research help. Malkia Lydia enriched my understanding of the Free Culture movement, introducing me to key players. Marcia MacNeil worked with me during the most intense period of writing, editing, and manuscript preparation, and I am extremely grateful for her dedication and good spirits throughout the process. Regan Carver provided indisensable assistance during the copyediting phase.

I thank the MIT Press for publishing my work. Special appreciation goes to Katherine Innis, who saw the promise of this book early on, to Robert Prior, for his steady support throughout the long time it took to turn that promise into a reality, and to Mel Goldsipe for shepherding the manuscript through the editing and publication process.

I never could have written this book without the support of my family. My parents, Robert and Ellen Rose, brother Rob Rose, and sister Julie Jones were always there for me as I labored through the long research and writing process. My sister Patricia Harriman read drafts of several chapters in their early stages, giving me valuable feedback and cheering me on to completion. My husband, Jeff Chester, has played so many roles in my life that I cannot even count them. He has been my professional and political partner, my writing collaborator, my supportive spouse and coparent, and my conscience. Finally, my daughter, Lucy Chester—my Digital Generation child—has been my inspiration, my joy, and my expert consultant as I explore the terrain of her media culture.

1 At the Center of a Cultural Storm

In September 2003, as high school and college students returned to their classes for the fall, the Recording Industry Association of America (RIAA) made an unprecedented move. Frustrated by the rising tide of illegal music file-sharing on the Internet, the trade group filed lawsuits against hundreds of young people and their families, charging infringement of the 1998 Digital Millennium Copyright Act (DMCA).[1] With threats of fines as high as $150,000, the high-profile legal actions prompted a flood of press coverage. The news media carried dozens of stories about teenagers charged with piracy for downloading their favorite songs in the privacy of their bedrooms or college dorms. Though the RIAA had given fair warning of its plans for months, many parents were caught off guard. Suddenly thrust into the public spotlight and faced with thousands of dollars in penalties and legal fees, they were forced to admit they had no idea what their children were doing online. As one startled father put it: "They're in there doing their homework and listening to all these different songs. I don't know what the heck they did, but I know one thing; they're not going to be doing it anymore."[2]

The lawsuits also sparked a flurry of heated public debate, with some observers commenting on the irony of an industry suing its own customers. Brianna LaHara, a middle-school honor student whose picture was plastered on the front page of the *New York Daily News,* quickly became the poster girl for public backlash against the music industry's draconian measures. "It's a disaster, PR-wise," commented a lawyer for the nonprofit Electronic Frontier Foundation, an advocacy group fighting against the RIAA crackdown. "Putting a 12-year-old defendant on the front page is not a way to lure people into the record stores."[3] Even some musicians found the music industry's actions indefensible. "Lawsuits on 12-year-old kids for downloading music, duping a mother into paying a $2,000 settlement for her kid? Those tactics are pure Gestapo," rapper Chuck D charged.[4]

For its part, the RIAA countered that it had no choice but to take legal action against individuals. Since the launch of popular peer-to-peer file-sharing networks, the industry had suffered more than a 25 percent drop in album sales.[5] While copyright litigation against Napster had forced that company into bankruptcy, newer decentralized file services such as KaZaA and Grokster had quickly taken its place.[6] Despite a succession of strategies aimed at curtailing online file-sharing, the practice had continued unabated. A survey by the Pew Internet & American Life Project found that nearly 70 percent of Internet users who were downloading music did not care whether the music was copyrighted or not.[7] As *USA Today* columnist Kevin Maney observed, "The RIAA is lashing out because it has no idea what else to do. And there's a reason for that. Nothing like its situation has existed before."[8]

The controversy over music file-sharing is emblematic of the powerful, complex, and sometimes contradictory roles that children and youth play in this era of the Internet. Since the advent of the World Wide Web in the early 1990s, the "Digital Generation" has been at the epicenter of major tectonic shifts that are transforming the media landscape. Coming of age at the beginning of a new century, these young people are leading the way into the uncharted territory of cyberspace, avidly embracing not only the Internet, but also a host of wireless devices and digital products—from video games to cell phones to iPods. Never before has a generation been so defined in the public mind by its relationship to technology. Pollsters, market researchers, and journalists closely track their every move, inventing a gaggle of catchy buzzwords to describe them—"N-geners," "Webheads," "Keyboard Kids," "CyberChildren," and "the MySpace Generation," to name only a few.[9] As active creators of a new digital culture, these youth are developing their own Web sites, diaries, and blogs; launching their own online enterprises; and forging a new set of cultural practices. From time to time, members of the Digital Generation have placed themselves on the front lines of seminal policy battles over the future of the new media, taking to the Web to organize their peers on issues such as intellectual property, privacy, and content regulation. More often their role has been symbolic, with politicians on the right and the left invoking the welfare of children in their high-stakes public-policy battles.

In our collective effort to make sense of this new generation, the public discourse often has been contradictory, reflecting an ambivalent attitude toward both youth and technology. As our children have ventured fearlessly into cyberspace, seizing upon all manner of digital gismos and gadgets, the public has responded with a mix of wonder, fear, and perplexity. In many ways, young people are treated with a certain level of fascination. Numerous books and articles pay homage to their role as pioneers. "The Net Generation has arrived!" Don Tapscott announced in his influential 1998 book, *Growing up Digital*. "Today's kids are so bathed in bits that they think it's all part of the natural landscape. To them, the digital technology is no more intimidating than a VCR or toaster."[10] The "Net-geners," Tapscott predicted, "are breaking free from the one-way centralized media of the past and are beginning to shape their own destiny. And evidence is mounting that the world will be a better place as a result."[11] Youth "are at the helm of the largest sociological shift in a generation," proclaimed *Brill's Content*. "Raised on email, instant messages, and Internet time, teenagers are developing into young Turks of technology."[12]

These "young Turks" have earned a reputation for leap-frogging over adult authorities in the realms of business and politics, making their mark on the world in ways unprecedented by previous generations of young people. Michael Lewis's *Next: The Future Just Happened* chronicled the forays of 13-year-old Jonathan Lebed, who quickly discovered how easy it was to set up an online brokerage account and make a killing in the stock market, much to the consternation of Wall Street and his own parents.[13] Teen "netpreneur" Shawn Fanning, a young, self-taught computer guru from Massachusetts, parlayed an avid interest in programming into a multimillion dollar business with the invention of peer-to-peer music-sharing software—Napster—throwing the music industry into turmoil and triggering an ongoing controversy over who controls content on the Internet.[14] Young people also have seized the power of the Internet as a tool for activism, deploying cell phones, laptop computers, and radio scanners to wage what Howard Rheingold calls "smart mob" campaigns against the powers-that-be.[15]

Even the youthful outlaws of the Digital Age have captured the public imagination. A cover article in the *New York Times Magazine* explored the hidden, underground lives of teenage hackers, some of whom have become folk heroes for wreaking havoc on the Internet, releasing digital worms and viruses that disrupt service, destroy files, and force companies to lose millions of dollars. Interspersed with starkly lit photos of some of the infamous players in this new game of disruption, the article asks: "Are they artists, pranksters or techno-saboteurs?"[16]

But much of the press coverage of young people and the Internet has focused on their vulnerabilities in a dangerous cyberworld that entices and ensnares them.[17] In the wake of the 1999 massacre at Columbine High

School, where two troubled teenagers murdered and terrorized dozens of people, news accounts reported that the boys were immersed in the dark shadows of a new digital-media subculture, spending hours acting out virtual murders in ultra violent video games and posting haunting, hate-filled commentary on their own personal Web sites.[18] More recently, a front-page story in the *New York Times* uncovered "the secret life of a teenager who was lured into selling images of his body on the Internet over the course of five years." The honor student had used his own Webcam to take pictures of himself "undressing, showering, masturbating and even having sex—for an audience of more than 1,500 people who paid him, over the years, hundreds of thousands of dollars."[19] A story in the *Washington Post* revealed that some teen girls are flocking to the thousands of Web sites that promote anorexia and bulimia as "lifestyle choices" and offer visitors a full complement of special features designed to normalize and support these life-threatening behaviors, including " 'thinspirational' pictures of extremely underweight women, menu suggestions, and discussion boards."[20] Another article told an eerie story of a college freshman who left an ominous "away message" in his instant–messaging account, telling his classmates "Goodbye" before jumping to his death from a dormitory balcony.[21]

These heightened concerns over online dangers are in sharp contrast to widespread beliefs about the positive role of technology in children's lives.[22] A survey by the Annenberg Public Policy Center found that most American families "are filled with contradictions when it comes to the Internet. Parents fear that it can harm their kids but feel that their kids need it."[23] Such conflicted responses echo public debates over such earlier "new media" as film, radio, and television. "Each new media technology," media scholars Ellen Wartella and Nancy Jennings observe, "brought with it great promise for social educational benefits and great concern for children's exposure to inappropriate and harmful content."[24]

But something is distinctly different about this new media culture and the role that young people have played, and continue to play, in its development and expansion. Beginning in the early 1990s and continuing into the early part of the twenty-first century, a powerful combination of technological, social, and economic trends has placed children and youth at the center of digital politics, commerce, and culture.

Coming to power during the same year as the launch of the World Wide Web, the Clinton Administration presided over the dramatic transformation of the Internet from an obscure government-run research network to a privatized and commercialized mass medium. As "the education president," who had campaigned on behalf of children, Bill Clinton placed the Digital Generation at the center of the official vision of a gleaming technological future in which the powerful new "Information Superhighway" would connect everyone in an electronic global village, bring to all "the vast resources of art, literature, and science," and make the "best schools, teachers, and courses" available to "all students, without regard to geography, distance, resources, or disability."[25]

On the eve of a new millennium, the Clinton Administration's 1993 Agenda for Action called for all schools, libraries, and hospitals to be connected to the Internet by the year 2000. "Bridging the digital divide" quickly became the rallying call of policymakers and industry leaders. Parents, educators, and librarians embraced the goal of wiring every classroom and library, which promised to be the magic bullet for ameliorating many of society's long-standing inequities and putting all children on an equal footing for full participation in the new economy. The Department of Commerce's National Telecommunications and Information Administration (NTIA) funded a series of high-profile demonstration projects around the country to showcase "how access to the Internet creates opportunities to tap the creativity of children and youth, to nurture their artistic talents, to engage them in civic enterprises, and to create bonds across generations."[26] New-media corporations teamed up with state and local governments around the country to connect schools and libraries to the Internet.[27]

The Internet's educational promise also helped drive its rapid penetration into American homes. In 1993, only 3 million Americans were connected to the Internet.[28] By the end of the decade, that figure had risen to over 150 million. A U.S. Department of Commerce report noted in 2001 that "family households with children under age 18 are far more likely to have computers than families without children: 70.1 percent, compared to 58.8 percent." By age 10, the study noted, "young people are more likely to use the Internet than adults at any age beyond 25."[29] A 2005 Pew Internet & American Life study found that 80 percent of parents with teens were going online, compared to 66 percent of all American adults. And 87 percent of youth aged 12–17 were connected to the Internet.[30]

As the Internet continued its expansion into the mainstream, children were thrust into the center of an ongoing series of "culture wars." The 1990s witnessed the first major Congressional rewrite of the nation's telecommunications laws since the New Deal, sparking a wave of regulatory and legal actions, as industry lobbyists, politicians, and interest groups competed over future control of the electronic media system. Throughout the deliberations, legislative proposals to protect children from indecency and pornography on the Internet took center stage, sparking widespread debate and sensational press coverage. Though the 1996 Communications Decency Act was struck down on First Amendment grounds, legislative efforts to shield children from Internet harms have continued, each triggering contentious legal battles that have dragged out for years.

These simultaneous tumultuous shifts—in the media and the regulatory landscape—have taken place during a particularly stressful era for families. Today's generation of young people is growing up amid economic and social changes that have triggered major dislocations in family structure, further complicating the job of parenting. The Annie E. Casey Foundation reports that 31 percent of children are living in single-parent families, and 18 percent are living below the poverty line.[31] Experts and social critics express alarm over high divorce rates and their impact on children's well-being.[32] More and more, mothers have found it necessary to work outside the home; many cannot afford the cost of childcare, raising concerns about young "latchkey children" being left home alone.[33] These trends have "shaped, reshaped, distorted, and sometimes decimated the basic parameters for healthy development," observes Patricia Hersch, author of *A Tribe Apart*. "All parents feel an ominous sense—like distant rumbles of thunder moving closer and closer—that even their child could be caught in the deluge of adolescent dysfunction sweeping the nation, manifesting itself in everything from drugs, sex, and underachievement to depression, suicide, and crime."[34]

As parents struggle with the stresses of modern-day life, they are faced with a rapidly expanding and immersive media culture far different from that of their own youth. A 2005 Kaiser Family Foundation survey of children and their families found that members of "Generation M"—U.S. children between the ages of 8 and 18—spend an average of six and a half hours per day with media. While television maintains a central place in their lives, it has been joined by a growing array of new digital technologies that young people are embracing and integrating into their daily activities. Two thirds of children own a portable CD, tape, or MP3 player; half of them have handheld video-game players. Though the study noted that overall today's youth are "largely happy and well-adjusted," it also found that "those who are least content or get the poorest grades spend more time with video games and less time reading than their peers."[35] These trends worry many parents who are concerned that sitting in front of a computer or television for extended periods of time can lead to weight gain, or that endless instant messaging can interfere with children's ability to form face-to-face relationships.[36]

This explosion of "new media" is occurring at a time when society has not yet resolved many of the issues related to "old media." Long-standing concerns about the impact of violent TV programming have been compounded further by the growing popularity of violent videogames such as *Grand Theft Auto,* in which children can act out the murder and mayhem themselves.[37] Parents are inundated with "how-to" books, Web sites, filtering software, and other tools to help them navigate their way through the new media environment. But many of them feel behind the curve, only learning about their children's favorite media pastimes from the front page of the newspaper or evening newscasts. While experts instruct parents to limit their children's "screen time" and keep the television out of a child's bedroom, most families seem to have a tough time making or enforcing such rules.[38] Feeling guilty, frustrated, and overwhelmed, parents lament their lack of control over a seemingly limitless media culture that surrounds and engulfs their children. "There comes a point," *Washington Post* reporter Bob Thompson commented, "if you're a parent with some pretensions to helping shape your family's value system, where you realize that fighting popular culture is a finger-in-the-dike proposition. There's a raging ocean of it outside your door, pumped out by the Great American Entertainment Machine, and it floods the malls, washes over the schools and seeps through modems and fiber optic cables straight into your living room."[39]

The driving force behind this exploding media culture is the exponential rise in children's spending power during the past several decades, prompted in part by the very changes in family structure that are confounding the role of parents. Never before have children and youth played such a powerful role in the marketplace. The Internet emerged as a new mass medium in the midst of an already highly commercialized children's culture. An enormous advertising and market-research industry was ready and waiting to adapt its strategies to the Digital Age. With the promise of e-commerce profits, "Echo Boomers," "the New Millennials," and "Generation Y" became prized consumers in the growing new economy. "The Generation Y tsunami is already gathering force," *BrandWeek* announced in 1999. "As young people become acculturated and socialized through the Internet, they are beginning to look for entertainment, or at least entertaining content, on their computers. Eyeballs glued to a screen are audiences; where there are audiences, there is advertising; and where there is advertising, there is commerce."[40]

Quintessential "early adopters" of new technology, members of the Digital Generation have eagerly embraced an endless supply of new technological gadgets, becoming both consumers and producers of a flood of digital content. One trade publication commented that young people have not simply adopted the Internet, they have *internalized* it.[41] Their engagement with digital media is ushering in a new set of behaviors, values, and expectations that this generation will carry with them into adulthood. "Pay close attention to Millennials," a market research report advised, "as their usage of media influences other demographic groups and they literally represent the world to come."[42]

The turbulent changes in the media system have created particularly fertile ground for a renewal of long-standing debates over the role of media in the lives of children and youth. From time to time, the shifting and conflicting images of children—as prescient gurus, innocent victims, powerful consumers, or active citizens—have forced a rethinking of what it means to be a child in the twenty-first century.

The story that unfolds in the following pages traces some of the major developments in the formative period of the Digital Age, focusing on the heightened role of children, teens, and young adults in the politics, commerce, and culture of the era. Chapter 2, "Digital Kids," shows how a confluence of historical trends throughout the twentieth century made children and teens a particularly valuable target market during the early commercialization of the Internet, triggering a burst of growth and activity in an already burgeoning youth-market research and marketing industry. Chapter 3, "A V-Chip for the Internet," documents the highly publicized political battles in the early 1990s about indecency and pornography on the Internet, describing how concerns over children prompted policymakers and industry leaders alike to devise strategies for ensuring a "family-friendly Internet." Chapter 4, "Web of Deception," is a case study of consumer-group intervention in the children's online marketplace, which ultimately led to the passage of the first law protecting children's privacy on the Internet. It also shows how the campaign for e-commerce safeguards in the United States became part of a much larger international battle for consumer protections in the global market. Chapter 5, "Born to be Wired," focuses on teenagers as the key "defining users" of new digital media, linking their uses of interactive technologies to their fundamental developmental needs of identity development and social interaction. As the chapter illustrates, digital marketers have found ways—largely under the radar of parents and policymakers—to tap into these psychological needs, creating sophisticated new market-research and profiling strategies that have become a pervasive and powerful presence in the lives of adolescents. Chapter 6, "Social Marketing in the New Millennium," profiles three contemporary public-education campaigns that have modeled their efforts on the cutting-edge strategies of digital marketers in order to change health and social behaviors among youth. Chapter 7, "Peer-to-Peer Politics," chronicles the myriad new ways that youth have seized the power of the Internet as a political tool, including the unprecedented mobilization of young voters during the 2004 presidential election, and the efforts of a new generation of young "cyberactivists" who are waging battles with the music and media industries over control of cultural expression in the Digital Age. Finally, chapter 8, "The Legacy of the Digital Generation," assesses the multiple roles and impacts of this first generation of young people growing up in the age of the Internet, and offers recommendations for policy makers, corporations, scholars, and parents that will help ensure an open, diverse, and equitable digital media system for the future.

2 Digital Kids

The Yahoo! campus sat on a large acreage of land at the edge of the San Francisco Bay in Sunnyvale, California, the heart of the famed Silicon Valley. Its sleek, modern buildings of glass and steel were surrounded by broad stretches of bright green, immaculately manicured lawn. The design and color of the place mirrored those of the Yahoo! Web portal, giving the corporate headquarters a fantastic, otherworldly, cartoon-like quality. A ribbon of purple carpet greeted and guided visitors into this digital Oz. Whimsical chairs and sofas in curved shapes and brilliant hues of yellow, purple, and chartreuse dotted the interiors of the buildings. In this corporate culture that encouraged long work hours, domestic accoutrements were part of the package, creating an all-accommodating home-away-from-home for Yahoo! employees. A dry cleaners was positioned in the middle of the campus, along with a grocery store, volleyball courts, a swimming pool, and a glass-enclosed 24/7 gym, where workers could be seen lifting weights, pumping stair climbers, and trotting on treadmills.

On this particular summer morning in 2003, the campus featured an added element of domesticity. A half dozen teenage bedrooms had been transplanted from their nearby suburban homes to the Yahoo! headquarters. Everything but the walls had been dismantled and reassembled inside one of the company's buildings. No detail had been overlooked in this re-creation of adolescent life. Clothing was strewn on floors; beds were unmade; posters were plastered everywhere; and rooms were filled with an array of technological paraphernalia—from CD players to TVs to computers. To complete the picture, the actual inhabitants of the bedrooms were there in the flesh, lounging on their beds, blasting their downloaded music, chatting with friends on their cell phones, and surfing the Web.

This elaborate, real-life display was part of a one-day summit called "Born to Be Wired." Yahoo! had assembled about fifty representatives from market-research firms, major advertising corporations (McDonalds, Frito-Lay, Walmart, Toyota, Pepsi Cola, etc.), and media companies to hear the latest research on youth and digital media. The morning keynote speaker was Neil Howe, coauthor of the book *Millennials Rising* and noted expert on the youth generation. Market researchers from Teenage Research Unlimited and Harris Interactive, armed with their most recent study of the "First Wired Generation," had arrived to impart their expertise about the ever-changing habits of this valuable, but elusive, demographic group. To get a firsthand look at what makes young people tick, participants were invited (and rewarded with a contest and prize drawing) to stroll through the bedroom exhibit and sit down and chat with the teenagers, who eagerly engaged in the conversations.[1]

The Yahoo! summit was only one of dozens of seminars, conferences, and trade shows organized by the youth-market-research industry every year. The catchy names for these events—"Digital Kids," "Teen Power," "Kidscreen Summit"—pay homage to the value and influence of the children and youth demographic. Eager marketers fork out thousands of dollars to attend these meetings, and tens of thousands more for the steady stream of specialized trade publications and reports offering strategies, secrets, and success stories from the front lines of the children's marketplace. Each event tries to outdo the others with the latest, cutting-edge techniques for cashing in on this lucrative age group. "Teen Power 2003" promised to "help you unravel all of the mysteries and secrets that occupy the minds and lives of today's teens." Session topics ranged from "Hip Hop Nation: Decoding the Urban Mystique through a Teen's Eyes," to "Teens and Technology: Using Electronic Media to Connect with the New 'Influencers,'" to "The Teen Brain: How a Teen's Brain Develops Has a Direct Impact on How You Communicate with Him/Her." At the Youth Marketing Mega-Event 2003, one of the hottest preconference workshops was a field trip into the streets of Los Angeles, billed as "an ethnographic immersion experience," inviting participants to "leave your business attire at home and join two cultural anthropologists on a voyage of discovery of the fast pace and often eccentric lifestyle of Los Angeles—home of the hot, the new, the fringe, the trendy, and the famous . . . one of the best ways to find new ideas is to hit the street, but not just any street. The streets we want are those places where people are living tomorrow's trends today."[2]

The youth-market-research industry has a long-established history in the United States, and has grown steadily during the past several decades, as children and adolescents have become an increasingly valuable demographic group. Hundreds of advertising agencies, market-research companies, and trend-analysis firms are involved in probing the behaviors, aspirations, and anxieties of children, "tweens," and teens. In recent years, these companies have tapped the expertise of a diverse array of specialists in sociology, psychology, and anthropology to explore youth subcultures and conduct motivational research, all in the search for knowledge and insight about this profitable market segment.[3]

In the 1990s, as the Internet swiftly moved into American homes, the youth-marketing enterprise intensified and expanded its efforts even further. The dramatic growth of the Internet occurred at precisely the time when children and adolescents had become a highly prized target for advertisers. When the World Wide Web was launched in 1993, children already were positioned in the center of a burgeoning media marketplace, with a full array of "brands" tailored exclusively to their needs—from specialized TV channels to magazines to music.[4] In the ensuing dot-com boom, the value of youth in the new digital marketplace became even greater. All the ingredients were in place to create a highly commercial digital-media culture, with unprecedented access to the child consumer. The dramatic crash of the overhyped online market did little to stop the flow of the new media into young people's lives. As a consequence, the Digital Generation has become the most heavily researched demographic group in the history of marketing.

Consumers-in-Training

The seeds of this digital commercial culture were sewn as early as the 1930s, when marketers began tailoring products and services to the special needs of the child consumer. By embracing and popularizing some of the earliest research on child development, the marketing industry played a role in framing the notions of childhood in the popular mind. Drawing on the work of French theorist Jean Piaget, a series of magazine articles and books began to popularize the scientific literature into user-friendly terms that could be understood and applied in the business world. For example, *Child Life Magazine* ran stories on children's color preferences and other behavioral patterns, as did the 1938 book *Reaching Juvenile Markets: How to Advertise, Sell and Merchandise through Boys and Girls*. These efforts marked the beginnings of what sociologist Daniel Thomas Cook calls a "distinct children's consumer culture."[5]

By the mid-1940s, the intersection of social changes and market developments had created a new category of childhood. As historian

Lizabeth Cohen explains, "Although 'adolescence' had been labeled since the nineteenth century to signify a developmental stage, and 'youth culture' referred to a select group of eighteen-to-twenty-four-year-olds in the 1920s, it was not until a majority of teens were graduating from high school in the 1940s that the teenage period of life—from ages thirteen to eighteen—took on the attributes of a mass cultural experience."[6] The United States's entry into World War II served as a catalyst, transforming this segment of the youth population into an identifiable and viable target market. According to Thomas Hine, the word "teenager" (spelled "teen-ager" until 1945) was coined anonymously during the war. As adults left their jobs to join the war effort, adolescents were able to enter the workforce, "earning real money for the first time in their lives." It was during the war that *Seventeen Magazine* was launched, the first publication uniquely tailored to the desires and anxieties of adolescent girls. As Hine explains, *Seventeen* catered to the "millions of young people looking for acceptance, competency, fun, and sex and, therefore, the right clothes, cosmetics, clear skin, great shoes, new music, and all the latest things."[7]

By the 1950s, an abundant postwar economy, the expansion of the middle class, and the introduction of new media technologies helped place adolescents at the center of a compelling new teen culture. Television's rapid intrusion into mainstream America caused both radio and movies to shift their strategies, creating new opportunities for these media to begin catering more directly to the teenage market. Radio stations across the country developed rock-and-roll formats, fueling the rise of a lucrative record business. Television was siphoning off adult audiences from movie theaters, so teenagers—with their increased spending power, freedom, and mobility—became a key target. New teen idols such as Elvis Presley, Sal Mineo, and Natalie Wood captured the hearts and minds of the teen audience. All of these events created what Thomas Hine calls "an outburst of teenage culture."[8]

The 1950s also saw the emergence of one of the first gurus of the teen market. After graduating from Northwestern University, Eugene Gilbert founded his own marketing company and migrated to New York City, where he became one of the most prominent ad men to promote the advantages of marketing to adolescents. Gilbert Youth Research also was the first to involve young people themselves in the market-research enterprise, recruiting 5,000 teenagers around the country to poll their peers about their buying habits, brand preferences, and other interests. Gilbert did not just provide the information they gathered to his clients; he also publicized it widely to the American public through his own syndicated column, which ran in more than 300 newspapers, and in his 1957 book, *Advertising and Marketing to Young People*.[9]

Identifying teens as a distinct market was part of a larger trend in the advertising and marketing industry: market segmentation. In order to define and effectively target these groups, marketers needed to understand what motivated them. Enlisting the help of psychologists, psychiatrists, and sociologists from leading universities, the advertising industry began to assemble a full arsenal of expertise and research tools—from in-depth interviews to Rorschach inkblot tests to thematic apperception tests—in the new enterprise of "psychographics," which combined demographic and psychological factors to develop more precise definitions of target markets.[10]

These new scientific tools were at the heart of a growing enterprise of specialized market research directed at children and teens, which became more formalized and prominent in the 1960s. The postwar baby boom and rising family incomes were expanding the children's market. New ad agencies devoted exclusively to children were founded.[11] Academic researchers, focusing on the unique aspects of children's developmental stages, helped marketers develop strategies for communicating directly with children, both in their focus groups and in their ads. When conducting focus groups with children age 5–12, advertisers were advised to try to establish a "one-to-one" relationship with the child, which only could happen when the mother was absent. In television commercials, they were instructed to rely more on visuals (including animated characters such as Kellogg's Tony the Tiger) because they made it easier for children to recognize and identify with the brand.[12]

In 1969, Texas A&M marketing professor James McNeal wrote an influential article in *American Demographics* that helped lend legitimacy both to market research on children and to the concept of the child as a consumer. Referring to the ages of 5–13 as the period of "consumer apprenticeship," McNeal translated psychological and developmental concepts into popular language by laying out a hierarchy of needs that characterized particular developmental stages.[13] Through his books and frequent press appearances, McNeal became one of the most prominent and visible proponents of children's marketing. By defining children as "consumers-in-training," he promoted a concept of childhood deeply rooted in the values and imperatives of the marketplace. As Cook notes, the tools of the trade of the market research enterprise—focus groups, surveys, and ethnographic studies—were presented by McNeal and others as "morally appropriate and ethically palatable" methods for "giving voice" to an otherwise unheard segment of the consuming population.[14]

Market research on children became even more sophisticated in the 1970s as the Saturday morning "kidvid" time slot became synonymous with children's television, and advertisers found they could use TV to market more directly to the child audience.[15] Communications scholars produced a wealth of new research examining the process of "consumer socialization" and the relationship between television advertising and children.[16] As Cook points out, these scholars "sought to uncover the social psychological dynamics behind 'how children learn and buy,'" measuring a variety of variables "relating to children's media exposure, information processing, and parent-child communication (including 'parental yielding' to children's purchase requests.)"[17]

The "Kids-Only" Network

By the 1980s, the growth of cable television and the introduction of more specialized channels had begun to alter the media landscape. As a heated public debate waged over the highly commercialized and violent Saturday morning kidvid programming on the broadcast networks, the cable industry offered Nickelodeon—a commercial-free network, and the first TV channel designed exclusively for children. Developed by Warner Amex, Nickelodeon quickly began winning awards from Action for Children's Television, the National Education Association and the National PTA.[18] Cable-system operators paid the network a fee of ten cents per subscriber to include the new network in their basic cable packages, which helped the companies win valuable exclusive franchises from local governments. To promote its service, Nickelodeon ran ads in the trade publications urging the cable operators to "Be a hero on your block for 10 cents per subscriber."[19]

But the network did not operate ad-free for very long. Within a year of its launch, Nickelodeon had decided to take corporate underwriting, arguing that like its public television counterpart it needed this added income to support the acquisition and development of new programming. These ten-second IDs acknowledging the underwriters quickly morphed into full-fledged commercial spots. The shift to a commercial business model proved very profitable for the network. Nickelodeon went on to become one of the most successful enterprises in children's television, not only making large profits, but also continuing to win accolades from its own peers as well as educators and parents groups. Its advertising revenues rose significantly from $13 million in the mid-1980s to more than $60 million in the 1990s, part of an expanding children's TV advertising market.[20]

Nickelodeon (along with Viacom's MTV) had a powerful influence on children's television programming, pioneering a new aesthetic with wild, colorful, trendy graphics; creating in-your-face, "kids-only" content; and breaking many of the rules (e.g., programs for girls cannot succeed) thought to govern kids' TV. Critically acclaimed, the channel also demonstrated the financial viability of children's content and helped trigger a proliferation of television programming and services for children.[21] By the 1980s, a new generation of children's TV networks had sprung up, and they continued to grow throughout the 1990s. Fox Kids Network, for example, became a profitable unit of the News Corporation's Fox Broadcasting, and Time Warner's Cartoon Network transformed the Warner Brothers animation library into a highly successful cable enterprise. In 1998, Fox bought the Family Channel, converted its daytime lineup to children's shows, and announced plans for two new networks: the Boyz Channel and the Girlz Channel. Discovery launched the Discovery Kids, and the Disney Channel created Toon Disney.[22] "Nickelodeon," Lawrie Mifflin of the *New York Times* observed, "was the motor that generated this prolific marketplace."[23]

Nickelodeon also was the first to build an ongoing relationship with children into its market research. Its "Nickelodeon Kid Panel" linked children electronically in thirteen markets around the country and encouraged them to communicate among themselves and with corporate researchers. through their home PCs.[24] By using commercial online services to connect the families, the online panel predated widespread adoption of the Internet, where such ongoing data collection operations would become routine. After Viacom purchased the company in the mid-80s, Nickelodeon teamed up with the market research firm Yankelovich to launch the Nickelodeon/Yankelovich Youth Monitor, which quickly became a major provider of market research to other businesses, selling its studies for as much as $26,000 each.[25] The Youth Monitor also served as a clearinghouse for public information about children and youth, regularly distributing data to the media on the habits, opinions, and pastimes of young people. A 1999 study, for example, found that "the top worries of 9- to 17-year-olds include not doing well in school (65 percent), not having enough money (59 percent) and getting cancer (52 percent)."[26] Nickelodeon pioneered such unconventional practices as the "weekend research retreat," where market researchers employed participant observation methods to monitor children over several days of slumber partying and other play activities. "Finally," enthused one of the Nick staffers, "there is a method of research for Kids which lets them feel comfortable in a natural setting and provides group interaction as well as enough variety to keep the children's attention and interest."[27]

The use of the word "Kids" was no accident. It was part of a conscious and deliberate effort to promote a particular view of what it meant to be a child. Nickelodeon executives hardly ever used the word "children" in press materials or public statements. The network's mantra, "let kids be kids," promised a carefree and safe playground in the media landscape, created especially for children. This "kids-only" philosophy was key to the network's success. As media scholar Linda Simensky observed: "By the 1980s, Nickelodeon had honed its 'big idea'—a philosophy that could be summed up in 'us vs. them.' The basic idea was that kids lived in a grown-up world, and that it was tough to be a kid when you had to follow all the grown up rules. Either you were part of them . . . or part of us. . . . Nickelodeon positioned itself as understanding kids, being for kids, giving them what they wanted to see, and giving them a place where they could be kids."[28]

Rise of the Big (Little) Spenders

The dramatic expansion of the children's-media and advertising market could not have happened without the steady rise in the spending power of children and teens, which increased dramatically in the latter decades of the twentieth century, doubling between 1960 and 1980 and tripling in the 1990s. By the turn of the century, 31.6 million teenagers were spending more than $155 billion of their own money per year in the retail market, and children aged 4–12 controlled or influenced the spending of nearly $500 billion.[29] Marketing experts attributed this surge in spending to several demographic, social, and economic changes during the previous two decades. As the baby boomers began having their own children, the "Echo Boom" generation was born, placing more and more children into the society, and thus into the marketplace. Significant changes also occurred in the American family. The percentage of working mothers more than doubled, so that by the early 1990s, 70 percent of U.S. mothers were working full-time or part-time.[30] Dual-income households became the norm. A rising divorce rate also created more single-family households. During the same period, teens and even younger children began to take more part-time jobs. A report in 2000 by the U.S. Labor Department found that "half of 12-year-olds do babysitting or lawn work, and the numbers go up from there: 57 percent of 14-year-olds and 64 percent of 15-year-olds have jobs."[31]

The adjustments that many parents were making in order to balance family and professional life became a boon to marketers, who quickly seized upon these new opportunities for targeting children and teens. In families where both parents worked, children began to assume more responsibilities for family grocery shopping. As marketers Selina Guber and Jon Berry noted, "Millions of children have taken on chores from buying groceries for dinner to cleaning the house, to cooking themselves after-school snacks, to even getting dinner started for their working parents."[32] Companies began introducing special versions of their popular brands designed especially for kids. As James McNeal explained, "usually these kids-also products differ by being scaled down and funned up. For example, Dial For Kids, an antibacterial soap, is formulated with mild, non-irritating cleansers to be gentle to kids' skin. The fun is added by packaging it in bold colors and producing it in such fragrances as screaming strawberry and power purple."[33] In divorced families, where children often were shuttled from one home to another, parents and grandparents began giving their children what the marketers labeled "DWI gifts," extra money and toys to help their children "deal with it."[34]

In the early 1990s, food manufacturers launched a new generation of "fun food" designed to take advantage of children's increased spending power and independence.[35] "[Kids] want to create their own identity," one marketer explained to *Brandweek*. With their own products, "they can create a sense of ownership and difference." But the move also was spurred by the bottom line. With the saturation of their traditional markets, another industry representative explained, corporations "saw an easy way of shoring up sales by going to a new market. And that new market is kids." Promoting food products directly to kids was a bargain, since buying time in children's TV programming was much less expensive than placing commercials in prime time.[36]

Working with "color consultants" and other children's marketing experts, food manufacturers added garish colors, playful elements, and entertainment cross-promotion strategies to create products that would be especially appealing to kids. Heinz E-Z Squirt injected "yucky" colors such as green and purple into its new kid-oriented ketchup, encouraging children to use the squeeze tube to draw pictures. When the company launched its new "Blastin' Green" brand in 2000, the demand was so high that factories had to crank up production to an around-the-clock schedule to keep it in stock. In a partnership with the Walt Disney Company, Kellogg's Mickey's Magix cereal offered "Mickey marshmallows and twinkling oat stars that turn milk blue." The company's Buzz Blasts included "aliens and rocketship graham cracker pieces that orbit around Buzz Lightyear of *Toy Story* fame." Combining advertising and product-placement strategies, Kellogg aggressively promoted its Pop-Tarts brand to teens. "Kids grow so fast and we need to keep Pop-Tarts relevant to them so we play up color and music," one marketer told the press. "Pop-Tarts lets kids express their rebelliousness and it provides a refuge for them."[37] These new "fun foods," however, had more in them than fun. "The trend in product innovation," Juliet Schor explains in her book *Born to Buy*, "has been to increase calories, and especially sugar content."[38]

Children were becoming more effective at influencing their families' consumer decisions, partly because of a shift from authoritarian to more permissive parenting styles. "Baby Boom and later generations of parents," Schor explains, "have been far more willing to give voice and choice, to see consumer decisions as 'learning opportunities.'"[39] Kids developed an arsenal of different styles of nagging, closely tracked and encouraged by the marketers, using the "nag factor" to prod their parents into buying toys, candy, and cereals.[40] Through their "kidfluence," children even began having an impact on such "big ticket" purchases as cars, hotel, and airlines, often responding to the advertising campaigns directed at them, including such kids-clubs as Camp Hyatt, Best Western Young Travelers Club, and Delta Airlines Fantastic Flyer Program.[41]

To take advantage of the growing children-and-teen market, ad agencies and market-research firms set up special divisions such as Griffin Bacal's KidThink, Saatchi and Saatchi's KidConnection, Doyle Research Associates' Kid2Kid, the Wonder Group, Teenage Research Unlimited, the Geppetto Group, and the Zandl Group, to mention only a few.[42] Retailers created new divisions especially for children and teens, including GapKid, Kid's Foot Locker, and Kids R Us.[43] As the *Washington Post* observed, "With the number of children in America larger than at the peak of the baby boom, and their purchasing power growing faster than economists can measure it, a vast service industry of market researchers, public relations firms, newsletters and ad agencies has sprung up to lead corporate America to young hearts, minds and piggy banks."[44]

Children Getting Older Younger

One of the products of this burgeoning market-research enterprise was yet another category of childhood—"tweens"—a concept invented by the advertising industry and embraced and promoted by the press. The word began to appear in both the marketing literature and the mainstream press during the 1980s, referring to the 27 million children age 8–12 (though some market researchers included even younger ages) that quickly were becoming a powerful demographic group.[45] As "the largest number in this age group in two decades," *Newsweek Magazine* observed in 1999, "they are a generation stuck on fast forward, children in a fearsome hurry to grow up. Instead of playing with Barbies and Legos, they're pondering the vagaries of love on *Dawson's Creek* . . . they're computer savvy, accustomed to a world of information (and a social life based on e-mail) just a mouseclick away."[46]

Marketers frequently cloaked their descriptions of the "tween" category in scientific terms, borrowing from existing child-development research and adapting it to fit the imperatives of the consumer culture. "Our studies of the scientific literature," the authors of *The Great Tween Buying Machine* wrote, "combined with our extensive first-hand observations, reveal a number of markers that clearly and consistently support the existence of a tween market segment that is distinct from kids and teens." If Piaget were writing today, the book suggested, he "might have considered the *tween* segment" as "the Period of Concrete Operations."[47] Marketers also invoked more recent academic research suggesting that children, particularly girls, were entering puberty at earlier ages than before. There was not a strong consensus, however, within the scientific community to confirm that this was a real trend, and a number of observers were quick to point out that too much was being made of what was still a quite tentative set of findings. "While earlier physical maturation may play a small role in defining adolescence down," Kay Hymowitz commented, "its importance tends to be overstated."[48] Despite the inconsistencies, the promoters of the "tween phenomenon" repeatedly touted the mantra that "children are getting older younger."

"Tweens are the brightest star in the consumer constellation," James McNeal gushed at a "Marketing to Tweens" conference in the late 1990s, where he offered practical strategies for tailoring products exclusively to this age group.[49] The marketing guru presented his own interpretation of this newly created developmental category. "They are at a stage in their life when they are becoming very concerned about their position in society. They are asking how conspicuous they are." But, at the same time, he explained, "they are told that they are still children. They are going back and forth between toys and clothing." Marketers could be most effective, he advised, by consciously tapping into the unique anxieties and aspirations of this age group.[50]

According to child psychologist and media critic Susan Linn, the creation of the new demographic category had less to do with developmental psychology and more to do with social changes in children's lives during the 1970s and 80s. As she explains, with more and more parents in the workplace, increasing numbers of children were left to fend for themselves. "These kids did not go unnoticed by Madison Avenue, and a new marketing demographic sprang from the vulnerability of children alone at home, unsupervised," she writes in her book *Consuming Kids: The Hostile Takeover of Childhood.* "Of course, the phrase 'latchkey kids,' referring as it did to kids who were fending for themselves and taking on household responsibilities, is rather downbeat, so the advertising industry created 'tweens.'"[51]

Entire product lines—from specialty stores to television channels to food—were developed exclusively for the tween demographic. Kraft Foods developed "Kraft Easy Mac," a macaroni-and-cheese product "in snackable form," designed for 9–13-year-olds. By running ads in popular tween shows such as *Dawson's Creek, 7th Heaven,* and *Buffy the Vampire Slayer,* the product doubled its sales in the first week of the campaign.[52] All the kid-vid cable networks—Nickelodeon, Fox Family, Disney Channel—created special tween programs.[53] "We always had the [kids] demo," explained a senior vice president for marketing at Nickelodeon. "But we wanted to create a place where we could program to tweens [and] market to them, a place for them to call their own."[54] A Disney Channel executive told *Multichannel News* that tweens were so important to his network that they were doing weekly research on them, using "every method you can imagine"—from focus groups to telephone surveys to stopping kids in shopping malls to interview them, aka "mall intercepts."[55] The network already had hit upon a successful programming formula with its popular live-action tween series *Even Stevens* and *Lizzie McGuire,* the magazine explained, and had "packed its tween-targeted content into a slot that stretches from mid-afternoon Friday, when that closing school bell rings, through Sunday."[56]

The proliferation of children's marketing sparked increasing concerns from public-health professionals and consumer groups. In the 1970s, Action for Children's Television and the Center for Science in the Public Interest had urged government regulators to impose some limits on children's TV commercials. But while they succeeded in establishing a few safeguards, they were unable to stop the steady progression of commercialism into children's lives. (Chapter 4 tells the story of these campaigns.) By the 1980s and 90s, as marketers became even more aggressive in targeting young people, there were renewed outcries from a growing number of organizations.[57] The American Psychological Association and the American Academy of Pediatrics issued warnings about the harmful effects of advertising on children, citing a large body of research documenting children's vulnerabilities to persuasive and manipulative commercial messages.[58] Consumers Union's 1997 report *Captive Kids* documented the alarming expansion of marketing into schools.[59] The Center for Commercial-Free Public Education mobilized students across the country to fight Channel One, a school-based television network offering free equipment to school districts that agreed to air the channel, along with commercials, in their classes.[60] But despite this rising tide of opposition, commercialism had become so pervasive that it was impossible to fight it on all fronts. And as activists continued to take aim at marketing in television, movies, and schools, advertisers already were setting their sites on digital media.

Early Adopters

When families began moving online in the early 1990s, children, "tweens," and teens already were heavy and avid consumers of electronic media. A 1998 Roper survey found that 86 percent of children ages 6–17 had access to a VCR (23 percent in their own rooms), 70 percent had a video-game system at home (32 percent in their own rooms), 50 percent had a TV in their own room, 40 percent had their own portable cassette/CD player, and 35 percent their own stereo system.[61] While parents and educators were concerned about the role of all this media in children's lives, many were hopeful that computers would provide an educational alternative to the highly commercialized entertainment content that dominated children's lives. Computers were becoming commonplace in many schools around the country, and academic experts such as MIT's Seymour Papert wrote extensively about how they would revolutionize education.[62] According to a survey by the Annenberg Public Policy Center, 84 percent of parents with home computers connected to the Internet believed that such access helped their children with schoolwork, and 70 percent regarded the Internet as a place for children to learn "fascinating, useful things."[63] By 1999, families with children represented one of the fastest-growing segments of the population using the Internet. Among these households, computer ownership had become almost as common as cable television, and Internet access had nearly caught up with newspaper subscriptions.[64] The new medium also held great promise for business. As the dot-com bubble began to expand, investors rushed into the high-tech and Internet market, banking on the potentially huge profits that "e-commerce" would generate. The low barriers of entry to this new medium allowed any individual or institution to launch a site on the Web, which had grown at a staggering rate. Stories of overnight millionaires abounded. As journalist John Cassidy recalled, "a group of college kids could meet up with a rich eccentric, raise some more money from a venture capitalist, and build a billion-dollar company in eighteen months."[65] In 1993, two graduate students from Stanford University started an online list of their favorite Web sites. By 1994, the site had generated tens of thousands of users. By the end of 1995, the site—Yahoo!—had attracted $1 million in advertising revenues.[66]

Although advertising had been forbidden by the unwritten rules of "Netiquette," only two years after the introduction of the World Wide Web it had become one of the keys to the Internet's future profitability. The Internet, with its static black-and-white images, had been less appealing to advertisers anyway. But the Web had the flash that Madison Avenue craved, and commercial Web sites soon began to crop up in exponential numbers. By April 1995, of the more than 30,000 Web sites on the Internet, 10,000 were commercial, according to *Interactive Age*.[67] To ensure the growth and success of advertising in the new interactive media, the ad industry launched its own aggressive campaign, forming the Coalition for Advertising Supported Information and Entertainment (CASIE), led jointly by the American Association of Advertising Agencies and the Association of National Advertisers. CASIE was created because of concern within the ad industry that "there would be no advertising in cyberspace." The role of the coalition would be to promote existing ad-supported programming and to encourage providers of new services to rely on advertising as a key source of revenue. CASIE already was involved heavily in lobbying efforts in Washington, both on Capitol Hill and at the various federal agencies, including the Federal Trade Commission.[68]

As early adopters of new technologies, children and teens were becoming one of the most valuable segments of the exploding online marketplace. "Kids are hip to technology as early as age 2," *Advertising Age* reported, noting that children already collectively spent a billion dollars of their own money on a variety of electronic products—from CDs to video games to TVs.[69] "For a glimpse at the hottest, fastest growing market for interactive-computer technology," the *Hollywood Reporter* wrote, "feast your eyes on the true future of multimedia: your children."[70]

Hundreds of eager digital "content providers" rushed online to take advantage of this new hot market—from new Web-based entrepreneurs to established media conglomerates such as Disney and Nickelodeon. The interactive market was developing so rapidly that a new monthly trade publication, *Digital Kids,* was created to track the swift changes underway. Games began emerging as one of the "killer applications." Time-Warner, cable giant TCI, and Sega of America joined to launch the Sega Channel, an interactive service of around-the-clock video games. The newly created Microsoft Network announced plans to offer an online service for kids called "Splash Online, created by Splash Studios, a developer of children's CD-ROMs and videos. Included in the mix would be SplashKids, a game that children would pay around $5 to play, and an online store that would sell Splash products and other paraphernalia."[71] "Recent moves indicate that major companies are positioning to launch children's sites on the Web," one trade publication announced in early 1995. Among the companies that had recently requested Internet domain registrations were: Burger King, Hasbro, Kellogg, and Tonka.[72]

Targeted at children aged of 3–12 and billed as the prototype "next-generation interactive channel for kids," da Vinci Time and Space was described as a "virtual world populated with characters whose ongoing stories are intertwined with entertaining as well as unique forms of interactive advertising."[73] Nine leading product brands and six agencies were working with da Vinci to design "places" or "interactive advertising components" that would "entice each child to interact" and track "every interaction between a kid and an advertisement."[74] Because this is "the generation that spends the most time glued to a computer monitor," trade publication *Selling to Kids* explained, "online marketing is going to be more important for this group than any previous."[75]

As the Internet moved further into children's lives, academic research on the new medium's relationship to children was slow in coming, with scholars paying no attention to the burgeoning online commercial enterprise. Researchers in the field of children and media were focusing most of their work on television (with hardly any critical research on advertising). The handful of university researchers investigating other aspects of the Internet often were forced to operate under government restraints, requiring elaborate paperwork before engaging in research with any individuals, adult or children. Government funding agencies such as the National Science Foundation still were struggling with ethics and guiding principles for online research. But market researchers, who had no such rules, were amassing a wealth of information on children's relationship to digital technologies. In addition to the large network of market-research companies already heavily involved in advertising to young people, new firms formed that specialized in the online environment. Some were developing unorthodox methods for penetrating the children-and-youth subculture, including hosting slumber parties for teen girls to "meet children on their own terms" and to "build trusting relationships with girls."[76] Others were finding that the Internet itself could be a powerful research tool and were beginning to communicate directly with children and teens on specially designed Web sites. The new Internet market research industry soon became an influential source of information for the press, which was avidly following the rapid growth of the high-tech marketplace.[77]

Market-research firms and trend-analysis companies began offering expensive industry seminars, conferences, and trade shows to cater to the rush of interest in the hot new "cybertot" and "cyberteen" demographics. At its first "Digital Kids" conference in 1995, Jupiter Communications assembled executives from the biggest advertising and market-research firms to share their latest research on children and digital technologies. A hundred young, hip dot-com entrepreneurs had flocked to the downtown Manhattan hotel to hear an executive from Saatchi & Saatchi's Kid Connection unveil the findings from its new report.[78] Invoking the mantra of "marketing to kids from the inside out," Saatchi had hired a team of researchers to probe the "cultural and emotional drivers" involved in children's relationship to the Internet. Cultural anthropologists had spent more than 300 hours observing children and their families while they were online, following up with a series of one-on-one psychological interviews designed to uncover "other hidden motivators that maybe weren't so obvious in the observation."[79]

As the audience members eagerly clicked away on their laptops, the expert explained to them why the Internet is "a medium for advertisers that is unprecedented." Drawing on psychological research, the study had determined, for example, that "interactivity is a natural response to four very powerful themes of childhood: attachment/separation, attainment of power, social interaction, and mastery/learning." Industry psychologists also had found that when kids go online, they quickly move into a kind of hypnotic "flow state," creating the perfect environment for advertisers. "There is nothing else that exists like it to build relationships with kids, and there's probably no other product or service that we can think of that is like it in terms of capturing kids' interest." Since advertising in the new digital media was "all about relationships," the key to success was getting to know your customers on a "one-to-one" basis. One way to do this was to create online "product spokescharacters"—like Ronald McDonald, the Gummi Bears, and Snap, Crackle, and Pop—who would engage in interactive play with children and develop ongoing personal relationships with them.[80]

The Clickerati

The dot-com boom produced its own high-tech gurus, who coined a variety of clever terms to define and celebrate the generation of children growing up with the Internet. "As consumers, N-Geners already have greater disposable income than previous generations of youth," Don Tapscott explained in *Growing up Digital,* his popular 1998 book. While issuing some warnings about the blurring of advertising and content on the Internet ("on the Net, the boundaries between ads and content melt like the cheese on a Whopper"), Tapscott enthused about the future of online marketing and children. "The new formula," he predicted, "will be N-Gen + the Net = electronic commerce. The net is becoming a new medium for sales, support, and service of virtually anything, as tens of millions of Net-savvy purchasers come of age."[81]

Idit Harel, a high-spirited PhD and former professor at MIT's Media Lab, used the label "clickerati" to describe the most active of the new Digital Generation. Founder of her own dot-com company, MaMaMedia, Harel aggressively promoted her business and her ideas in the trade and mainstream press, and she was a frequent booster of the online market at industry association meetings. "These clickerati kids thrive on and accept change," she told one trade publication. "They think both linearly and non-linearly. They're looking for tools, not for answers. And they're looking for ways they can create their own media and not just consumer media."[82] "High tech is now my tech," Harel told a 1998 gathering of researchers and new media producers. Today's children, she explained, have an entirely new set of relationships with media compared to their parents. They have new expectations, new attitudes, and new behaviors in connection with the new media environment, concerning both what that environment should provide and how much control it should yield.[83]

Her educational Web site, mamamedia.com, which launched in 1997, blended her passion for constructionist theory (she was a protégée of Seymour Papert) with her zeal as an online entrepreneur. Her enterprise billed itself as a company dedicated to "creating unique activity-based learning media for kids 12 and under, their families and educators."[84]

Children entering the Web site were immersed in a graphically rich, brightly colored, interactive playground, bathed in psychedelic colors of hot pink, lime green, and lemon yellow. The company's promotional materials described the various activities on the site:

> Romp—creatively designed directories and themed activities let kids explore the Web while learning word meanings and classification; Surprise—kids develop their own digital stories, build multimedia towns, and create digital drawings complete with audio and animation; Zap—kids express their own creativity as they custom design their interfaces and build unique digital signatures that say, "this is who I am today on MaMaMedia"; Buzz—here kids showcase their own creations and exchange projects and ideas with other MaMaMedia kids.[85]

The site earned recognition for its innovative use of new technologies and received a Computerworld Smithsonian Award from the Smithsonian Institution. "E-commerce is going to be integrated into any family and household," Harel told reporters in 1999, predicting that children soon would be able to make online purchases throughout the Web with "virtual allowances, coupons, and gift certificates."[86]

The company's New York headquarters also boasted its own room for conducting focus groups with children, and its Web site wove advertising and marketing into its educational content, forging partnerships with children's product brands. On the MaMaMedia home page, for example, an icon invited children to "Zap your friends with Wacky DigSig cards sponsored by Fruit Gushers and Fruit Roll-ups." This "sponsored activity" was designed to drive traffic to the General Mills children's Web site, You Rule School.[87] In return, packages of Fruit Gusher snacks carried the Internet address for the MaMaMedia site.

In the heady period of the late 1990s, online entrepreneurs, seeking to cash in on the lucrative cybertot and cyberteen demographics, rushed to the Web with a variety of imaginative, sometimes outlandish, new ventures. At trade shows and industry conferences, they could be seen, often with their venture-capital partners on their arms, bubbling with enthusiasm to reporters and handing out branded freebies such as mouse pads, screen savers, and toys. These events often took on a circus-like quality, as characters from popular television shows and new Web sites mingled with the other participants. At the Digital Kids 1999 conference in San Francisco, sponsored by Internet research firm Jupiter Communications, promoters for the kids' Web site Headbone Interactive, sporting beanies with stuffed bones on top, handed out cards and flyers touting the "empowering" features of their online community. As Bob Thompson, a reporter for the *Washington Post* assigned to cover the event, observed:

clusters of enthusiastic conference goers from Nickelodeon, Hasbro, Disney, the Learning Company and Discovery Communications mingled with equally jazzed cadres from small, but ambitious outfits that are not yet household names: Zeeks.com, MaMaMedia, Headbone Interactive, Curiocity's FreeZone. Cell phones sprouted in the lobbies and on the stairwells after every session, and snatches of earnest networking (That's why I'd like to get you into a meeting with this individual I'm talking about . . .") were overheard.[88]

In the two days of PowerPoint presentations and discussions, participants were introduced to an array of new ventures, innovative online business models, and a flood of statistics, all promoting the limitless profit-making opportunities in the "kidspace" of the Internet. The headlines that flashed on the screen suggested a cornucopia of opportunity in the online market: "Online: A Marketer's Paradise" . . . "As Schools Get Wired" . . . "and Kids Spend More Time Online" . . . "and Online Becomes Infectious" . . . "the Inhibitions to Buy Are Less for Kids Than Adults" . . . "Online Shopping: A Ticket to Ride"[89] Analysts predicted that children and teenagers would be spending $1.3 billion online three years from then, a number, Thompson noted, "that was cited, mantra-like," throughout the conference.[90]

Branded Environments

Advertising and marketing were quickly becoming a pervasive presence on the World Wide Web. The forms of advertising, marketing, and selling to children that were emerging online departed in significant ways from the familiar thirty-second spots that punctuated children's TV programs. The interactive media were ushering in an entirely new set of relationships, breaking down the traditional barriers between "content and commerce," and creating unprecedented intimacies between children and marketers. The idea of advertising as something separate from programming was beginning to disappear as the Web blended them into seamless "branded environments." In many ways this was a throwback to the earliest years of television when sponsors created and packaged programs, often weaving their ads into the storylines. But on the Web, the two often were indistinguishable.[91] This model opened up endless possibilities for children's Web sites.

Mattel's Barbie.com site positioned itself as a community for girls. With links to high schools and other ".orgs" the Barbie site offered a variety of online activities designed to appeal to girls, such as sending e-postcards, receiving newsletters, entering contests, and "voting" for their favorite

Barbie. The site profiled many of the popular doll's new personas (including Soccer Barbie, Pet Lovin' Barbie, and Vintage Spring in Tokyo Barbie) as well as many of the classic styles. Following on its successful Barbie Fashion Designer CD-ROM, the site also gave girls the opportunity to design their own personalized Barbie by choosing from an inventory of physical features, clothing styles, and personality traits. "You get to choose all kinds of stuff about her," Mattel's Nancie Martin told her marketing colleagues, "you get to decide what her face or her skin color is, her eye color, her hairstyle, her hair color . . . and you get to choose what she's wearing and her accessories—because of course, without accessories we're nothing—and you get to make up a story about her." (The personalized Barbie could be purchased online for $39.95 and custom made for that child.)[92]

The development of "brand loyalty" among children had become axiomatic among marketers, a core element in the strategy of "cradle-to-grave marketing." Researchers knew that children began developing brand preferences in early childhood, before they entered school. "And not just for child-oriented products," James McNeal explained, "but also for such adult-oriented things as gasoline, radios, and soaps." This tendency is strongly rooted in the developmental needs of children, he added. "The belonging (affiliation) need, which causes us to seek cooperative relationships, is very strong among children. . . . Also, children are looking for order in their lives. There are so many new things to encounter that some order is necessary to cope with them all. A trusting relationship in which satisfying acquisitions can always be expected helps give order to an increasingly complex life."[93]

On the Internet, branding could be taken to a new level. "Marketers here have an unparalleled opportunity to get kids actively involved in brands," one market researcher explained. Online-content developers were encouraged to create playgrounds where children could interact with and manipulate various brand "elements"—"brand characters, brand logos, brand jingles, brand video"—cutting and pasting them, and authoring their own creations, which could then be posted in public online places for other children to see."[94] Another industry executive advised: "[I]f you create an ad that's as much fun as the content," such as "games that kids can play that involve the products. . . . then there'll be a reason for kids to click on the ads and interact with them and enjoy them."[95] Products such as Crayola, Lego, Barbie, Coke, and Fritos soon had their own branded environments. At the 1999 Digital Kids conference, participants spoke proudly of branded communities for teens—Web sites built around products—invoking the slogan, "love my community, love my brand."[96]

New dot-com ventures—with names like Rocketcash, IcanBuy, DoughNET, and CyberMoola—were launched to enable children to make online purchases. When market research surveys suggested that many parents did not want their kids buying products on the Internet, the companies came up with strategies designed to overcome these concerns. Some created "digital wallets" that would allow parents to use a credit card to deposit limited amounts of money into a child's online account. Others offered barter or other kinds of non-cash transactions as a first step.[97] Beenz.com offered "virtual cash" that children and adults could collect online by visiting participating sites and taking part in surveys, promotions, and other marketing enterprises.[98] DoughNET let kids "play with money" by investigating how the stock market worked and enabled them to donate real-world money to nonprofit causes.[99] For teens who did not have money to spend online, DoughNET let them earn "DoughPoints" by answering polls by HarrisZone, a division of Harris Interactive, which billed itself as a "full service research firm, harnessing the power of the Internet to gather the market intelligence businesses need to compete in the new economy."[100]

Dot-com Crash and the Aftermath

But the promise of a burgeoning online media Mecca—fueled by the seemingly limitless coffers of discretionary children's spending—was short-lived. The highly inflated dot-com market, which reached its peak during a twelve-month period in 1999 and 2000, went into a dramatic slide at the dawn of the new century. The economy of the Internet plummeted, as venture capital quickly dried up amid projections of less-than-expected revenues. Hundreds of commercial Web sites folded, both well-known and obscure online ventures, including a number of highly touted kids and teen sites.[101] Other dot-com companies that had built up their staffs quickly during the boom were forced to cut them drastically. MaMaMedia laid off 30 of its 150 employees.[102] A visit to MaMaMedia's New York headquarters in 2001 found the spacious, trendy suite of offices largely empty, no receptionist at the front desk to greet visitors, and the tiny group of workers huddled in a small space at the back, struggling to keep the operation alive.[103] The ambitious plans for da Vinci Time and Space's interactive children's television network never materialized. DoughNET, IcanBuy, and

CyberMoola went under, along with other kids' e-commerce ventures. The only surviving teen credit card was Visa Buxx, Visa Company's credit card for teens.[104]

To many it may have looked as if the new online children's marketplace had vanished altogether. But, in fact, this was not the case. Though the hopes of a thriving Internet e-commerce market may have been dashed, the Internet and other digital media already were becoming a lasting presence in the lives of children and teens. The dot-com boom era had served as an intense period of experimentation, when online entrepreneurs, newly freed from the conventions and constraints of traditional media, hatched innovative business models that were designed to take advantage of the ease with which children and teens navigated the digital environment.

As the dot-com world recovered from the shock of its dramatic downward slide, marketers and content developers regrouped for the post-crash digital era. The companies with significant offline "bricks and mortar" businesses were able to survive better, as were some of the "leaner and meaner" of the Web-only ventures. Big media conglomerates, which were growing larger through a spate of mergers and acquisitions, also were well positioned to prosper after the crash. All continued with their Web ventures, adopting and adapting many of the innovative practices that had been invented by the failing dot-coms. "The current kid Web site climate has become strikingly Dickens-esque," one trade publication wrote. "For brand extension entities such as Foxkids.com and Nick.com, it's the best of times, as the companies refine their business models and relaunch their sites to position themselves for long-term financial health. But for many of the independent players, the last six months have been the worst of times, with financial uncertainty looming large, and the prospect of more closures very real."[105]

Network programmers for the successful children and teen channels already had begun preparing for the eventual convergence of television and the Internet. Closely monitoring the online behaviors of the Digital Generation, they were experimenting with "cross-platform" strategies for encouraging young viewers to stay involved with programming over long periods of time. "Multitasking," a computer term that described a machine's ability to run several programs at once, quickly entered the Internet-age lexicon to describe the particularly unique teenage phenomenon of being able to attend to several different media at the same time. As Katie Hafner explained in a 2001 article in the *New York Times,* "now that more than half of American households have computers, the younger generation has steadily been sharpening its skills. While many adults are perfectly adept at, say, toggling between e-mail and a spreadsheet, perhaps with a phone conversation or two thrown into the mix, teenagers can multitask many of the elders under the table, juggling a half-dozen activities without missing a beat."[106]

As the Disney Channel's Rich Ross explained to the press, "[Kids have] come to expect that, when they watch a show on Disney, there's an online component equally as exciting as what they see on TV. Kids are savvy, so the site has to give them something extra: more information about a show or its characters, games with 'collector cards,' or other things. Otherwise, they'll get bored with programming." The Cartoon Network's executive vice president added, "The merging of TV and the Internet is one of the most interesting changes that's happened in Kids TV," describing his network's "Total Immersion" cartoons, "where we run a show or a special promotion on TV, and online content is offered for a simultaneous experience."[107]

The Web offered multiple ways to forge deeper emotional relationships between viewers and TV programs. Some cross-platform strategies blended the fictional and the real, inviting fans to experience the inner worlds of their favorite TV characters. NBC's *Passions* allowed online navigators to read one of the cast members' diaries. DawsonsDesktop.com, created by Sony Columbia Tri Star as a companion to the popular teen television drama *Dawson's Creek*, combined fan culture with online product pitches.[108] Launched during the second season of the show, the Web site was designed to replicate the personal computers of the show's four main characters, with a different desktop featured each week. "Ever wish you could see inside someone's computer?" the site asked,

> Someone like one of your favorite characters on Dawson's Creek? Well, here's your chance! Each week one of four different desktops appears here—Dawson's, Joey's, Pacey's or Jen's! Between the episodes, you can delve into their journals, emails, Instant Message chats—even their trash cans! Just click "Go" to enter. Once you're on the "desktop" you can click on any of the icons, or just click "Begin." Like on your own desktop, there are many ways to navigate. And since Dawson and his friends are often online, there's something new every day![109]

In the words of the *Wall Street Journal*, the site is "a voyeur's paradise where every day brings a new tidbit about the show's lead characters. . . . It offers original content as well as plot lines and character development that run parallel to the show. That mix is key, since fans have to shuttle back and forth from show to site to make sure they don't miss anything. The more fans get involved in the site, the more they watch the show."[110]

Early in the third season, the episode "Homecoming," which aired on the WB network on October 6, 1999, featured a scene in which Dawson presented a necklace to Joey, his ex-girlfriend and current best friend, as a symbol of their friendship. Shortly thereafter, the weekly e-mail newsletter featured the following announcement: "Own Dawson's Necklace! You asked for it and now it is available. . . . Dawson gave his necklace to Joey last week and now you can own the exact replica. Preorder yours today." Directed to the Web site by the advertisement in the newsletter, fans were presented with the opportunity to purchase their own $12.99 replica of Dawson's necklace.[111]

The Internet of the 1990s had served as a test bed for more sophisticated approaches that would take full advantage of the capacity of digital media. "As we enter the second decade of the Internet Explosion, it is finally safe to say that lessons were learned," explained an online brochure by Claria, an Internet market research company. "Most advertisers looked at the Internet consumer and saw just another big, broad demographic. They entered the Internet market relying on traditional advertising campaigns, trying to adapt them for the Internet and hoping—fingers crossed—for results. Most were disappointed." Interactive technologies called for a new model: "behavioral marketing." Its key strategies were: "1) Market to the individual, not to a mass demographic, based on his or her unique Web-surfing behavior; 2) develop insights into consumers' preferences, based on anonymous profiling; and 3) blend promotional advertising with brand awareness campaigns." Behavioral marketing, its proponents argued, "encompasses a more refined approach to Internet advertising, one that treats the Internet consumer as an individual, not as part of a mass demographic."[112]

In the coming years, this "more refined approach" would shape the content not only on the Internet but also on an array of "platforms" in the expanding media culture of the Digital Age. For younger children, the model of behavioral marketing would not develop completely. A little-known law passed in 1998, the Children's Online Privacy Protection Act, would curtail and limit some of the data-collection practices and personalized communications that were at the heart of this new system. But for teenagers, the system would emerge full-blown.

3 A V-Chip for the Internet

The image was haunting. A very young child sat at the computer keyboard, his face awash in the sickly green glow of the screen, his eyes and mouth wide open in a hypnotic gaze straight out of a horror film. Looking half-doll/half-human, the boy on the July 3, 1995, cover of *Time* appeared almost computer generated himself.[1] The magazine carried an exclusive story about "cyberporn," a new term entering the popular lexicon that year. According to a study at Carnegie Mellon University, with just a click of a mouse, children were being plunged into a dark underworld of sordid images, pornographic material, and pedophiles on the Internet. Some children had even received unsolicited e-mail with pictures of couples engaged in "various acts of sodomy, heterosexual intercourse, and lesbian sex."[2]

Within days of its publication, the *Time* story sparked a storm of protest. Civil libertarians and academics attacked both the reporter and the university researchers for allegedly shoddy work and for causing undue alarm among the public. Line-by-line critiques of the study's methodology were posted immediately on the Web. But despite these efforts to discredit the research, the story generated enormous media coverage and public concern. It also helped frame a growing policy debate over children and the Internet. The image of an innocent child confronting the vast dangers of cyberspace was to become one of the most compelling and ubiquitous visual symbols of the digital era, appearing over and over again in the next few years—on television screens, in magazines, and in newspapers.[3]

As more and more families juggled work and home responsibilities, it was increasingly difficult for them to monitor the various media their children were consuming. With the growth of satellite and cable television, more and more channels were being brought into the home, often offering direct access to sexually explicit and violent programming. Added to these worries were new concerns over sexual lyrics in music and graphic video games.[4]

The Internet was promoted as an educational alternative to these other media. In his speeches across the country, Vice President Al Gore repeatedly invoked a glowing technological future where a school child in his tiny hometown of Carthage, Tennessee, would be able to "come home, turn on her computer and plug into the Library of Congress."[5] Now, however, children were stumbling into the dark, dangerous caverns of cyberspace. And in some cases, the harmful tentacles of cyberporn were reaching out to them, invading the safe havens of their online journeys. The problem of cyberporn, the *Time* article commented, was the "flip side of Vice President Al Gore's vision of an information superhighway linking every school and library in the land. . . . When kids are plugged in, will they be exposed to the seamiest sides of human sexuality? Will they fall prey to child molesters hanging out in electronic chat rooms?"[6] As *Newsweek* reporter Steven Levy wrote: "Until now parents have believed that no physical harm could possibly result when their progeny were huddled safely in the bedroom or den, tapping on the family computer." But news reports of endangered children "have triggered a sort of parental panic about cyberspace. Parents are rightfully confused, faced with hard choices about whether to expose their children to the alleged benefits of cyberspace when carnal pitfalls lie ahead." A *Newsweek* poll in 1995 found that 85 percent of Americans were concerned that pornography was too easy for young people to access through the Internet, and 80 percent expressed worries about "virtual stalking" through unwanted messages.[7] But there was no consensus on what to do about it. A *Time*/CNN poll had shown people sharply divided on the policy remedies for harmful Internet content, with 42 percent of respondents favoring Federal Communications Commission–like control over sexual content on the computer networks and 48 percent against it.[8]

All this was taking place during the mid-1990s, when the Internet was still in its infancy as a mass medium. Less than 15 percent of Americans were online, and many of those were connected through popular proprietary gateway services such as America Online, CompuServe, and Prodigy.[9] But already legislative action to reign in the new frontier of cyberspace was gaining momentum.

When the *Time* cover story hit the stands in the summer of 1995, Congress was deep into deliberations, negotiations, and political deal-making over the rewrite of the Communications Act of 1934. The rationale for the proposed new law was that only by unleashing unfettered competition in the media marketplace would the nation be able to take advantage of the enormous and powerful changes taking place in telecommunications. In September 1993, the Clinton Administration had released a 46-page report outlining the establishment of a National Information Infrastructure (NII) "that would create a seamless web of communications networks, computers, databases and consumer electronics that will put vast amounts of information at users' fingertips. . . . [That] can help unleash an information revolution that will change forever the way people live, work and interact with each other."[10] But while the rhetoric for the future was high-minded and visionary, the actual terms of the policy debate were much more complex, and the various industry players each had their own political agendas and hoped-for outcomes.[11] The sweeping "telecommunications reform" legislation created one of the highest stakes lobbying battles of the era, with every major communications industry interest at the bargaining table.

In the midst of the historic rewrite of the nation's telecommunications laws, a flurry of culture wars was beginning to erupt over the role of the media in the lives of children. Interest groups on the right and left—with competing visions of the digital future—were mobilizing to fight it out on Capitol Hill, at the White House, and in the courts. With a national election looming, Republican and Democratic politicians would engage in a political tug-of-war, each side laying claim to the powerful symbols of this public debate. Before the final law would pass, both the new medium of the Internet and the old medium of television would be forced into the public-policy crucible. The legislative remedies that would emerge from this intense political activity would do little to solve the problems they were supposed to address, instead sparking even more controversy. The government's attempts to shield children from harmful content in an increasingly complex media landscape would be deeply flawed. At the same time, the public furor over sex, violence, and indecency would deflect attention away from the critical policy decisions that would determine the media power structure of the new Digital Age.

A Clash of Visions and Values

Months before the watershed *Time* article, the controversial amendment to the telecommunications bill had been introduced in the Senate, aimed at curbing pornography and indecency on the Internet. Known as The Communications Decency Act (CDA), the legislation would outlaw obscene content and impose fines of up to $100,000 and prison terms of up to two years on anyone who knowingly made "indecent" material online available to children under 18.[12] The proposed law followed a series of related regulatory, legislative, and court decisions during the previous two decades, all of which had been highly contentious. In the 1970s, conservative media-watchdog group Morality in Media had filed a complaint with the Federal Communications Commission (FCC) against the radio broadcast of a recording of comedian George Carlin's "Seven Dirty Words" on the New York station WBAI, part of the progressive Pacifica Network. After the FCC decided in favor of the complaint, the radio station challenged the decision in court. Ultimately, the case reached the Supreme Court, whose historic 1978 decision concluded that the material in the recording was indecent and had to be restricted to broadcasting during hours when children were not likely to be in the audience. As newer communication technologies came along, Congress continued to struggle with ways to legislate protections for children from indecent content. In the 1980s, confronted with a burgeoning new industry of "900 number" companies that charged fees to let callers hear sexually explicit messages, lawmakers passed a law to ban telephone indecency. When the Supreme Court struck that law down for being too broad, Congress passed a more narrowly tailored law requiring 900-number companies to create mechanisms to ensure minors could not access explicit material.[13] This ongoing tug-of-war—involving conservative groups, civil-liberty organizations, Congress, and the courts—continued well into the next several decades.

The chief sponsor of the Communications Decency Act was Nebraska Senator James Exon, who had served in the Senate since 1978. Though Exon himself was a conservative Democrat, he enjoyed the support of a number of Republicans, who had gained control of both houses of Congress in the landslide 1994 election. With Republicans in the majority, conservative religious groups were in a particularly influential position. These groups had waged pressure campaigns and lobbied government bodies in the past over the content of American mass media. With the advent of the Internet, they renewed their efforts, finding sympathetic supporters on both sides of the political aisle. One of the most powerful of these groups was the Christian Coalition. Founded in 1989 by the Reverend Pat Robertson, the organization had played a very prominent role in national politics, actively lobbying Congress and the White House on a variety of issues, and organizing grassroots campaigns around the country to support a conservative policy agenda.[14]

In May 1995, Ralph Reed, Executive Director of the Christian Coalition, released a ten-point "Contract with the American Family" laying out a "pro-family" legislative agenda designed to follow on the "Contract with America," which Republican House Speaker Newt Gingrich had successfully used to frame the election-year debate the year before. The first item in the Christian Coalition's treatise was a call for Congress "to enact legislation to protect children from being exposed to pornography on the Internet."[15] Joining the Christian Coalition in its fight against Internet indecency were a number of conservative groups dedicated to moving family values to the top of the U.S. public-policy agenda. These included the Family Research Council, the National Law Center for Children and Families, the Eagle Forum, the American Family Association, and the Traditional Values Coalition.[16] To these groups, the Internet's global reach and unfettered openness posed formidable threats to the sanctity of the family. As this new medium penetrated further into American homes, conservatives sought to place the same kinds of protections against indecency upon it that they had successfully instituted in television.

Once conservatives began the push for Internet indecency legislation on Capitol Hill, a broad coalition of groups mobilized to oppose the measure, including established organizations such as the American Civil Liberties Union (ACLU) and the American Library Association (ALA), as well as newer groups created specifically to deal with the policy debates of the new digital era. Many of the activists involved in these organizations had been early cyberspace pioneers, avid users of the Internet long before the introduction of the World Wide Web. As enthusiastic members of some of the first online communities, they had heralded the Net as an unprecedented participatory medium that held the key to a more democratic future.[17] One of the pioneering groups was the Electronic Frontier Foundation (EFF), founded in 1990 to promote a vision of the Internet as a "new frontier for free speech and society."[18] The Electronic Privacy Information Center (EPIC), established in 1994, dedicated itself to focusing "public attention on emerging civil liberties issues and to protect privacy, the First Amendment, and constitutional values."[19] By 1995, EPIC already had become a prominent advocate in Washington for Internet freedom and privacy. The Center for Democracy and Technology, another of the new Internet policy groups, aimed to "develop and implement public policies to protect and advance individual liberty and democratic values in new digital media."[20]

In their campaign against the Communications Decency Act, these organizations argued that the Internet was a far different medium from television, and that because of its unique nature it deserved to remain unregulated. Any attempt to impose restrictions on it constituted a severe threat to the First Amendment and essentially would "transform the vast library of the Internet into a children's reading room, where only subjects suitable for kids could be discussed."[21] Unlike television, they argued, where government regulation had been required because there were limited airwaves, "bandwidth on the Net is unlimited, and the government's permission is not required to attach a server to it in the same way as a radio or television station. Content providers on the Internet are publishers, not broadcasters."[22]

While those who opposed the legislation framed their arguments in terms of legal principles, the CDA's supporters made the claim that they represented the interests of children. They invoked compelling language and imagery designed to resonate with the fears and frustrations of modern-day families, wrapping their positions in a call for a return to family values. "I want a country where children come first again and where virtue is honored," the Family Research Council's Gary L. Bauer proclaimed at a meeting of Christian broadcasters, "a place where values matter and the American dream is still real."[23] "We are talking about our most important and precious commodity—our children," wrote Senator Exon in an op-ed. "We cannot simply throw up our hands and say a solution is impossible or the First Amendment is so sacrosanct that we must stand idly by while our children are inundated with pornography and smut on the Internet." Drawing on the metaphor of the Information Superhighway, Exon argued that the Communications Decency Act was designed to "make this exciting highway as safe as possible for kids and families to travel. Just as we have laws against dumping garbage on the interstate, we ought to have similar laws for the information superhighway." He also addressed the stresses of modern family life, asking: "Does anyone really think parents can stand over their children's shoulders and monitor them all of their waking hours every day?"[24]

In the media and on the Hill, rhetorical weapons and powerful visual symbols were able to trump the principled arguments of the bill's opponents. The news media seized upon the cyberporn issue, covering it incessantly. As media scholar Henry Jenkins observed, the haunting image of the innocent child that had first appeared on the cover of *Time* magazine had ultimately framed the debate in the public mind, "rendering the arguments about the First Amendment beside the point."[25] To help ensure crucial votes for the CDA's first passage in the Senate, Senator Exon had a friend download some of the most graphic Internet pictures, compiling them into a blue folder, inviting his colleagues to stop by his desk on the floor of the Senate to look at them. The technique proved to be highly successful. The end of the debate was carried live on C-SPAN, shining a public spotlight on the controversial proceedings. As *Time* noted, "few Senators wanted to cast a nationally televised vote that might later be characterized as pro-pornography."[26]

As a last-ditch attempt to thwart passage of the amendment, civil-liberty activists staged an online "Internet Day of Protest" in December 1995, sending more than 20,000 e-mail messages to lawmakers.[27] In what the *New York Times* called "the largest organized protest on the Internet," hundreds of Web sites switched to black backgrounds in an orchestrated demonstration labeled by one of the protesters as "a thousand points of darkness."[28] But these efforts failed to stop the momentum. The Communications Decency Act was passed as part of the telecommunications bill in early 1996.

Supporters of the CDA considered it to be an effective remedy that would protect children from harmful content on the Internet. It imposed fines and possible jail sentences on anyone found guilty of transmitting obscene or indecent material knowing that minors (under 18) could receive it. But its vague language suggested a failure to understand fully the nature of this new medium. Calling the new law "a masterwork of internal confusion," Marjorie Heins, one of lawyers who opposed the CDA, explained in her book *Not in Front of the Children:*

> The "send" and "transmit" provisions of the law were dubious enough—applying, so it seemed, not only to one-on-one email but to group messages or online discussions among hundreds of people, if even one minor is present. But the provision criminalizing "display in a manner available" to minors was the truly loose cannon in the CDA. It applied to all of cyberspace—Web sites, archives and libraries, discussion groups, and "mail exploders" or "listserves" (emails sent to multiple recipients). And it did not require an identifiable young reader, merely the possibility of one.[29]

These flaws in the legislation gave the activists an opening to challenge the CDA on legal grounds, which they immediately set out to do.

For the online industry, the intense CDA debate, culminating in its successful passage, was a critical wake-up call. With the growing popularity of the Internet, companies such as America Online and CompuServe began launching new features that allowed customers to venture into cyberspace without forcing everything through their branded, proprietary services.[30] But these same companies found themselves caught in the middle of the raging cyberporn controversy. In September 1995, the FBI arrested a dozen people who had been using AOL to lure children and teens into sex and to distribute images of child pornography.[31] Company executives were cooperating with the government, but the whole situation was raising disturbing child-safety issues. If the Internet were to become the profitable mass medium they expected it to be, then remedies would need to be developed to assure wary parents that it was safe for their children to enter cyberspace.

President Clinton signed the Telecommunications Act into law on February 8, 1996, at a special event at the Library of Congress, where actress Lily Tomlin, in a reprise of her TV role as an old-fashioned telephone operator, made a visit by satellite. Using an electronic pen on a computer screen to instantly post the new law on the Internet, the president proclaimed: "Today with the stroke of a pen our laws will catch up with our future. We will help to create an open marketplace where competition and innovation can move as quick as light."[32] In the slightly less than 2,000 words of that speech, the president said nothing about the controversial Communications Decency Act. He did mention another provision of the new law, lobbied in by education and library groups and supported by the White House, that created a federal fund to ensure affordable access to the Internet for schools and libraries—a policy later called the "e-rate".[33]

In fact, during the entire congressional CDA debate, the Clinton Administration said very little about it. When the amendment was being considered in the Senate the year before, the White House had issued a cautionary statement about the proposed legislation, urging lawmakers to move slowly on it but stopping short of opposing it outright. "The President thinks that this issue deserves thoughtful discussion," a spokeswoman told the press. "The administration abhors obscenity, in whatever form it is transmitted," she explained. But "there are important First Amendment issues that need to be addressed before legislation is rushed through. We ought to have a serious approach—such as hearings—to find the best solution."[34] The entire matter had put the administration in a difficult position. In a crucial reelection year, President Clinton could not openly endorse a proposal that raised such critical constitutional issues. Nor did the White House want to align itself with right-wing proposals. Supporting the law also would alienate the online industry, which had given financial support to Clinton in his 1992 election.[35]

Cyberporn was just part of a host of "family value" issues that began to take center stage in the 1996 presidential campaign. As the Christian Right continued to keep its political agenda in the limelight, polls suggested broad public support for many of the issues it raised. For example, a *USA Today* survey found that 73 percent of respondents agreed about the importance of "family values," which included "strengthening the family and encouraging young people to abstain from sex until marriage."[36] In his bid for the presidency, Republican Senator Bob Dole inserted media and family values into the campaign, attacking Hollywood for "offering nightmares of depravity."[37] Some press reports suggested that there was internal conflict within the Clinton administration over how to deal with the Communications Decency Act. The *Wall Street Journal* reported that the Justice Department—which later would have to defend the CDA in court challenges—was worried even before the law passed that it was too broad, but noted that the agency had "avoided voicing its criticism for fear of being branded pro-pornography."[38]

On the same day that Clinton signed the CDA into law, a coalition of nonprofit and educational groups led by the ACLU filed suit in federal court to challenge the act's constitutionality.[39] The controversy was far from over.

A Simple Thing

While the president never mentioned the CDA in his Telecommunications Act signing speech, he spoke at length about the V-chip, one provision of the legislation mandating that all new television sets 13-inches and larger would have to be equipped with a new technological device to enable easy blocking of certain TV programs. The V-chip would be designed to work with a new television-program ratings system, to be developed voluntarily by the industry. "I am very proud that this new legislation includes the V-chip," he announced, displaying one of the tiny chips to the audience. "It's not such a big requirement," he explained, "but it can make a big difference in the lives of families all over America." In a strong appeal to morals and values, Clinton added, "I thank the Congress for reducing the chances that the hours spent in church or synagogue or in discussion around the dinner table about right and wrong and what can and cannot happen in the world will not be undone by unthinking hours in front of a television set . . . if every parent uses this chip wisely, it can become a powerful voice against teen violence, teen pregnancy, teen drug use, and for both learning and entertainment."[40]

The White House had played a particularly proactive, hands-on, public role in promoting the V-chip. The summer before, President Clinton had endorsed it at Vice President Al Gore's annual "Family Reunion" conference in Nashville before an audience of child advocacy, education, health, and family groups. Decrying what he called "indiscriminate sex and violence on television," the president had hailed the V-chip as a user-friendly,

First Amendment–friendly tool that families could use to protect their children from harmful television content. "It's a simple thing," he explained. "This is not censorship, this is parental responsibility."[41] Six months later, as Congress was about to pass its historic telecommunications legislation, Clinton inserted the V-chip into his State of the Union speech, making a special appeal to Congress to pass the requirement so that "parents will be able to assume more personal responsibility for their children's upbringing."[42] (The mention of the V-chip had prompted one industry trade reporter to comment on the "unusually specific telecom policy references in the speech.")[43]

By embracing the V-chip, Clinton had found a family-values issue that could work with his Democratic base while giving him ammunition in the battle over media culture that already had become a prominent part of the campaign. Criticizing the media was not new for Bill Clinton. Though he enjoyed close relationships with Hollywood, and had received strong industry support for his first presidential campaign, he had not been shy even in 1992 about calling the entertainment industry to task over its content. He commented in a press interview that "makers of films and TV programs could, without undermining their artistic integrity, have a major new impact on the way people view the world . . . there's no question the cumulative impact of this banalization of sex and violence in the popular culture is a net negative for America."[44]

In the 1996 campaign, the Clinton-Gore ticket could wield its V-chip victory as a powerful symbolic weapon in the fight for the hearts and minds of American voters. The White House worked with its own internal pollsters, who conducted surveys and focus groups to identify key voter segments and the issues that would resonate with them.[45] The polls identified "soccer moms" as a key constituency.[46] The message of "parental empowerment" was designed to resonate with these voters, and to address their frustrations and fears of living in a media-saturated world.

But despite the rhetoric, the V-chip was far from a "simple thing." The fight to pass the new requirement had been as polarized and contentious as the controversy over the CDA. The law itself was a complicated mix of government regulation and industry self-policing, creating a system that combined mandated hardware with voluntary software. The wording was purposefully constructed to walk a very thin legal tightrope, designed to avoid a slip into dangerous unconstitutional territory. Its passage had required a buy-in from several of the most powerful lobbying organizations in Washington that had spent decades resisting any government intervention into programming content. To secure the industry's approval, deals had to be struck, cards had to be traded. And even as the ink was drying on the new V-chip law, there was trouble brewing over its implementation.

Chipping away at TV Violence

As the country moved swiftly into the Digital Age—with all the visionary talk about the Information Superhighway—it seemed a little anachronistic for old debates about television to become such a prominent part of the policy deliberations for the future of media. Even more curious was that, for the first time in its fifty-year history, the television industry had been forced into a position it had so successfully avoided in the past. Political battles over TV violence dated as far back as the early 1950s, when Congress first took up the issue in hearings on juvenile delinquency. By the 1960s, social scientists in the growing field of mass communication had begun to accumulate studies documenting the nature and extent of the harms to children and youth posed by media violence. Over the years, several generations of academic experts testified before congressional committees, and dozens of bills were introduced aimed at reducing television violence, but none became law. In the private sector, boycott campaigns were waged against the TV industry by a wide spectrum of interest groups—from the National Parent Teacher Association to the American Medical Association to the National Federation for Decency. But while these efforts generated widespread press coverage, their impacts had been transitory. Under heightened pressure, the TV industry would reduce violent content, but as soon as the pressure lifted, programming would return to business as usual and violence would rise to its previous levels.[47] After decades of congressional debate, millions of dollars in research, and repeated industry promises of self-regulation, no workable remedy had emerged for the long-standing problem of television violence.

The violence issue had so dominated the research, public debate, and policymaking about television that it overshadowed and obscured many of the deeper structural and policy developments that were having a far-reaching impact on the quality of the media culture.[48] Throughout the 1980s, with dozens of new cable channels competing feverishly for viewers' attention, television continued to push the envelope of acceptable content, triggering greater public concern. Yet during this same period, industry mechanisms that had successfully deflected outside criticism in the past either were weakened or eliminated altogether. The National Association of Broadcasters' "code of good practice" was dismantled after the courts found its restrictions on advertising time to be a restraint of trade.[49] During the Reagan Administration, the FCC had scrapped many of the public-interest rules for broadcast television. Corporate takeovers and media mergers during this deregulatory era had triggered a wave of cutbacks in TV network standards and practices departments, whose job had been to maintain programming standards and deflect outside criticism.[50]

All of these changes prompted outcries from politicians, parents, and advocacy groups. Action for Children's Television and a coalition of education and health organizations went to Congress in the 1980s to restore some of the children's television requirements that the FCC had eliminated, and they successfully passed a new law in 1990 that required TV broadcasters to air educational programming for children as a condition of license renewal. That same year, Congress passed the first legislation aimed at addressing the problem of TV violence. Introduced by Sen. Paul Simon (D-IL), the law suspended antitrust provisions for the television industry for three years. Its intent was to enable (and encourage) industry leaders to join together to work out a self-regulatory plan for curbing excessive violence in their programming.[51]

There was much skepticism that such a waiver would prompt the industry, which had become more and more competitive, to work together to draft a single set of voluntary guidelines on television violence. As *Electronic Media* reported on the passage of the new law, "few believe such an agreement will be forged."[52] Many in the television business hoped in vain that the law and its deadline soon would be forgotten.

By the time the law was scheduled to sunset in 1993, lawmakers launched a new spate of congressional hearings on television violence. The TV industry tried several desperate tactics to hold off further legislative action. In July, the four broadcast networks announced a new system of "advisories" to warn parents of violence in an upcoming program. But this move failed to satisfy anyone. Producers called it censorship. Conservative parents' groups, such as Americans for Responsible Television, called the advisories a joke.[53] And when network executives were summoned to hearings on Capitol Hill, they were met with highly skeptical lawmakers. The networks did not "deserve one kind word" for their advisory system, commented Rep. Jack Bryant (D-TX). Instead, industry leaders should be condemned for "sitting on their hands for three years."[54] The following month, at a high-profile television-industry summit in Los Angeles, Senator Simon warned that if the TV leadership did not do something within sixty days, Congress would legislate.[55] Almost immediately, several new anti–TV violence bills had been introduced.[56]

Amid the various congressional proposals was a novel idea for using technology to solve the problem of television violence. Tim Collings, an engineering professor at Simon Fraser University in Canada, had just unveiled a new invention he called the "V-chip," which stood for "ViewControl chip."[57] Through this tiny electronic device installed in television sets, parents could block out programs with objectionable content. The system worked on the same principle as closed captioning, using the vertical blanking interval in the television signal to send and receive a special code in the programming that indicated the show's score according to a simple numerical rating system for violence, sex, and language. The V-chip would read the rating and shut off the television program if the score fell outside the range the parents approved.[58] The chip had been embraced already by policymakers in Canada, where a series of recent violent crimes had prompted a widespread public debate over violence in the media.[59] Keith Spicer, chairman of the Canadian Radio-Television and Telecommunications Commission (the Canadian counterpart to the United States's FCC), had been making the rounds in both Hollywood and Washington to promote the "Canadian solution" as a "workable compromise between the extremes of draconian regulation and anything-goes permissiveness."[60]

The V-chip was perfectly suited to the politics of the digital era. It was a new tool for overworked, stressed-out parents who did not have time to monitor or control every minute of their children's viewing. It also was less likely than other legislative remedies to run afoul of the U.S. Constitution. Rather than the government passing laws to restrict the time periods when TV violence could be shown, the V-chip would impose no government constraints on the programming itself. Instead, parents would be able to choose which programs were acceptable (or not) for their own families. When Congressman Ed Markey (D-MA), chair of the powerful House Telecommunications Subcommittee, introduced the first V-chip legislation, he told the press that parents "will be given the power to send a message directly to the industry. The government will not be involved."[61]

But implementation was a little more complicated than that. For the system to work, someone would have to rate all the programs on television. This was not something the government could do. And the television industry still had no intention of developing a ratings system. It was a small task to slap a few advisories on the most violent shows, quite another to evaluate and assign a rating to every TV show produced and broadcast. Though pay cable channels such as HBO already put ratings on their programs, other sectors of the industry remained adamantly opposed to a uniform rating system. Broadcasters were particularly vocal, raising First Amendment and economic concerns. Any attempts to rate and block programs could seriously undermine their revenue from advertisers, they warned legislators.[62]

As one of the most powerful lobbying forces in Washington, the National Association of Broadcasters (NAB) enjoyed particularly privileged access to congressional leaders, federal regulatory agencies, and the White House.[63] Because it represented local media in every congressional district, the NAB could wield more than monetary influence. One of its key lobbying tactics was to bring the station managers from a congressperson's district to meet with the lawmaker to plead in person regarding whatever particular case was on the table. The encounter carried an unspoken threat that failure to cooperate could result in lack of TV coverage for the politician.[64] Broadcasters also had other important business before congress in the mid-1990s. As debates were beginning on the telecommunications-reform legislation, the TV industry was seeking relaxation of the ownership rules that had restricted the number of stations a single company could own, as well as requesting longer license periods. Broadcasters also wanted assurances that they would be awarded valuable parts of the electromagnetic spectrum. Concerned that the FCC might permit competitors to apply for digital licenses, the NAB was pressing Congress to award this spectrum exclusively to "incumbent" broadcasters, arguing that they needed it in order to make the transition to high-definition television. But the industry already was making other plans for this digital real estate, whose value was estimated to be as high as $70 billion. New technological processes made it possible for a small portion of the new spectrum to be sufficient for digital TV broadcasting, which would allow the remainder to be diverted for a number of other potentially lucrative uses, such as data transmission and multicasting. Thus "spectrum flexibility" became the new buzzword.[65]

By 1993, the call for a V-chip was coming not only from congressional leaders in the Democratically controlled Congress, but also from high-level officials in the Clinton Administration. On the Hill, the issue quickly became enmeshed in the debates over the telecommunications reform legislation. Congressman Markey warned broadcasters that if they wanted the digital spectrum, they would need to accept the V-chip. "It is difficult for broadcasters to claim that they will use the new spectrum for the public interest," he charged, "when they are unwilling to use a scintilla of spectrum for the V-chip."[66] Framing the issue as a public-health problem, U.S. Surgeon General Jocelyn Elders told a congressional committee that a rating system would be an important ingredient in an overall violence-prevention plan for the country, since some economically disadvantaged children watched as many as thirteen hours a day of television, much of it violent fare.[67] Approaching violence as a crime issue, Attorney General Janet Reno informed another panel of lawmakers that it was constitutionally permissible to regulate violence on television, sending a strong message to the industry to develop a meaningful ratings system to avert government action.[68] During his confirmation hearings in fall 1993, Clinton-appointed FCC Chair Reed Hundt promised the Senate Commerce Committee that one of the commission's top priorities would be television violence.[69]

When the Democrats lost both houses of Congress in 1994, there was considerable speculation in Washington about how this sudden reversal of political fortune might affect government initiatives to regulate TV violence. *Washington Post* reporter Ellen Edwards observed that many people within the industry were anxious that congressional efforts would broaden beyond concerns over violence into areas of sexual content and language. While some people speculated that "the V-chip may be history," others were concerned that it might become "the S-chip" (for sex) instead. As one industry insider put it: "When Republicans talk about violence they also talk about sex. 'Why can't we get that filth off TV?' They don't distinguish between the two."[70]

Over the next two years, as the telecommunications bill worked its way through Congress, the tandem issues of TV violence and Internet indecency generated much controversy and coverage in the mainstream press. Debate over the digital spectrum, however, was confined largely to the trade press and inside-the-Beltway discussions among policy wonks. Media scholar Jim Snider documented the dearth of television news coverage of the issue along with a lack of serious treatment of other topics of interest to broadcasters, such as the provisions to relax ownership rules.[71] When the bill passed the Senate in summer 1995, the trade publication *Communications Daily* published an unusually strong editorial:

> It seems to the Senate that studios, networks and cable operators are so cavalier about the threat of television violence that the V-chip is the only solution. But, far removed from the violence issue, the legislators pushed through a telecommunications bill that would allow that same small cluster of corporations to create a virtual media monopoly. . . . Not surprisingly, news of the V-chip got reasonably good play

in the nation's press and on the television networks. Not surprisingly, the telecommunications act rewrite, which passed the Senate Thursday and could radically change the landscape of U.S. communications, did not.

The entire legislative process had "happened without much of the public's knowledge," the editorial continued:

The race toward industry deregulation is one of those supposedly welcome signs these days the government bureaucrats are letting the private sector flower. Violence in the media, on the other hand, is in the spotlight. In survey after survey, Americans claim to be mightily upset about it. So are senators and representatives. And so, the government is there to help. It is guarding our morality at the same time that, more quietly, it has sold its own.[72]

By the time the bill became law in 1996, broadcasters had received much of what they had asked for, including official guarantees from Congress that existing broadcasters would be exclusive recipients of the digital spectrum, with assurances of flexibility. But they also got the V-chip. Last-minute pressure from the White House, and from key Congress members, played a clear role in convincing industry leaders to finally agree to accept the controversial device and to develop program-rating system.

In the complex negotiations over the legislation, amid various kinds of deal-making efforts between Republicans and Democrats as well as alliances across the aisles, what emerged was a slightly morphed version of the V-chip in line with what some had augured in the wake of the 1994 election. The new mandate called for a technology and accompanying rating system designed to block out not only violence but also sex and any other "indecent material."[73] As legal scholar Monroe Price notes, the law "was carefully crafted to respect an imaginary line between unconstitutional coercion and accepted coerced voluntariness."[74] While it technically did not require the industry to rate its programs, if programmers chose to do so, then the ratings signal would have to be transmitted. Any guidelines the industry developed would be submitted to the FCC, which was required to consult with "public interest groups and individuals in the private sector," in order to determine whether the guidelines were acceptable.[75] If the industry did not come up with an acceptable ratings system within one year, the FCC would convene an advisory group to develop one.[76]

The final Telecommunications Act was nothing less than a complete restructuring of telecommunications industries. Under the mantra of "competition," the law essentially was a deregulatory policy. Among other things, the law removed government controls of cable rates and lifted long-standing ownership limits in television and radio that had been designed to prevent concentration of media control in a democratic society.[77]

Several prominent media critics were quick to pounce on the V-chip legislation. "In the annals of dumb solutions to serious problems, history will have a ball with the V-chip," observed *New York Times* columnist Frank Rich:

> Far from making television safer for children, the V-chip will merely postpone and confuse the issue until well into the next century—even as it provides politicians with convenient cover. . . . If you look at the bigger picture, this law is also about a mammoth expansion of mass culture—more media, more outlets—and a rapid expansion of power for the handful of mega-corporations that control it all, from TV, movies, music and publishing to both print and electronic news. It was perfectly symbolic that on the day Mr. Clinton signed the bill, Disney got its official Federal approval to swallow up ABC. Into this vast new universe of omnipotent media goliaths comes the tiny V-chip, designed to help parents block the coarse outpourings of an exploding digital universe. Common sense alone dooms this gizmo to failure.[78]

Tom Shales, TV critic for the *Washington Post*, was similarly derisive of the legislation. "Perhaps it can be thought of as a television condom," he quipped. "Its intended goal, after all, is safe sets." Writing of the political deal-making that had produced the controversial amendment to the bill, he noted that "[M]any industry observers think the ratings and the chip constitute a virtual quid pro quo in exchange for huge new chunks of the spectrum at virtually no cost." He also warned that the ratings would have a harmful effect on quality and creative expression in television, concluding that "a skeptical citizen might well wonder whether there's really anything on television as frightening and obscene as the latest crop of proposals to clean it up."[79]

On February 28, 1996, just a few weeks after the Telecommunications Act became law, a White House summit with representatives of the major production studios, cable companies, and broadcast networks, convened to devise the required rating system to work with the V-chip. Though press reports noted that the industry representatives looked a little stressed after coming up with a "voluntary" agreement, the president was buoyant. "We're handing the TV remote control back to America's parents so that they can pass on their values and protect their children," he announced. The new system was to be developed and implemented by the Motion Picture Association of America (MPAA), modeled on the movie industry's own rating system, which had been in place for twenty-seven years.[80]

The sequence of events that led to this moment was becoming a familiar pattern on Capitol Hill. In the 1980s, the music business found itself under public and government pressure and finally resorted to a system of parental-advisory stickers in order to placate parents' groups and Congress.[81] (See chapter 7 for a detailed account.) More recently, video-game producers had been on the Congressional hot seat. In 1993, some of the same lawmakers who were waging war against TV violence had taken aim at the growing $5.5 billion video-game industry.[82] In ritualized, public shaming events, executives were summoned to testify and called to account for their actions, as gory scenes from *Mortal Kombat* and *Night Trap* were screened for national TV audiences. Facing impending government legislation, the industry had agreed to begin rating its own products. By the following year, two separate rating systems had been created. A new Entertainment Software Rating Board regulated console video games. Its age-based scheme was based loosely on the MPAA movie ratings. The computer industry came up with its own "content-based" system, developed by the Recreational Software Advisory Council.[83] These divergent approaches to labeling were later at the heart of the debates over both Internet and television content.

As part of the legislative agreement on TV ratings, the industry was to hold a series of meetings with industry groups, academic experts, and health professionals as it developed the new system for labeling programs. The entire process was rife with problems and complications. Though the major trade associations had agreed to the legislation, many Hollywood writers and producers remained strongly opposed to having to rate programs at all. Dick Wolf, creator of such award-winning crime drama series as *Law and Order* and *New York Undercover*, launched a personal crusade against the chip, warning that ratings would dumb down television and prompt advertisers to pull their commercials from some of the most successful prime-time shows. "Who's kidding whom?" Wolf exclaimed in the press. "Anything stronger than a G-rating will be an all-points bulletin for politicians, special interest groups, pundits, academic grant seekers, critics, and navel lint gazers of every stripe."[84] Richard Cotton, general counsel at NBC, the network most critical of the ratings, expressed similar concerns: "We're moving toward a world in which the only thing that will pass muster on TV is 'Sesame Street.'"[85] By fall, the *Los Angeles Times* reported that the television industry had broken into "squabbling factions." MPAA President Jack Valenti portrayed the TV business as "a mutually antagonistic, factious industry where everybody wakes up hoping everybody else has failed the night before." With all the grumbling and suspicion, he feared there was a "real danger that no single, unified ratings system would emerge, tempting FCC bureaucrats to step in."[86]

Unhappiness with the V-chip ratings came from outside the industry as well. The ACLU opposed it on First Amendment grounds, but most of the group's efforts were taken up with challenging the Communications Decency Act. Conservative media critics were skeptical of the idea of parental blocking technologies, worrying that the labels might just give cover to producers who wanted to push the envelope on sex and violence. The American Family Association's Donald Wildmon, who had waged a number of high-profile boycott campaigns against network television, told the *Washington Post* that he hated the V-chip because it "institutionalizes violence."[87]

Even the child-advocacy groups that had been urging lawmakers for years to pass TV-violence legislation were becoming increasingly dissatisfied. In supporting the V-chip, organizations such as the National PTA and the American Psychological Association (APA) had sought a system similar to the one that V-chip inventor Tim Collings had proposed for Canada, with indicators for levels of violence, sex, and language. Surveys had shown that parents found "content ratings," such as the one already in use by HBO, to be preferable to "age based" labels like the MPAA's movie ratings.[88]

By summer 1996, it was clear that the V-chip—that "simple thing" that President Clinton had hailed only a few months before—was embroiled in controversy. Valenti told a TV-industry meeting that he expected the new ratings plan to be "pilloried and assaulted from all sides," warning that if Congress tried to step in and meddle with it, "we'll be in court in a minute" to challenge the constitutionality of the law.[89] Sometime during this process, an aide to the FCC chairman approached the Center for Media Education and urged the group to organize a political intervention before the whole matter ended up in the hands of the FCC. This was uncharted constitutional territory into which the government agency did not want to venture.[90]

Founded in 1991, CME had become the leading organization waging media-policy battles in Washington on behalf of children. It had just completed a four-year campaign to strengthen the FCC's rules on the 1990 Children's Television Act. Leading a large coalition of child-advocacy, health, and education groups, the center had negotiated with FCC staff, the White House, and the broadcasting industry on the final agreement on children's educational programming requirements, which were set to take effect that fall. The group's role regarding the TV violence issue, however, had been minimal. Although it had participated in some of the meetings with congressional leaders and other advocates prior to passage of the V-chip legislation, for the most part CME had remained on the sidelines during the early stages of this debate.[91]

In early December 1996, Valenti conducted one of his last consultative meetings, this one at the NAB's Washington headquarters, where representatives from more than a dozen educational institutions and nonprofit advocacy groups had been invited. Though the meeting ostensibly was called to solicit "input" from the groups, a story leaked to the *Washington Post* ran that same morning, revealing that the industry already had come up with its final plan for the new, age-based system.[92] When confronted with the article, Valenti vehemently denied that the system had been finalized, insisting that he was there to listen to ideas from the participants. But from his remarks during the remainder of the meeting, it was obvious that the industry had no intention of changing its plans. Though he referred to the industry proposal as "content-driven ratings," the *Post* article made it clear that they were only a slight alteration of current movie ratings (TV-G, TV-PG, TV-14, etc.). The MPAA had made some modifications for television, adding TV-MA for "mature audiences" and TV-Y for children's programs. A special category had been created to label violent children's shows (TV-Y7, for children 7 years and older), but the industry had gone out of its way to avoid any explicit reference to violence in any of the designations. When participants asked Valenti about labeling content, he adamantly pronounced that content labels were "off the table."[93]

With the new guidelines scheduled for release a few weeks later, Valenti invited the groups to endorse the ratings, promising them "free advertising" in the brochure.[94] But the individuals he had assembled that morning were not happy. Following the meeting, the Center for Media Education and the National PTA began organizing a protest against the proposed guidelines, urging their colleagues at other organizations to refuse their support and instead sign an open letter to Valenti opposing the age-based system and calling for a rating system that would rate the content in the individual programs. On the day of the industry's press conference announcing its new ratings guidelines, the advocacy coalition held a "counter press conference" in another room at the National Press Club, with Congressman Markey and other influential lawmakers present. "These groups have every right to be heard, but they have no way of forcing the industry to listen," Markey told reporters. "Parents have been cast in the role of Oliver at the orphanage. If parents criticize the industry for serving this gruel of a ratings system they are told they are ungrateful. If they have the temerity to ask for 'More,' they are told to sit down and shut up."[95]

By the time the FCC approval process was underway in early 1997, the controversy over the V-chip was raging in the press. Pressures continued to rise for the industry, with opposition coming from multiple directions. More than two dozen major newspapers ran editorials against the age-based ratings.[96] In Congress, lawmakers dusted off old antiviolence bills that were languishing on their shelves and readied them for reintroduction. At a Senate hearing in February, Senator Dan Coats (R-IN) announced that he planned to introduce a bill to prohibit the FCC from granting or renewing TV-broadcast licenses unless the applicant demonstrated a plan for implementing a "program-specific, content-based ratings system." Senator Kent Conrad (D-ND), who had worked closely with advocacy groups to develop legislation on TV violence, threatened to introduce a new resolution clarifying congressional intent on the V-chip law. Former Senate Commerce Committee Chair Fritz Hollings, a powerful Democrat from South Carolina, teamed up with colleagues with plans to reintroduce his "safe harbor" legislation, which would restrict violent programming during the "family hours," when children were likely to be watching. Frustrated, Jack Valenti told a Congressional committee that he was "bewildered by the ferocity" of the criticism.[97]

By April, the parents' groups had formally urged the FCC to reject the proposed ratings, arguing that the system was purposefully vague and failed to tell parents anything about the content of TV programs.[98] Congressman Markey, along with fifteen other representatives and seven senators from both sides of the aisle, sent a letter to the FCC saying that the system was unacceptable.[99] A month later, the Hollings safe harbor bill had cleared the Senate Commerce Committee. Committee Chairman John McCain (R-AZ)—who had earlier opposed the bill on constitutional grounds—decided to support it this time because the law would only impose a time-channeling requirement if the industry refused to develop a content-based rating system. The V-chip technology, he explained, "puts parents, not the government, in the rightful role of determining what children will watch."[100]

By early summer, the FCC had extended the deadline on its review process, hoping for some resolution of the conflict. Industry leaders had begun to soften their positions somewhat and were making public comments indicating a willingness to discuss some adjustments in their system, which Valenti now was calling a "work in progress." Sometime in June,

Congressman Markey called together representatives from some key advocacy groups to inform them that industry representatives were willing to begin meetings to discuss possible changes in the proposed system. However, he made it clear that persuading the industry to scrap the system altogether was impossible; rather, some modifications could be made in the form of additional "content descriptors."[101]

In an unprecedented development, the heads of the three major TV-industry lobbying organizations—the National Association of Broadcasting, the National Cable Television Association, and the Motion Picture Association of America—agreed to sit down with representatives of ten advocacy groups to work out a compromise.[102] Throughout the month-long negotiations, lawmakers continued their threats to legislate, forcing industry representatives to remain at the table. The amended "TV Parental Guidelines" finally were unveiled at a White House event on July 10, 1997, and the new content labels began appearing on television programs later that fall.[103]

Although the public-interest advocates succeeded in getting content descriptors added to the age-based ratings, what finally emerged from this highly politicized process was a confusing set of labels, probably only fully understood by the fourteen participants present at the negotiations. The written text of the agreement described each of the categories, but unwritten agreements also were embedded in the system that was finally adopted. Some of the highly technical issues had been argued over for weeks. In the middle of the process, another letter was tossed into the already-confusing alphabet soup. In addition to the "S" for sex, "L" for language, and "V" for violence, industry negotiators insisted on adding a "D" for sexual innuendo. It was implied that introducing this additional label might give NBC, which had refused to participate in the process, some incentive to agree to run the content descriptors. (The "D" label could be used for sitcoms such as *Friends*, which derived much of their humor from jokes about sex, without requiring the networks to use the more controversial "S.") But NBC never was persuaded to join the other networks in using the content descriptors, though it did use the simple age-based ratings. (Cable network Black Entertainment Television, BET, refused to participate in the ratings system altogether.)

The label for violent children's shows, "TV-Y7-FV," epitomized the convoluted and highly contentious negotiation process. When advocacy groups called for adding a "V" to children's program labels, the industry insisted on a modifier. After much haggling over which word to choose, advocates reluctantly agreed on "FV" for "fantasy violence." This purposefully ambiguous label caused confusion as the ratings system was implemented, with some parents and reporters thinking the letters stood for "family values."[104]

But no matter how flawed the final set of guidelines, the concept of technological solutions to "empower" parents to protect their children from the potential harms in an increasingly complex media world had caught on with policymakers. The development of the V-chip allowed Congress and the White House to declare victory in the cultural wars over television. Working out the details for how this seemingly simple remedy actually would function was left to the industry and the parents' groups.

Sometime during this controversy, regulators in Canada quietly decided to drop the content-based rating system that had served as the model for advocates in the United States.[105] As one press report noted, the Canadian authorities realized that "a system that isn't compatible with the U.S. system would be a logistical nightmare because three-fourths of Canada's prime time lineup comes from the U.S."[106]

A Family-Friendly Internet

Just as V-chip negotiations were getting underway, the U.S. Supreme Court declared the controversial Communications Decency Act unconstitutional.[107] Civil-liberty groups were jubilant. "Today's opinion defines the First Amendment for the next century," declared EPIC legal counsel David Sobel, who had served as cocounsel in the court case, Reno v. ACLU. "The Court has written on a clean slate and established the fundamental principles that will govern free speech issues for the electronic age."[108]

Anticipating the decision, the White House already had begun formulating a new official policy for handling Internet indecency, modeled on the notion of "parental empowerment tools" the administration had used for television. "We can and must develop a solution for the Internet that is as powerful for the computer as the V-chip will be for television," President Clinton told the press. "With the right technology and rating systems—we can help ensure that our children don't end up in the red light districts of cyberspace."[109]

This time, though, instead of government regulation, the Clinton Administration was taking a clear hands-off approach to regulating content on the Internet. Part of the reason was the Supreme Court decision itself, which had pronounced the Internet a uniquely open and democratic medium, unlike any that had come before, and thus deserving of the highest First Amendment protections.[110] But this new stance on Internet regulation was part of a broader administration policy designed to allow the growth of e-commerce to proceed unfettered by government involvement. Online sales in 1995 had been $200 million in the United States alone and were expected to soar to more than $1 trillion by 2010.[111] The White House had just released a "Framework for Global Electronic Commerce," outlining the government's official policy regarding a wide range of Internet issues, including copyright protection, tariffs, trade, and content regulation.

President Clinton, a self-described "technophobe," told the press that he was getting more into the Net since his daughter Chelsea had gone off to college. But the president made it clear that this free-market approach to government policy on the Internet "does not mean indifference when it comes to protecting children." Within a few weeks he began meeting with parents' groups, industry leaders, and consumer organizations to develop "an equivalent to the V-chip" for the Internet.[112] In a highly publicized White House event on July 15, 1997, the president and the vice president stood on the dais with industry leaders and parents' group representatives to unveil a "virtual toolbox" that would "empower" parents to make the Internet safe for their children. "Our message to parents is clear," America Online president Steve Case announced. "You don't let your kids take a trip in a car without a safety belt. You shouldn't let your kids travel in cyberspace without this Internet toolbox." Vice President Gore assured parents that the tools were user-friendly, "much easier than programming your VCRs."[113]

At the heart of the toolbox strategy was a whole array of new filtering software and blocking technologies that had sprung up amid the ongoing debate over Internet indecency. During deliberations over the Telecommunications Act, several enterprising companies stepped up to deliver technological solutions to the problem of online indecency. SurfWatch, one of the first products on the market, debuted in 1995. Developed by a Silicon Valley company, the software promised to "filter out" any inappropriate cyberspace content by matching Internet addresses with a list of forbidden Web sites. "This is the kind of software that can offer the individual choice as opposed to censorship," Surf Watch's vice president Jay Friendland told *Newsweek* magazine.[114] Other companies followed suit with similar products. Their names—NetNanny, CyberPatrol, CyberSitter, SafeSurf, etc.—carried with them the promise of security and protection, the ideal high-tech tool for today's harried, overworked parents. As promoters demonstrated the new devices at congressional hearings, conferences, and trade shows, online industry leaders and government policymakers embraced them.[115]

Most were software packages for parents to install on their home computers. Ads for the products began to appear on the Internet and in parents' magazines, each company claiming to have the most effective features for screening out harmful content. The systems operated on a keyword basis, equipped with long lists of objectionable terms that, when encountered on the Web, would block the user's attempt to enter a site. It quickly became clear that none of these products was a perfect solution. News accounts turned up embarrassing problems with the screening systems, some of which blocked out informational Web sites on breast cancer or kept students from doing legitimate research online. The programming codes developed to block content often were shrouded in secrecy, prompting some to charge companies with promoting ideological agendas in their choice of forbidden Web sites.[116] For example, CyberSitter marketed itself as a protection against Web-based pornography, obscenity, gratuitous violence, hate speech, and criminal activity. But, as the *Toronto Star* pointed out, "an increasing number of investigative Net journalists also claim CyberSitter, without fanfare, blocks access to Web sites based on political criteria." One online magazine reported that CyberSitter had blocked entry to the National Organization for Women's Web site.[117]

Bennett Haselton, an 18-year-old college student at Vanderbilt University, decided to wage an online campaign against the filtering companies. At the time of the court challenge to the Communications Decency Act, he later recalled, "no one was representing youth on this issue."[118] A computer whiz and math major who had entered college at the age of 16, Haselton created his own Web site, PeaceFire, whose purpose was to expose the inner workings of filtering software and instruct children and teens on methods for dismantling the tools. Hazelton's campaign was one of the early notable efforts by young people to seize the Internet as a powerful political weapon.[119] He used his site to publicize the secret list of forbidden Web sites blocked by CyberSitter, leaking the information to reporters at *Wired.com*. The story's appearance on the Web generated more press coverage, prompting the company that manufactured CyberSitter to threaten a lawsuit against Haselton, charging that he had "engaged in illegal criminal copyright violations to further his juvenile teenaged political agenda, and reduce the effectiveness of our product." The company also tried to pressure PeaceFire's ISP to shut down the site, though it never followed through on its threat to sue, nor was it able to force PeaceFire off the Web (although it did add the site to its own list of blocked Internet addresses).[120]

The groups that had challenged the CDA in the courts had based their case, in part, on the emergence of these new filtering software products. One of the core arguments in the court challenge was that these new tools would enable parents to protect their own children from harmful Internet content, thus making a law restricting such content not only unconstitutional, but also unnecessary.[121] But even as the civil-liberties lawyers prepared for the case against CDA, some had begun to have serious doubts about the legal and political implications of offering filtering software and blocking technologies as a defense. As Marjorie Heins recalled, it soon "became clear that this particular 'less burdensome alternative' entailed massive censorship risks of its own." Though the argument always was that using filters was voluntary, she noted, "as the preliminary injunction hearing in Reno approached, the major corporations that increasingly controlled Internet access were already beginning to introduce rating and blocking systems directly onto computer browsers."[122]

In the wake of the Supreme Court decision, the White House, along with some of the leading Internet companies, prepared to promote the widespread adoption of this practice. In a closed-door meeting on the day of the July 1997 White House Summit, President Clinton solicited and received the cooperation of five Internet companies—CNET, Excite, Infoseek, Lycos, and Yahoo!—that agreed to the idea of incorporating filters directly into their search engines and browsers. The companies released a statement at the Summit announcing this plan for a "family friendly" Internet.[123]

The goal of reigning in and rendering "family friendly" such a huge, uncontrollable global network as the Internet—with more than a million sites—may have seemed a bit unrealistic. But already new Internet ratings schemes and technologies were being developed to make the daunting task manageable. Net Shepherd Family Search launched a search engine "designed to make the Internet a friendlier, more productive place for families . . . through filtering out Web sites judged by an independent panel of demographically appropriate Internet users, to be inappropriate and/or objectionable to average user families."[124]

The Recreational Software Advisory Council (RSAC) offered a plan for individual Web sites to rate themselves. As the CDA controversy grew, RSAC launched a spin-off ratings system, "RSACi," designed for the World Wide Web. Unlike software products, which imposed outside criteria onto the Web, RSACi labels were generated from Web site operators themselves.[125] With icons that looked like a group of thermometers, each representing a particular kind of content, the "temperature" would reveal the amount of that type of content in the Web site.[126] The system was designed to allow parents to choose the levels of sex, language, or violent content that was appropriate for their own children, and then to program their browsers to read the icons. To help browsers read online ratings, the World Wide Web Consortium (W3C), the key organization developing standards for the Web, introduced a new Platform for Internet Content Selection (PICS), a rather elaborate system designed to read either self-ratings like RSAC or "third party" labels.[127]

The momentum toward comprehensive Internet filtering systems raised alarm bells for civil-liberty groups. Labeling these schemes "censorware," EPIC, the ACLU, and other groups began to voice their strong opposition to the White House's and industry proposals. Barry Steinhardt, associate director of the ACLU, argued in a public statement that systems like RSACi and PICS threatened "to change the very architecture of the Internet to incorporate universal content ratings. . . . Linked together, the various schemes for rating and blocking could create a regime of private 'voluntary' censorship that puts what the Supreme Court called 'the most participatory form of mass speech yet developed' at risk." Painting an ominous scenario if such plans were put into place, he predicted that "What we are likely to see is some self-rating and a handful of third-party systems all essentially reflecting the same moral construct. These few ratings schemes will come to dominate the market and become the de facto standard for the Internet." With increasing pressure from the government, they will be "built into Internet software as a default," threatening to block any speech that has not been rated.[128]

Despite these concerns, the White House moved forward swiftly with an initiative for large-scale promotion of its toolbox. Plans were underway by midsummer 1997 for a White House Summit on Children and the Internet. Organizers of the upcoming event comprised a rather strange mix of nonprofit groups and corporations, including companies such as America Online and Disney Online, which already were investing heavily in children's Internet content, as well as several filtering software vendors such as Microsystems Software, NetNanny and SurfWatch. America Online played a particularly prominent role, working closely with nonprofit groups from both sides of the CDA legislative and court battles. The National Law Center and the antiporn group Enough is Enough, which both had lobbied strongly for the CDA, were part of the steering committee, working alongside organizations such as the Center for Democracy and

Technology (CDT) and the ALA, which had fought the law in the courts. A strong proponent of filtering software, CDT worked closely with the Internet industry to promote the adoption of such systems.[129]

The coalition of child advocacy, health, and education groups that had participated in the TV-ratings negotiations—which included the National PTA, the National Education Association (NEA), the American Psychological Association (APA), the Children's Defense Fund, and CME—was absent from the early planning of the White House Summit. The groups had been invited to participate but had expressed reservations about the narrow "child safety" focus of the event. While most of these groups, with their large constituency of parents, shared the concern about shielding children from harmful online content, they also believed that a high-profile media event focussing only on this issue would divert public attention away from a host of other critical areas that would have to be addressed if the Internet were to serve the needs of children and families. The NEA, for example, had helped lobby in the e-rate provision of the Telecommunications Act, which guaranteed affordable access to the Internet for schools and libraries, and remained committed to solving the problem of the "Digital Divide." The National PTA and CME already were involved in a campaign at the Federal Trade Commission to ensure safeguards against excessive commercialism and invasion of children's privacy in cyberspace. (Chapter 4 is a case study of that campaign.)

For the online industry, however, the child safety issue was the most urgent. Conservative groups already were working with members of Congress to introduce new versions of the Communications Decency Act, redesigned to pass constitutional muster. Republican Senator Dan Coats, allied with the same conservative groups that had supported the CDA, introduced new legislation that would require any commercial Web site containing material judged harmful to minors to keep children from accessing it or face criminal charges.[130]

By late summer, with preparations for the December event already well underway, CME and several of the largest mainstream organizations—including the NEA, APA, and National PTA—informed the White House that they would boycott the summit if their demands for a broader agenda were not met. Administration officials hastily organized a meeting with the summit's steering committee to discuss expanding the agenda. A few concessions were made, including a promise by the White House to hold three additional summits over the next two years in order to adequately address the range of issues related to children's use of the Internet.[131] After heated discussions and insistent prodding, planners also finally agreed to add a panel debating the pros and cons of filtering software and blocking technology, inviting some of the most vocal civil-liberty groups to present their objections to the Administration's "toolbox" proposal.

The summit was becoming a much more contentious affair than originally planned. Rather than a chorus of supporters promoting the joint industry–White House plan for a "family-friendly Internet," the event had sparked a cacophonous public debate, with a host of competing voices jockeying for attention in the press. "[T]oday dawned in the capital with the public-relations machinery already running full speed to churn out positions," Dan Brekke wrote in *Wired News* on the opening day of the summit. As America Online, Disney and other companies announced a public-education campaign to promote the use of Internet filters and ratings, civil-liberty groups attacked the initiatives. The Electronic Frontier Foundation (EFF), ACLU, American Society of Newspaper Editors, and seventeen other organizations announced the formation of a new "Internet Free Expression Alliance," renewing their efforts to fight any attempts to control online content.[132] EPIC, whose general counsel David Sobel reluctantly had accepted the invitation to speak at the event, released "Faulty Filters," a report documenting major flaws in some of the plans for search-engine filters.[133] The ACLU's Barry Steinhardt, also a reluctant participant in the summit, released a statement expressing "grave concerns about the so-called digital tool box that is the centerpiece of this event," adding that "the uncritical acceptance of many of these tools does a great disservice both to free expression and to the real interest of parents to choose the values they wish to impart to their children."[134] And conservative groups, the Family Research Council, the Christian Coalition, and Enough is Enough expressed skepticism about weak, self-policing measures for regulating cyberporn, vowing to continue lobbying for new laws.[135]

In the five months since it had been announced, the White House's original plan for a V-chip for the Internet faced significant challenges, forcing both the Clinton Administration and the online industry to devise other strategies for rendering the new online medium family-friendly. Though the analogy to television worked well as a sound bite, rating the Internet, with its international reach, exploding growth, and wide-open architecture, was rife with problems.[136]

The three-day event finally opened on December 1, 1997, under the banner "Internet Online Summit: Focus on Children." Held in a Washington hotel, the meeting attracted more than six hundred participants representing a diverse set of interests and agendas, from online companies, filtering-software vendors, government, and a spectrum of nonprofit groups, all watched closely by the press. Though most of the panel discussions went smoothly, the session on filtering and ratings flashed into heated debate. But these fault lines were overshadowed by the dominant theme of the conference, which clearly supported the idea of parental-empowerment tools for the Internet. An ongoing "technology tool display" in a hotel ballroom housed elaborate booths by software vendors, search engines, and others, all touting the benefits of such empowerment tools. Conference participants wandering into the room were greeted by enthusiastic sales representatives eager to demonstrate the unique features of their products.

Public-relations firm Fleishman Hillard produced a promotional video featuring the summit's headliners and issued daily press releases announcing dozens of new initiatives emanating from the event, all aimed at assuring the public that the Internet could be a safe place for families and their children. The U.S. Department of Education offered a "Parents' Guide to the Internet "with instructions on "how to safely navigate the Internet," as well as a glossary of Internet terms and advice on what families should consider when buying a computer or selecting an ISP. The nonprofit National Center for Missing and Exploited Children debuted a "Cybertips Line," billed as the "911 for the Internet," where "individuals can report illegal Internet activity related to child pornography and predation." Private industry and police teamed up for the "Law Enforcement Internet Safety Forum."[137] America Online promised a half-dozen new programs, including "a permanent parental control button" on the AOL Welcome screen, a "Neighborhood Watch area," and a "Notify AOL" alert button to "report inappropriate content or behavior directly to AOL's Community Action Team." At Disney, the Three Little Pigs made interactive cyber-safety public-service announcements aimed at parents and children. Time Warner drafted Batman and Wonder Woman to host a national teach-in, and it loaned out Fred Flintstone and Scooby-Doo for America Online's online-safety program for children.[138]

To ensure that the American public got the message, Vice President Gore announced the launch of a year-long national public-education campaign, "America Links Up." Using the slogan "Think, Then Link," the centerpiece of the effort was an "Internet Teach-In" coinciding with the start of the school year. It included town-hall meetings in schools, libraries, and community centers across the nation. PSAs ran on television, pointing viewers to an 800 number and a Web site where parents could download a toolkit and other resources for their families.[139]

The timing of the event turned out to be somewhat unfortunate for the Clinton Administration. The September 1998 launch of the "America Links Up" campaign and the back-to-school "Kids Online Week" took place only a few days after the release of the infamous Starr Report, which documented in sexually explicit detail the White House trysts between President Clinton and intern Monica Lewinsky. Commenting on this ironic twist of events, columnist Lawrence Magid wrote:

> About a year ago, the President of the United States met with leaders of the online industry to ask them to come up with voluntary procedures to protect children from inappropriate material in cyberspace. Little did the President know that he would be the subject of what will undoubtedly become the Internet's most widely read X-rated document. That document was first posted Friday, September 11th on Congress's Web site—the very body which, two years ago, passed a law that would have made criminals out of anyone who "makes, transmits, or otherwise makes available any comment, request, suggestions, proposal, image, or other communication which is obscene, lewd, lascivious, filthy, or indecent."[140]

In the final analysis, the orchestrated effort to promote a family-friendly Internet was unable to preempt further attempts to legislate access to "indecent" content in cyberspace. Within the next two years, new laws replaced the defunct Communications Decency Act. Both the 1998 Child Online Protection Act (COPA) and the 1999 Children's Internet Protection Act (CIPA) prompted further constitutional challenges and heated public debates that continued well into the first decade of the next century. Congress held more hearings. An official commission was set up through the National Academies of Science to work through some of the complicated challenges posed by the Internet indecency legislation.[141]

The official America Links Up campaign evolved into GetNetWise, an ongoing public education initiative sponsored by many of the industry and nonprofit participants in the first White House Summit. Its Web site contains a list of tools for families, featuring links to more than seventy filtering-software products, blocking technologies, and ratings services. The cornucopia of names on the drop-down menu reveals a dizzying array of products to suit every possible need and taste—from NetNanny and CyberPatrol to ChatWatch, Covenant Eyes, and CrayonCrawler.[142]

As for the V-chip, the complicated icons (e.g., TV-14, DSLV; TV-Y7, FV) continued to appear during the first fifteen seconds of most television programs. Not surprisingly, subsequent studies showed that parents remained confused about what these symbols meant.[143] Neither the broadcasters nor the TV manufacturers did much to educate the public about the V-chip or the ratings. Under pressure from the FCC and advocacy groups, broadcast and cable networks produced a handful of PSAs to explain the ratings to parents. Though these spots appeared from time to time—often in the middle of the night—they probably were seen more frequently in congressional and FCC hearings than they were on TV.[144] The television ratings did not homogenize content, as some in the industry had predicted; on the contrary, a new generation of even more provocative programming tested the boundaries of public acceptability as never before. According to several industry insiders, while labeling the programs caused little disruption or concern, no one in the business wanted to encourage parents to *use* the V-chip to actually block any programming.

In the years that followed, even in the midst of an exploding of digital culture, television would once again find itself at the center of a raging controversy over media content, and political expediency would force industry leaders to revisit their troubled relationship with the V-chip.

4 Web of Deception

As parents shuddered in horror of all the dangers lurking in cyberspace, many were comforted by a series of news articles that began appearing in the mid-1990s, highlighting "safe zones" for kids on the Internet. These Web sites, designed specifically for children, often touted educational benefits. "If you're tired of all the whining about dirty pictures on the Internet—and want to protect your children at the same time," wrote Joe Kilsheimer in the *New Orleans Times Picayune,* "it's time to go looking for kid-friendly cyber-playgrounds. In fact, there are a lot of sites on the World Wide Web that children will find fun and engaging."[1] One of these "safe" sites was KidsCom.

Launched in 1995, KidsCom billed itself as "a communications playground for kids age 4 to 15."[2] Taking advantage of the Internet's interactivity and the World Wide Web's compelling graphic interface, the site was filled with fun activities—educational games, chat, and even opportunities for kids to post their own stories on the Web. *USA Today* praised KidsCom, calling it "a great site for children to get some education while having fun."[3] Approved for classroom use by school districts across the country, the site won accolades for its educational content.[4] As KidsCom developer Jorian Clarke proudly explained to the press, "We work to make it an edutainment site, where kids can hopefully learn something while they're having fun."[5] As of January 1996, more than 15,000 children had registered at the site.[6]

But KidsCom was much more than a virtual playground. Probing the depths of the Web site by following the links to its parent company, SpectraCom, or consulting recent articles in the marketing trade literature revealed KidsCom's true purpose. The Web site was essentially an online market-research tool, designed to elicit a wealth of demographic, behavioral, and preference information from children. SpectraCom described itself as "a strategic planning, interactive on-line marketing, research and communications company located in Milwaukee, Wisconsin, and serving clients nationwide."[7] The company was at the forefront of online research, harnessing the power of the Internet to glean detailed information from one of the most valuable target markets in the Digital Age. One of its specialties was creating Web sites that could be used for marketing and market research, in order to provide its clients with a variety of services. These included surveys on "brand awareness" and "usage habits" as well as "database development detailing demographics and psychographics," "customer satisfaction and opinion studies," and "lifestyle monitoring for popular culture trends in areas such as music, television, film and celebrity assessments, food fashion, [and] life outlooks." SpectraCom also offered "tracking studies," "usage testing," and social-issue research on "race and ethnicity, education [and] cohort behavioral tracking."[8] The company served a prominent group of clients, including Blockbuster Video, Levi Strauss & Co., Pepsi Cola, and Kraft, Inc.[9]

SpectraCom's approach to Internet e-commerce was emblematic of the wave of new businesses seeking to cash in on the opportunities in this promising communications frontier. Its business model—conducting online market research that could be sold to other companies—was to become a staple in the new digital economy. Like many other commercial children's Web sites, KidsCom had created a variety of mechanisms to encourage kids to provide valuable information about themselves. In order to "play" on the KidsCom site, children had to register, which meant answering a detailed questionnaire that asked for such personal information as name, sex, birthday, grade, number of family members, e-mail address, city, state, and zip code. The mandatory registration also required children to provide the names of their favorite TV shows, commercials, and musical groups, as well as the names of the friends who had referred them to the KidsCom site. Children were questioned about their dreams and aspirations, what they would like to do when they grew up, in which part of the world they lived at the moment, and where they would like to live. Every blank in the survey had to be completed before a child was allowed to enter the KidsCom site.[10]

While the rapid growth of marketing and advertising on the Web was apparent to anyone in the industry, it was a well-kept secret to the rest of the world. Many parents still did not know what the World Wide Web was, let alone the nature of the new online commercial environment. With Congress in the final stages of debating the controversial Telecommunication Act, both policymakers and the public were preoccupied with fears about pornography, indecency, and other dangers to children on the

Internet. Hardly anyone was paying attention to the commercial Web sites for kids that were flooding the online landscape.

In 1995, the Center for Media Education began to investigate the emerging marketing practices targeted at children on the Web.[11] Over the next four years, the group would lead a coalition of child advocacy, health, and consumer organizations in a press and public-policy campaign aimed at establishing rules to govern children's marketing in the new digital media. The goal of forcing government intervention in the digital marketplace would come in direct conflict with the Clinton Administration's agenda for promoting e-commerce. But both the U.S. government and the online industry already were embroiled in a much larger, international public-policy debate that ultimately would help the children's advocates in their efforts. At the same time, attempts to curb commercialism in children's online media would be shaped by the intense public obsession with online safety. Though the final outcome of the campaign would be a narrower victory than the advocates wanted, it would have a significant impact on the emerging commercial practices targeted at children in the digital media.

Legacy of the "Kidvid Wars"

In many other countries, children's advertising had been strictly regulated, especially in television, where government-run broadcasting systems established restrictions and, in some cases, complete bans on TV advertising aimed at this vulnerable group of consumers.[12] In the United States, however, advertisers have enjoyed considerable freedom to target children, with little interference by the government. American broadcasting's emergence in the 1920s coincided with a period of rapid growth and power in the advertising industry. Commercials quickly became an accepted staple in American radio and subsequently in television. Throughout the twentieth century, powerful media and advertising-industry lobbies had weathered numerous attacks from public-interest and consumer groups. These organizations became part of the power structure in Washington politics, armed with a full arsenal of political weapons designed to deflect, disarm, and undermine outside criticism. The lobbyists were familiar personalities in Washington policymaking, marching into regulatory hearings armed with statistics and legal documents to defend the hundreds of billions of dollars that were at stake in the global advertising business. Like other powerful industry interests, the ad lobby enjoyed easy access to regulatory agencies, Congress, and the White House.[13]

When child-advocacy and consumer groups tried to convince the government to regulate television advertising to children during the 1970s, the battle became a quintessential case of Washington special-interest power politics. Drawing on a wealth of government-funded research that documented the effects of television advertising on children, Action for Children's Television and the Center for Science in the Public Interest petitioned the Federal Communications Commission and the Federal Trade Commission to take regulatory action.[14] Both agencies launched rulemaking procedures, proposing a range of restrictions as well as the possibility of an outright ban on advertising to young children.[15] But these regulatory moves came at a time when the business of children's television was well established, with a large infrastructure of industry stakeholders in place, a healthy market, and an eager following of child viewers. Attempts to regulate children's advertising aroused the ire of the television and advertising industries, including the large number of food, toy, and beverage companies that marketed to children. A press war erupted. The industry warned that without advertising revenue, all children's programming would dry up.

At the FCC, advocates achieved partial success. Despite forceful industry opposition, the commission passed a set of guidelines in 1974 to govern TV-advertising practices targeted at children 12 and younger.[16] But at the FTC, the ultimate outcome was disastrous. Industry lobbyists went directly to Congress to thwart the FTC's efforts, pushing through legislation that not only forced the commission to terminate the rulemaking procedure on children's television advertising, but also stripped the commission of its powers to develop broad rulemaking procedures on any issue. The press played a pivotal role in the controversy. One of the final events that helped to crush the FTC's rulemaking proceeding was the now-famous *Washington Post* editorial, accusing the commission of trying to be the "national nanny."[17]

Though the advocacy efforts in the 1970s were met with powerful resistance, they also left a legacy that became the groundwork for renewed activism in the 1990s. The FCC established a set of guidelines in 1974 for commercial advertising in children's TV programming. While the rules were weak and poorly enforced, they were based on a significant body of social-science research documenting the special vulnerabilities of children to the powerful appeals of marketers. Studies had shown, for example, that young children have difficulty distinguishing between programs and the commercials that surround and interrupt them. As a consequence, "separators" were added. These little messages—"we'll be right back" or "now we return to our show"—became familiar moments in the kidvid landscape, though subsequent research suggested they made little difference.[18] Research also showed that young children found it hard to resist persuasive product pitches by program hosts. Like the teachers and parents in children's lives, these authority figures—whether real-life or animated—wielded special powers to influence young people. So in kids' shows, the hosts were forbidden from pitching products to their young audiences. But the guidelines failed to stop the growth of toy-based programs like the *Mighty Morphin' Power Rangers*, which some argued were nothing more than program-length commercials themselves.[19] When the FCC tried to abandon these rules in a deregulatory sweep in the 1980s, ACT successfully sued the commission for their reinstatement. Subsequent passage of the 1990 Children's Television Act codified some of these guidelines, such as time limits on commercials.[20]

The policy battles over children's advertising in the 1970s also spawned a new self-regulatory body: the Children's Advertising Review Unit. CARU was a special division of the National Advertising Review Council, an industry agency established in 1974 in the midst of a growing consumer movement. CARU vowed to ensure "truthful, non-deceptive advertising to children under the age of 12," and to respond to public complaints against errant advertisers. It published a booklet of lofty-sounding principles and guidelines that was widely distributed to regulators, consumer groups, and the press. But most of the principles were so vague as to be unenforceable. Nor did CARU have any actual power to crack down on violations.[21]

But CARU's primary purpose was to quell criticism and preempt any further attempts at government regulation. And it did that very well. The group established an advisory board that included some of the same experts whose research on the harmful effects of television advertising had been used in the FTC's rulemaking procedure. It sponsored high-profile conferences, inviting consumer groups, academics, and the media to debate pressing issues of the day.[22] When trouble brewed in Washington, CARU staff could be flown in to testify before congressional committees or regulatory agencies, assuring policymakers that the industry could police itself.

The most significant legacy of the 1970s policy debacle was a severely weakened regulatory infrastructure, particularly at the FTC, where congressional hand-slapping over the children's TV proceeding had left the agency powerless to protect children from a burgeoning advertising and marketing industry. Within the FTC, the experience remained strong in the institutional memory of many staffers. The strong deregulatory surge that swept through Washington during the Reagan Administration in the 1980s further weakened the government's role in the marketplace.[23]

The Whole Web Is Watching

By the time the Center for Media Education began its work in Washington in 1991, the kidvid advertising wars had become part of the folklore of the advocacy community, a cautionary tale about the hazards of taking on a powerful and organized group of industry stakeholders. However, the newness of the Internet created a unique window of opportunity for policy intervention. At that point, the online market was still a nascent medium. Regulatory action in the Web's earliest stages of development could influence how marketing and advertising to children would develop in the new digital media. But waiting to take action posed risks. Once marketing practices were in place, they would become fixed into the economic structure and business practices of the industry and would be staunchly defended if there were threats to regulate them.

However, while TV commercials had become an all-too-familiar part of the contemporary children's media environment, online advertising remained largely unknown. The first task for advocates was to make publicly visible in the press some of the new online marketing and data-collection practices, along with the principles that were guiding the growth of the digital marketplace. CME had used a similar tactic in its campaign for stronger FCC rules on children's educational programming. In 1992, the watchdog group released a report, based on TV-station license renewals, to illustrate the callousness of the industry to comply with its new mandate to serve children under the 1990 Children's Television Act. The study described numerous examples of lackluster response, cynical reporting, and irresponsible scheduling, by quoting directly from the industry's own statements. One of the most widely reported incidents was a broadcaster's imaginative description of the 1960s cartoon series *The Jetsons* as a program "specifically-designed" to educate children because it "teaches children what life will be like in the 21st Century." The press loved the report. The *New York Times* ran a front-page story featuring pictures of cartoon characters that the industry claimed were teaching children such valuable lessons as "good vs. evil." This public shaming in the press helped place the issue on the FCC's policy agenda.[24]

Researchers investigating the online children's marketplace found a gold mine of material, to date largely unknown to parents, policymakers, and the press. If parents thought the Internet offered an alternative to the highly commercialized world of children's television, they were in for a surprise. In the unregulated media environment of the Internet, advertisers were free from the legal and regulatory rules that had constrained them in television. Along with the hundreds of new dot-coms targeting children online were the largest children's-TV advertisers of food, toys, and candy that already had established Web sites.

In these virtual playgrounds, children could play with the familiar product-hawking characters of TV land, who had gleefully hopped into cyberspace, where they could interact freely and directly with children, without parental oversight. Nabisco, maker of Oreo cookies and a wide range of crackers and other popular snack foods, invited children to visit its Nabisco Neighborhood, described as "a place where you can play for hours on end and never worry about the sun going down or running out of quarters." Oscar Mayer's CyberCinema featured an online Sega contest and game; a narrated, interactive guided history of the Wienermobile; and a Super Bowl party and Cyber Halftime Show, with real-time audio. On the Kellogg site Snap, Crackle, and Pop were the official greeters at the Clubhouse, beckoning children to come inside and color pictures of the characters, download Rice Krispies Treats recipes, and do word-find puzzles. The Kellogg General Store sold licensed merchandise, including Tony the Tiger watches, Toucan Sam sweatshirts, and Snap, Crackle, and Pop T-shirts.[25] In the early 1990s, when the Frito-Lay company had tried to develop an entire television show around Chester Cheetah—the mascot for its popular Cheetos snack food—children's and consumer advocates had cried foul, forcing the company to scrap its plans.[26] But online, children could be invited into Chester's Closet to play with the character without their parents ever knowing.[27]

One of the most disturbing features of Internet advertising was that computer technology made it possible to collect large amounts of personal information from children. Online marketers were learning that children loved to answer questions about themselves and eagerly filled out questionnaires that asked for intimate details about where they lived, their likes and dislikes, the names of their friends, and in some cases their families' income. Some Web sites offered incentives, promising free gifts such as T-shirts, mouse pads, and screensavers, in exchange for children's e-mail addresses, street addresses, purchasing behavior, and preferences, as well as information about other family members. Others required children to complete registration forms in order to enter the site. The youth-oriented area of the new Microsoft Network promised to make children "Splash

Kids" if they cooperated and answered questions about themselves. To sweeten the offer, the site offered the chance to win tantalizing prizes, including a Sony Discman. On the Time Warner site for the movie *Batman Forever,* supplying personal information became a test of loyalty. "Good citizens of the Web," the site urged, "help Commissioner Gordon with the Gotham Census."[28]

The KidsCom site had one of the more imaginative reward systems for collecting personal information from children. Its Loot Locker featured a whimsical school locker exploding with goodies. Children could earn points, which could be turned into KidsKash. This virtual money could then be used to purchase "loot." Just by visiting the site, a child could earn one point per week. She could earn an additional five points for each friend, parent, or teacher she persuaded to register on the KidsCom site. But the fastest way to get rich was to participate in a survey (worth 10–20 points). Each survey asked a variety of questions, usually about a particular kind of product. For example, one survey asked what brand of athletic shoes the child wore most of the time, why she had chosen that brand, who bought the shoes for her, where they were purchased, and how often they were replaced. Another survey asked about what kinds of computers were being used at the child's school. Loot, which cost a minimum of 75 points, included such products as Power Rangers videos, Nabisco food products, Bubblicious gum, baseball cards, and CyberPuppy software. More expensive items—such as an Atari Jaguar video game system—required up to 600 points.[29] Staff at CME spent hours on the site, playing the games and answering the questions. Within a few weeks, the organization's shelves were lined with candy, gum, and toys.

These techniques for enticing children to volunteer information about themselves were only the most visible part of the online data-collection strategy. New technologies such as "cookies" made it possible to engage in covert data collection, tracking every move an individual made online. Compiled "clickstream data" created an elaborate and detailed profile of an individual's response and interaction with advertising.[30] The information could be the basis for online ads tailored to the psychographic and behavioral patterns of each child.[31] And third parties could buy it.

The practices that CME uncovered in its report were part and parcel of a new paradigm for marketing that had taken root quickly and spread in the growing dot-com sector. The strategy was based on the principle of developing unique, long-term relationships with individual customers in order to create personalized marketing based on individual preferences and behaviors. "Interactive technology," market research expert Don Peppers explained, "means that marketers can inexpensively engage consumers in one-to-one relationships fueled by two-way 'conversations'—conversations played out with mouse clicks on a computer, or touch-tone buttons pushed to signal an interactive voice response unit, or surveys completed at a kiosk."[32] As another marketing executive put it: "This goes beyond simple transaction processing and secure payment systems: it's about building relationships with customers online—knowing each customer by name, knowing their preferences and buying patterns, observing the customers over time, and using this data to sell more effectively to them."[33] At the heart of this system was the ongoing collection of personal information and tracking of online behavior. Through the data collected, marketers were able to create an irresistible package of ads and buying opportunities designed to "microtarget" the individual customer.[34]

These intrusive practices were troubling enough when applied to adult consumers. But with children, they were more serious. For the first time, advertising went beyond just manipulation; one-to-one marketing on the Internet posed significant privacy threats, not only to the children themselves, but also to their families.

Before releasing its report to the press, CME and its legal counsel made a visit to the FTC to alert the commission. Researchers brought screen shots and other documentation to illustrate the various kinds of abusive marketing and data-collection practices they had found. Well aware of the controversial kidvid rulemaking and its outcome, CME leaders explained to FTC staffers that they were not calling for a ban on all Internet advertising to children. Such a proposal would have triggered an aggressive response from both the new online industry and the well-established advertising industry and galvanized these forces in opposition to any form of regulation. All CME wanted from the FTC were some "rules of the road" for conducting business with children on the Information Superhighway. Despite this measured approach, the advocates were not greeted with a great deal of enthusiasm at the commission. The events from more than a decade before had left their mark on the agency. At the mere mention of children's advertising, several staff members practically recoiled in horror.[35]

CME released its report *Web of Deception* in March 1996, generating extensive exposure in major newspapers, TV news, and online publications.[36] "Kids Snared in Web of Ads," a *USA Today* headline warned.[37] Calling the new online marketing practices "unfair and potentially dangerous," children's media activist Peggy Charren told the *Christian Science Monitor* that "pitching to children on a one-to-one basis is like shooting fish in a barrel."[38] The study also caught the new online industry off guard. Major corporations had hired young twentysomethings to develop Web sites and were not paying much attention to the process. General Foods was about to launch its new Web site for kids, You Rule School, on the very day that the report hit the press. The site's launch was immediately called off. "Official word," *Advertising Age* reported, "is that agency Saatchi & Saatchi Interactive, New York, needed more time to implement new technologies, but we couldn't help noticing the move came shortly after the Center for Media Education blasted Web sites aimed at kids."[39]

For the advocates, the report's timing was fortuitous. It was the middle of an election year, with debates already raging over Internet indecency, TV violence, and moral values in the media. The prospect of personal and family information being collected online from unassuming youngsters also played right into rising public concerns about child safety in the Digital Age. Marc Klaas, whose 12-year-old daughter Polly had been kidnapped and murdered three years before, was waging a national campaign against database companies that sold personal information on millions of children. Charging that this sensitive material was getting into the hands of criminals, Klaas's group Kids Off Lists was seeking to ban the sale of children's personal information without parental consent.[40]

Though the Federal Trade Commission was limited in its authority to develop industry-wide rules on children's online marketing, it could act on individual complaints against "bad actors." CME partnered with the Institute for Public Representation, a public interest law clinic at Georgetown University, to develop a legal strategy for the complaint against KidsCom, one of the Web sites engaged in some of the most disturbing practices. The analysis concluded that the FTC's jurisdiction over "deceptive" advertising was the strongest legal tool available. In May 1996 CME filed a formal complaint charging that the Web site was deceptive and misleading and thus violated Section 5 of the FTC Act. CME already had identified KidsCom in its earlier report.[41] The highly publicized complaint caught KidsCom developer, Jorian Clarke, by surprise. Her reaction was described in a computer magazine several years later:

> Jorian Clarke was feeling good that spring morning in 1996. Driving back to her Milwaukee office, she was thinking about the great client meeting that she had just wrapped up, and the successful second quarter looming ahead for her company. The ring of her mobile phone distracted her from her thoughts for a moment. Clarke had barely said "Hello" when the reporter from National Public Radio got right to the point. "What do you think about this report that says you're using and manipulating children?" "I almost drove off the road," Clarke recalled.[42]

Ms. Clarke's response to CME's allegations suggests that she had not thought she was doing anything wrong in designing this Web site for children. Like many of her colleagues in Internet start-up businesses, she had not worked in children's television, so she was not familiar with TV's codes of conduct for advertising and marketing aimed at this special audience segment. When criticisms of KidsCom had first appeared in CME's report a month before, the press called her for comment. Maintaining that she "never gives out personal information to companies," the executive had commented that she agreed, "in theory" to some of the concerns raised in *Web of Deception*, but challenged: "who do they think is going to pay for these wonderful sites for children?"[43] Nor was she aware of how FTC rules against deceptive advertising might apply to the particular practices employed on the KidsCom Web site.[44]

Federal Trade Commission investigations always are conducted in private. Sometimes they take years and do not surface publicly until an agreement has been reached between the agency and the company in question. CME was never told outright whether or not the commission had decided to act on its complaint. But a week or so after the filing, the advocates walked into the office of one of the commissioners—a woman in her 60s or 70s who until this time had not yet ventured online—and found her hunched over her computer, intently engaged in an online game. "I've already earned 10 KidsKash points!" she gleefully announced.[45]

Transatlantic Pressures

While CME wanted broad rules to govern Internet marketing to children, it soon became clear that the immediate political opening was to push for privacy safeguards. Public-opinion polls showed a rising concern about privacy on the Internet. The press had begun to take up the issue, reporting horror stories of identity theft and other dangers posed by the emergence of digital commerce and global communications.[46] A growing public-interest digital-privacy movement was mobilizing, as new organizations joined established privacy-advocacy groups to serve as watchdogs for both industry and government and to campaign for new laws. One of the most important advocacy groups was the Electronic Privacy Information Center. Already engaged in the fight against the Communications Decency Act, the small organization also was becoming a visible and prominent advocate in the battle over Internet privacy. EPIC was press savvy and well connected to a network of academic experts from around the world. EPIC worked closely with Privacy International and a growing coalition of consumer organizations with a stake in both national and international data-protection policy.[47] Other U.S. nonprofits that were active on the privacy front included the American Civil Liberties Union, which had fought for years against government intrusion of privacy, Computer Professionals for Social Responsibility, and the Center for Democracy and Technology.

For years, regulation of privacy in the United States had been conducted in a piecemeal fashion, leaving an incomplete patchwork of federal and state provisions that were inadequately enforced. Unlike most other countries in the developed world, the United States had no federal agency devoted to developing and implementing privacy policy. "The approach to making privacy policy in the United States is reactive rather than anticipatory, incremental rather than comprehensive, and fragmented rather than coherent," explained privacy scholar Colin Bennett. "There may be a lot of laws, but there is not much protection."[48] Privacy advocates had been calling for years for the establishment of an independent privacy agency, but they had gotten nowhere. While the FTC had assumed some jurisdiction over online privacy, its authority was very limited. The agency could hold hearings and conduct research but had no broad powers to regulate information practices on the Internet. It only could prosecute deceptive or fraudulent practices that violated existing laws.[49]

Meanwhile, events were moving forward across the Atlantic that were creating new pressures on both the online industry and the U.S. government. After a long period of deliberation, the European Union finally issued its "Directive on the Protection of Personal Data and on the Free Movement of Such Data" in 1995. The document laid out a strong set of policies to govern privacy protection in the EU member states. But in a global economy, these policies also had a significant impact on the domestic policies of major countries around the world. One of the directive's most significant provisions forbade EU member states from engaging in "data transfer" with countries that failed to provide an adequate level of protection. These new provisions would have far-reaching implications for a wide range of U.S. and multinational corporate enterprises, from credit-granting and financial institutions, to hotel and airline-reservations systems, to the booming direct-marketing sector. "There is no doubt," explained Bennett, that the EU's directive "now constitutes the rules of the road for the increasingly global character of data-processing operations."[50]

By mid-1996 most European countries, and many others outside of Europe, had enacted their own comprehensive privacy laws designed to pass muster with the EU directive. The United States was one of the last holdouts, refusing to change its domestic laws to "harmonize" with the strong European regulations. But privacy advocates in the United States, armed with the EU policy, and working closely with the international privacy community, were mounting a campaign to force the United States to bring its standards up to par with those in the rest of the world.

The official position of the Clinton Administration was that Internet privacy, like the rest of e-commerce, should be governed not by laws but by industry self-regulation. The White House Information Infrastructure Task Force (IITF) had issued a report in 1995 that included a set of "principles" for companies to follow in order to protect the "privacy rights of citizens." But they were only recommendations to the private sector, and the industry had shown very little interest in implementing them.[51] With less than a year and a half before the EU deadline, the administration had put into place a set of strategies designed to shore up industry cooperation on the home front while undermining the EU's influence in the world regulatory debate. The White House assigned Ira Magaziner to head up the inter-agency electronic commerce working group and to serve as the key international negotiator on the issue. Magaziner, a "new age" management consultant, had gained notoriety in the early days of the Clinton Administration as director of First Lady Hillary Clinton's failed health-care-reform initiative. After the devastating demise of his complex, unwieldy government plan for national health care, Magaziner had reinvented himself as a free-market zealot whose new mission was to "keep government out of electronic commerce everywhere."[52] He was still working in the White House, this time drafting the United States's official white paper on e-commerce, which, among other things, outlined a policy designed to preempt actions by foreign governments that might limit the ability of U.S. companies to engage in international trade over the Internet.

The FTC, along with the U.S. Department of Commerce, was working with industry to encourage more effective approaches to self-regulation. The FTC had begun a series of workshops and public hearings on consumer protection and the Global Information Infrastructure, taking testimony from a variety of industry and consumer groups. By the mid 1990s, with the EU deadline less than a year and a half away, the agency began to focus more on Internet privacy. Consistent with the official administration policy, however, the commission took a softball approach to prodding the industry. At the opening of a June 1996 workshop on "Consumer Privacy on the Global Information Infrastructure," chairman Robert Pitofsky carefully characterized the event as "a fact-finding workshop, designed to provide a forum for discussion and debate. We are not here to lay the groundwork for any government rules, guidelines, or otherwise. Rather, we would like to learn more about industry and consumer initiatives that have emerged over the past year."[53]

On Capitol Hill, more serious threats loomed. Congressman Ed Markey, fresh from his victory on the V-chip law's inclusion in the 1996 Telecommunications Act, introduced an online-privacy bill aimed at protecting adult and child consumers. Arguing that the debate over online privacy had become wrapped up in technical detail, Markey told the press it was time to focus on basic rights. "Whether they are using a phone, TV clicker, a satellite dish or modem, every consumer should enjoy a Privacy Bill of Rights for the Information Age," he said. Urging Congress to "act swiftly because the current situation is utterly unacceptable," Markey warned that "the wondrous wire may also allow digital desperadoes to roam the electronic frontier unchecked by any high tech sheriff or adherence to any code of electronic ethics."[54] A separate piece of legislation was introduced by Representative Bob Franks (R-NJ), working with Mark Klaas and other advocates, that would prohibit the sale of personal information about children without parents' consent.[55] Dozens of other privacy bills would follow.[56]

Opting In or Out?

With the early stages of the dot-com boom already underway, the digital media held great promise for growth in the new economy. Above all, Internet businesses wanted to operate in an online environment that was free of any regulatory restraints. But faced with rising political pressures at home and abroad, the industry scrambled to devise self-regulatory regimes that would keep government off its back without killing the golden goose of e-commerce profits. In the meantime, lobbyists offered promises and platitudes. The Coalition for Advertiser Supported Information and Entertainment promised to develop a new set of goals on privacy for the advertising community.[57] Microsoft's Jack Krumholtz, testifying on behalf of the Interactive Services Association, offered his organization's view on the issue. "We believe at the ISA," he explained, "that the keys to privacy are what we refer to as the Two E's, education and empowerment."[58]

But the clouds of government oversight also offered a silver lining to the new online industry. When FTC commissioners had raised questions before about the possibility of unsolicited e-mail from online businesses, they had been assured that nothing of the kind could happen on the Internet. The well-understood rules of "Netiquette" ensured that anyone who dared "spam" others would trigger the wrath of the entire Internet community, whose members would immediately "flame" the perpetrator.[59] But these arcane customs, so integral to the pre-Web Internet, already were being overrun by the powerful engine of e-commerce. The public debate over online privacy presented an opportunity for marketers to completely wipe out the old model of Internet behavior and replace it with a new set of norms that would be more consistent with the business imperatives of the Digital Age.

The trade association with the most immediate stake in the new ecommerce marketplace was the Direct Marketing Association (DMA). Representing thousands of companies in the United States and abroad, the DMA had promoted and defended direct mail and database marketing for nearly a century, since its founding in 1917. It lobbied federal and state legislators, as well as numerous other federal agencies, including the FTC, the FCC, and the U.S. Postal Service.[60] Arguing that "the promise of technology and the social rules of the online marketplace have yet to be determined between consumers and business," the organization offered its own "self-regulatory principles" for addressing online privacy. Based on the core concepts of "notice and choice," the DMA proposed that online companies be encouraged to post their individual privacy policies on their Web sites and allow customers to choose whether to "opt out."[61] This "opt-out" concept, cloaked in the rhetoric of consumer freedom, was designed to ensure that data collection would become the default in online business transactions.

But even the marketing industry had to admit that where children were concerned, the picture was more complex. Simple ideas about "notice and choice" would not work, particularly with younger children, who had not developed the cognitive ability to make complex judgments. "It may take a whole village to raise a child, but it just takes one big corporation to exploit one," child psychiatrist Michael Brody from the Academy of Child and Adolescent Psychiatry told the FTC. "Not until they're adolescents do they even understand what personal information is . . . and develop a strong sense of privacy."[62] In their joint comments to the FTC, CME and the Consumer Federation of America (CFA) argued that "[O]pting out is not an effective mechanism for protecting the privacy of children."[63] Instead, the advocates proposed their own rules for the government to adopt in order to protect children's online privacy. Based on the "opt-in" principle, their goal was twofold: to make it very difficult for online marketers to collect personal information from children, and to ensure that parents were brought into all online business transactions involving their kids. If a commercial Web site wanted to collect such information from a child, it would have to get permission from parents in *advance*. Simply saying, "be sure to ask your parents before filling out this survey" would not be good enough. The CME/CFA proposal called for parental permission to be obtained *offline*, by fax or "snail mail."[64] Advocates hoped that this requirement—the last thing that industry wanted—would reverse the trends toward wholesale data collection in the emerging children's digital marketplace.

Over the next two years, as the Internet privacy issue played out in Washington and in the press, concerns about children continued to take center stage, with consumer advocates pressing for government intervention and industry leaders arguing that self-regulation would address the problem. The debate forced parents and policymakers to grapple with a whole new set of questions about the role of children in this new media marketplace. Who should be responsible for protecting children from exploitive practices on the Internet—the government or parents? How could the same families that rely on their kids for computer tech advice effectively police their children's behavior online? At what age were children responsible enough to make their own decisions about what information to give out online? Should marketers be required to stop collecting personally identifiable information from children altogether? How would that affect children's online experiences in this new educational medium?

The new content-blocking-technology industry quickly seized upon the privacy concerns, adapting its "empowerment tools," invented in the wake of the Internet indecency controversy, to address this most recent crisis. Representatives from CyberPatrol, NetNanny, and SafeSurf came to the FTC's 1996 workshop armed with PowerPoint presentations and new product displays, each promising the solution for protecting children's online privacy. CyberPatrol's Susan Getgood introduced her company's new Chat Guard feature, which allowed parents to program the software to keep their children from answering particular kinds of questions online. If a child tried to fill out a survey, the software would block certain responses. "It's hard for the kids to get around it," she claimed. "It's tamper resistant. I wouldn't say tamper proof because I'm asking for trouble."[65] But when the nonprofit Consumers Union, the group that publishes *Consumer Reports*, tested the software, the results were less than satisfactory. In a 1997 survey of child subscribers to the group's magazine *Zillions*, researchers concluded that CyberPatrol, CyberSitter, and NetNanny could not prevent a "determined, computer-savvy child" from circumventing the software.[66]

Consumer advocates argued that, even if filtering software were fool-proof, the whole idea of using technology to prevent children from giving out information was the wrong approach. "Parents' use of these products," argued Consumers Union's Jeff Fox, "doesn't relieve companies doing business on the Internet from the responsibility of respecting children's privacy any more than using a child safety seat eliminates the need to drive safely."[67] Another problem with the software approach was that the filtering tools were designed to protect children from accessing online adult content that was not intended for them. But threats to children's privacy were coming from Web sites designed especially for them. Parents should not have to set up elaborate screening software to protect children from their own online content areas. As *Privacy Times* editor and publisher Evan Hendricks explained, parents were being put in an untenable position: "It's either you let them have the information if your children use the computer. Or, if you care about any of this stuff, you can't let them use the computer. That's not the way to set national policies."[68]

Until the 1996 CME report, the industry had done nothing to advance self-regulation to protect children's privacy online. When a CME representative made a visit to the Children's Advertising Review Unit's New York headquarters a year or so before, she had found an underfunded operation with a one-person staff, and the computers had not even been hooked up yet.[69] CARU's Elizabeth Lascoutx admitted to the FTC in June 1996 that privacy was a "brand new area for CARU," assuring regulators that the group was in the process of putting out draft guidelines to the advisory board.[70] By the time the rules were finally released in April 1997, the debate over children's privacy was in full swing. Regulators gave the guidelines mixed reviews. FTC commissioner Christine Varney commended CARU for doing a "terrific public service," but as the *New York Times* reported, she stopped short of endorsing them, citing a different set of proposed guidelines submitted by a team from the Consumer Federation of America and the Center for Media Education. In sharp contrast to the CME/CFA proposal, CARU's proposed guidelines did not make specific requirements for marketers to obtain parental permission, but rather suggested that online companies make "reasonable efforts" to get children to ask their parents for permission before making a purchase or answering personal questions about themselves or their families on the Internet. As Varney explained, the two opposing proposals gave the commission a "good place to have a dialogue," with the challenge being to "narrow the gaps" between them.[71]

Of the five commissioners heading up the FTC, Christine Varney was the most visible official to speak out publicly on the issue of children's privacy. An attorney in her thirties, and the mother of two small children, she was particularly effective at articulating parental concerns about children's online activities, and she frequently challenged industry representatives to show how their practices would protect children. To child advocates, she appeared to be supportive of the idea of requiring parental permission before companies could collect personal information from children. In her public statements on the broader privacy issues, Varney also played a tough regulator role. For example, she warned the financial services industry in February 1997 that while the government had until then encouraged self regulation, Congress was likely to grow increasingly impatient over the next four years if no significant progress were made.[72] But Varney also had close ties with the Clinton Administration, having worked on the 1992 Presidential campaign and served as Cabinet Secretary and Assistant to the President prior to being appointed by President Clinton to the FTC. To overseas audiences, she frequently advanced the official White House position on self-regulation, arguing that no new legislation was needed to protect privacy and that the FTC could prosecute any parties trying to use the Internet for purposes that could be considered fraudulent under existing laws.[73]

"Let the Surfer Beware"

When the FTC held its long-awaited second privacy workshop in June 1997, industry groups came with a gaggle of brochures, software, and new technological tools, all designed to stave off government regulation. The World Wide Web Consortium unveiled its Platform for Privacy Preferences (P3P) project. "Web technology has suddenly made the privacy issue critical by providing at the same time a wealth of new threats and a wealth of new solutions," proclaimed consortium director and Web creator Tim Berners-Lee. Demonstrating the prototype software for commissioners, Berners-Lee showed how customers could engage in a "negotiation" with individual Web sites over the level of privacy protections offered.[74] But as some industry observers noted, these new tools could make privacy protection a complicated process. "If Internet groups get their way," commented Ira Teinowitz in *Advertising Age*, "dealing with privacy issues on the Web could be a lot like the haggling that goes on in car dealerships. Want to access a premium version of that news site? It'll cost you some personal data. Want one site to know where you live but not another one? You'll be able to negotiate that as well."[75]

Representatives from the Direct Marketing Association assured commissioners that it was "moving aggressively" to bring its members on board the self-regulatory train. The trade association had created a special section of its Web site called Privacy Action Now! with tools to help Web site operators draft privacy policies. The group also announced a new activity book, "Get Cybersavvy! The DMA's Guide to Parenting Skills for the Digital Age," as well as links to parental-control software sites.[76] Individual companies announced new corporate privacy policies with great fanfare. "Groundbreaking Policy Protects Cyberspace Privacy of Consumers," read the headline from the McGraw-Hill companies.[77] New corporate partnerships also were announced. As one trade publication observed, "Concern over Internet privacy has become so great that two sworn enemies—Microsoft Corp and Netscape Communications Corp—have joined forces to develop a universal standard for privacy software."[78] The online industry created a new nonprofit organization, TRUSTe, to address concerns over online privacy. Founding members of the new venture included computer industry guru Esther Dyson, AT&T, IBM, Oracle, Netscape, and other prominent U.S. corporations. TRUSTe was set up to serve as a kind of "Good Housekeeping seal of approval." Companies would pay between $500 and $5,000 to receive a seal documenting their adherence to certain privacy safeguards. The seal would be displayed prominently on the Web site so that customers could comparison-shop online vendors, choosing those companies that promised stronger protections of their data.[79] "If you can buy a sleeping bag from Eddie Bauer or L.L. Bean," Dyson explained to the press, "and one protects your privacy and the other doesn't, if you value your privacy, you'll buy from the one that protects it."[80]

But while some privacy-advocacy groups endorsed the self-regulatory proposals, others were sharply critical of the plans. When eight of the nation's largest consumer-database companies announced a joint agreement to limit their data collection practices, EPIC's Marc Rotenberg warned that the policymakers should be skeptical. "These companies have a bad record in the area of self-regulation," he told the press. Georgetown Professor Mary J. Culnan added: "There will be plenty of people out there who won't follow these guidelines. . . . Without a law, they can offer whatever they want."[81]

As the industry continued to promote its self-regulatory progress in the press, privacy advocates stepped up their own media efforts, supplying reporters with a constant stream of facts and illustrations to support their claims that privacy threats in cyberspace remained a major problem. EPIC's *Surfer Beware* report reviewed one hundred of the most frequently visited Web sites on the Internet and found that while almost half of them collected personal information from visitors, only seventeen had explicit privacy policies, and none met even the most basic standards for privacy protection.[82] Watchdog groups attacked America Online for its plan to provide its customers' telephone numbers to telemarketers, prompting the company to drop the controversial idea.[83]

CME and CFA released a study showing that collection of personally identifiable information from children on the Web was becoming even more widespread. Nearly 90 percent of the 38 largest child-oriented Web sites attempted to gather personal information, including names and addresses, according to the report. Forty percent used incentives, such as free merchandise and sweepstakes, to collect the information, while only 20 percent reminded children to ask their parents before providing information. The study had uncovered fresh examples of data-collection practices, ranging from the Colgate site, where the online Tooth Fairy instructed children to "Fill in the blanks below, get a good night's sleep, then check your e-mail tomorrow for a message from you-know-who," to Frito-Lay's DreamSite, where cartoon psychiatrist Dr. Dream urged kids to "pull up a couch, tell Dr. Dream about your nocturnal meanderings, and get ready for some instant analysis." The report also found that cookies were used by 40 percent of the children's Web sites.[84] "We have to keep online marketers out of the 'cookie jar,'" CME Executive Director Jeff Chester told the press. "Such 'Orwellian' practices to stealthily track every move made online, and share that information with other companies, should be prohibited."[85] Consumer Union's Charlotte Baecher added, "Web sites for children are more intrusive and manipulative than "the worst children's television."[86]

Hands-off the Internet

With the July 1997 release of its "Framework for Global Electronic Commerce," the White House articulated the official U.S. policy for the economic future of the Internet. The document was a free-market tract, predicting that e-commerce revenues could total "tens of billions of dollars by the turn of the century," but only if government stayed out of the Internet's business. The administration already had dropped its use of "Information Superhighway," substituting it with the less-colorful "Information

Infrastructure." The new nomenclature was consistent with the official position that placed the private sector, rather than the government, in control of both the development and governance of the Internet. "For electronic commerce to flourish," the white paper asserted, "the private sector must continue to lead. Innovation, expanded services, broader participation, and lower prices will arise in a market-driven arena, not in an environment that operates as a regulated industry." The White House also called on foreign governments to take a hands-off approach to the Internet, warning that for the potential of a robust online economy to be realized fully, "governments must adopt a non-regulatory, market-oriented approach to electronic commerce, one that facilitates the emergence of a transparent and predictable legal environment to support global business and commerce."[87]

In the midst of the growing global debate over Internet privacy, the paper acknowledged that e-commerce would not thrive unless privacy rights were protected. But it argued that industry self-regulation and technological solutions were the appropriate methods for addressing the problem. While promising to "continue policy discussions with the EU nations and the European Commission to increase understanding about the U.S. approach to privacy," the document conceded that it faced challenges in convincing American companies to develop effective self-regulatory mechanisms. "If privacy concerns are not addressed by industry through self-regulation and technology," the paper warned, "the Administration will face increasing pressure to play a more direct role in safeguarding consumer choice regarding privacy online."[88]

By the end of 1997, White House officials were beginning to hammer out agreements with the European Union on various policy issues related to regulating the Internet. In December, they reached an agreement that no new tariff duties would be imposed.[89] A few months later, the administration convened a "Global Summit" on Internet regulation.[90] But in the area of privacy, the U.S. government had a big problem on its hands. Despite assurances to European regulators that the online industry was developing an effective self-regulatory regime, it was clear that the United States had a long way to go before it would be able to comply with the EU policy. "So far, the industry's track record has been found lacking," *Business Week* reported in early 1998. A survey of the top hundred Web sites by the magazine found that only 43 percent displayed privacy policies. "Of the notices posted, some were difficult to find and inconsistent in explaining how data are tracked and used." And the widely hailed TRUSTe, which had promised to sign up 750 Web sites by March, had only gotten

10 percent of that goal in the nine months since its launch. "Such a lackadaisical approach has riled the government," the article noted, observing that among many privacy advocates patience also was running out.[91]

The debate over online privacy revealed a disconnect between what the industry wanted and what its consumer base expected from e-commerce. As Denise Caruso wrote in the *New York Times*, "Data vendors believe, as do most modern marketers, that one-on-one sales pitches are the future of commerce—and because the Internet links so easily to one user at a time, it is their dream machine."[92] One industry observer commented: "What the Internet has done is make explicit what used to be implicit—namely that there were dossiers on you that can be built up with great granularity."[93] Yet most online consumers wanted government safeguards. A 1997 Georgia Tech survey found that 87 percent of Web users believed they should have "complete control" over the demographic data captured by Web sites. More than 71 percent supported the idea of new laws to protect online privacy.[94]

Where children were concerned, the administration was in an even more difficult position. The White House had championed protections for children as a central component of its successful political strategy, so merely adhering to a voluntary, self-regulatory, private-sector seal-of-approval approach might appear contradictory, particularly as the children's privacy issue was being linked to online safety. As more and more learned about online data being collected from children, parents became increasingly alarmed. There also was a sense among the public that somehow this wholesale collection of personal information from youngsters was just unfair. A national survey in 1997 by Louis Harris & Associates found that 97 percent of parents whose children used the Internet objected to the practice of selling or renting the names and addresses of the children who visited companies online. Nearly three-fourths objected to the information being collected at all. And 64 percent thought companies should not be allowed to obtain e-mail addresses or usage patterns of children who visit their sites, even if it were done just to gather statistics. "As America's children rush onto the Internet, their parents greatly fear for their privacy and safety," explained an article in the *Pittsburgh Post-Gazette*. "Parents are afraid the widespread collection of information about their kids could have them targeted for advertising, or be used to lure kids into purchasing things online without permission. Worse yet, they fear a child's information could be obtained by pedophiles." The survey also found that 96 percent of parents believed companies violating their own policies about children's information should be held legally liable.[95]

Added to concerns over data collection by marketers were new worries about what children were encountering in online chat rooms. A third of the respondents in the Consumers Union survey had experienced problems with other users online, including use of pornography and profanity, as well as inappropriate advances. "We were checking out a chat room," one 11-year-old boy told *Zillions*, "and they were swearing and talking about drugs." A teenage boy reported that, because his e-mail address was visible in his profile, "Someone sent me a lot of pictures of little kids naked, or performing sexual acts." A 12-year-old girl commented that, "Sometimes I get an Instant Message with people asking me for my password, address, and phone number."[96]

As CME, CFA, and their allies pushed for government policies to protect children's privacy, the White House was realizing that some government action would be needed to quell parental fears. During the June 1997 FTC workshop, the administration already had begun making public statements that supported some kind of regulation along the lines of what CME and CFA had been seeking. "The Clinton administration will press for stricter rules on how information can be collected from children on the World Wide Web," reported the *Washington Post.*[97] Commissioner Varney suggested that marketers be blocked from obtaining information about age, sex, home address and e-mail address from children under 12 without parental consent. Varney also was reported to have begun pushing CARU "into toughening its new guidelines for online privacy."[98]

The Center for Media Education had lobbied Ira Magaziner's office to include special provisions in the official e-commerce white paper that would exempt children from the administration's free market policies. The White House consented, asking CME to draft language documenting the special vulnerabilities of children to advertising. The edited version, as it appeared in the final paper, reflected the government's conflicting positions on marketing and data collection. The section dealing with data collection, while promoting voluntary solutions, also created an opening for government regulation:

> The Administration is particularly concerned about the use of information gathered from children, who may lack the cognitive ability to recognize and appreciate privacy concerns. Parents should be able to choose whether or not personally identifiable information is collected from or about their children. We urge industry, consumer, and child-advocacy groups working together to use a mix of technology, self-regulation, and education to provide solutions to the particular dangers arising in this area and to facilitate parental choice. This problem warrants prompt attention. Otherwise, government action may be required.[99]

However, on the subject of marketing and advertising in cyberspace, the document left no such options. "A strong body of cognitive and behavioral research demonstrates that children are particularly vulnerable to advertising," the paper noted. "As a result, the U.S. has well-established rules (self-regulatory and otherwise) for protecting children from certain harmful advertising practices." But no further government actions would be needed to ensure that there were adequate safeguards in cyberspace. Instead, the paper explained, "[T]he Administration will work with industry and children's advocates to ensure that these protections are translated to and implemented appropriately in the online media environment."[100]

On July 15, 1997, one day before the Supreme Court struck down the Communications Decency Act, the FTC issued a ruling supporting CME's complaint against KidsCom. Its release to the press was timed to coincide with the high-profile White House meeting where Vice President Gore unveiled the "virtual toolbox" as part of the administration's new campaign to promote a family-friendly Internet. Both moves were designed to send a message that the administration was taking actions to protect children in cyberspace. "The President and the Federal Trade Commission Wednesday delivered a double-barreled attack against the exploitation of children on the Internet," wrote *USA Today*.[101]

While FTC staff had found that the KidsCom data-collection practices violated the commission's rules on deceptive advertising, it chose not to take legal action against the company. But the decision still was a major victory for the advocates. The letter established a framework for how all children's Web sites should handle data collection in the future. Applying the CME/CFA–proposed rules, the commission put operators of Web sites on notice that they should post clear privacy notices to parents, and that they should obtain parental permission before releasing any personal information to third parties.[102] As the *Washington Post* noted, the KidsCom decision was "the first time regulators have articulated a policy regarding the collection of information from children."[103]

These developments suggested that the United States was moving toward adopting a government policy on children's online privacy. But the White House still hoped to preempt legislation, even in the face of rising pressures at home and abroad. As Ira Magaziner continued to prod American industry into creating effective self-regulation policies, he also tried to assure regulators overseas that such efforts were working. Speaking at a November 1997 ad-industry event sponsored by CARU, he warned advertisers that they needed to fortify their self-regulatory policies. "What you are doing is essential to the Administration's strategy," he told industry representatives. "The tremendous economic benefits of the Internet will not work if we don't get efficient industry self-regulation on issues like privacy and content, especially in the children's area. . . . If you fail," he added, "we will have to go the legislative route." He also told advertisers that it was "essential that children's privacy be protected and children not be exploited, that information is not gained from children without parental permission." But at the same time that Magaziner was threatening advertisers to get into line, he was enlisting CARU in his overseas campaign to relax their restrictions. CARU representatives told the *New York Times* they were planning to urge the European Union to "follow the American *de minimis* approach to Internet regulation, rather than adopt Government strictures."[104]

A month later, just as the White House was to begin its summit on children and the Internet, Ty Inc., the company that sold the highly popular Beanie Babies stuffed toys, announced that it was making significant changes in its Web site, in compliance with a CARU directive based on its new guidelines. In addition to posting a privacy policy, the Web site would also require parental permission before collecting any personal information from children under the age of 17. "Self-regulation is a very constructive way for advertisers, as responsible corporate citizens, to create an environment that's safe and healthy, particularly for children," Ty's outside counsel, Ed Getz, told the *New York Times*. The attorney was careful to add that he "would not advocate mandatory regulations because one hallmark of the interactive medium is a measure of creativity, and regulations instill a rigidity in the process that is not in the best interest of the advertiser or the consumer." CARU's Elizabeth Lascoutx assured the press that this was only the first of many actions the group intended to take to curb practices that violated its guidelines. "If we don't protect kids, if we don't get it right, that will force regulation of the Internet," she explained. "But I'm very hopeful based on reaction to our guidelines . . . as long as we in the industry don't drop the ball, we'll be fine."[105]

But such public assertions of tough vigilance by the industry belied the fact that CARU only learned about the Beanie Babies problem because it was one of many Web sites whose data-collection practices had been exposed in a report released a few months earlier by the Center for Media Education and Consumer Federation of America. CARU had not been combing the Internet to rout out the bad apples in the new digital marketplace; CME and CFA had done much of the work for them.[106]

"Sweeping" Change

Throughout 1997, more privacy bills were introduced in Congress, including some to protect children's privacy. "The orgy of privacy bills from both Republicans and Democrats is a response to unprecedented levels of concern," wrote one reporter.[107] By early 1998, as eighty pieces of privacy legislation were pending on the Hill, it was clear to the administration that U.S. industry was still too slow in adopting adequate self-regulatory regimes to satisfy European regulators.[108]

In March, with only seven months until the EU deadline, the FTC announced it would do a random "sweep" of commercial Web sites—both adult and children's—to see how many of them had posted privacy policies. "Normally, such threats would raise hackles in the free-spirited realm of the Net," commented one reporter in *Business Week*. "But maybe not this time." A new *Business Week*/Harris poll had found that there was rising unease among the public about online privacy. A majority of respondents identified fears over privacy as the main reason they were not going online, rating this factor higher than cost, ease of use, or unwanted marketing messages. Even "hardcore Netizens" who were generally quite comfortable with the Internet were growing wary, according to the survey. More than three-quarters of them told researchers that they would use the Web more if privacy were guaranteed. "Perhaps even more striking," the article said,"50 percent of the computer users polled say that government should pass laws 'now' on whether personal data can be collected and used on the Internet."[109] Another reporter, writing in *Investors Daily,* noted that "the White House now sees how privacy concerns could hinder development of its Net pet."[110]

As Magaziner continued his overseas campaign to sell American-style self-regulation of the Internet, U.S. government agencies increased the pressure on industry to take action. By April, the White House efforts to enlist support from nations outside of Europe had begun to pay off. "The U.S. and Japan are poised to sign a breakthrough agreement on electronic commerce, locking the two world powers into a light-handed approach to regulating Internet trade," reported an Australian trade publication.[111] Meanwhile, at the Commerce Department, Secretary William Daley was turning up the heat, jawboning leaders of individual companies, and convening corporations, trade groups, and consumer groups in what one trade publication referred to as a "last ditch effort" to "come up with [an] online privacy program that will head off threat of government action." With a July 1 deadline to report back to President Clinton, Daley's pressure had put the "fear of God" into the online industry, sparking "feverish activity."[112] Another summit, this one sponsored by the Commerce Department, was scheduled for June 1998.

Even before the FTC announced the results of its Web-site-monitoring efforts, insiders knew the findings would be dismal. Expecting a further public-relations crisis, industry began to act. Companies and trade groups issued a flurry of press releases, unveiling new initiatives to address online privacy. The Better Business Bureau announced plans to launch BBB Online, another "seal-of-approval" program designed to "foster an ethical online marketplace."[113] The Interactive Services Association (ISA) (renamed the Internet Alliance) issued a " Privacy Tool Kit" to help online businesses develop privacy policies."[114] And a dozen high-tech trade associations wrote to President Clinton announcing an "industry-led, self-regulatory plan to address online privacy concerns." Included in the new group were: the American Electronics Association, the Business Software Alliance, the Software Publishers Association, and the Consumer Electronics Manufacturers Association.[115]

Christine Varney, who had recently left the FTC to work in the private sector, began heading up a new Online Privacy Alliance, a consortium of fifty American companies and trade associations, including America Online, AT&T, Bell Atlantic, IBM, Microsoft, and Netscape. Having been one of the most prominent government officials to criticize the industry, she became its chief advocate for self-regulation. Her reputation as a champion for children's privacy regulation gave the former commissioner added credence when she announced that the alliance had been working on guidelines for children's privacy that would be released very soon.[116]

Varney also tried to convince the Center for Media Education to join the alliance, but CME resisted this effort to coopt the advocacy community. At the time, CME was leading its own coalition of consumer, health, and parent organizations, including the Consumer Federation of America and the National PTA, in a national campaign for children's marketing and privacy regulations on the Internet. It was also working with a larger coalition of privacy groups, led by the Electronic Privacy Information Center. The privacy community remained committed to a national policy on Internet privacy and did not want to undermine that goal by signing on to industry plans aimed at preempting government intervention. CME was already in discussions with White House officials to secure support for new FTC rules on children's online privacy. At the same time, the children's advocates knew that the government would not act without getting a buy-in from some of the powerful companies in the children's media industry.

So CME began informal negotiations with Disney, America Online, and several other companies to forge an agreement on a regulatory framework.[117]

A few days before the FTC's scathing "sweep" report was to be released, America Online announced it would begin requiring written consent from parents before content providers could collect names or other personal information from children online.[118] This preemptive move did little to undercut the impact of the study's results. As expected, the commission had found an Internet e-commerce environment with very few protections for its consumers. Of the 1,400 Web sites surveyed, 85 percent collected personal information from visitors, but only 14 percent provided any information on their data-collection practices, such as what kinds of data they were gathering and how they were using the information.[119] Only 2 percent of Web sites surveyed provided a comprehensive privacy policy.[120] "The vast majority of online businesses have yet to adopt even the most fundamental fair information practices," the report concluded.[121] Children's Web sites were equally lacking in safeguards. Nearly 90 percent of more than 200 sites in the survey collected personal information such as names and addresses from children.[122] Though 54 percent of children's sites disclosed some privacy practices, few took "any steps to provide for meaningful parental involvement in the process. Only 23 percent even told children to ask their parents before providing information."[123]

The results of the FTC study, demonstrating such an obvious failure by the online industry to regulate itself, flew in the face of official U.S. government assurances that effective voluntary safeguards would be in place in time for the EU Data Directive deadline, which was then four months away. The commission needed to take some action to show that the United States was moving forward. But with considerable pressure coming from the White House not to call for legislation, the FTC was in a rather difficult position. Since the study's results had revealed serious problems with both children's and adult Web sites, the FTC should have called for comprehensive governmental policies to protect all consumers on the Internet. But instead, it chose a more narrow action. Arguing that it lacked the authority to develop government rules on its own, the agency called on Congress to pass a law to protect children's privacy on the Internet.[124] "The failure of self-regulation is particularly disturbing so far when it comes to children," FTC Chairman Robert Pitofsky noted. The survey's findings had produced a wealth of colorful illustrations to demonstrate the dangers facing young people online. Even with all the public controversy over children's privacy, commercial Web sites were still engaging in shady data-collection practices. For example, the "Young Investors" site asked children to supply an abundance of financial information, including any gifts they might have received in the form of stocks, cash, savings bonds, mutual funds, or certificates of deposit.[125] Such egregious examples provided ample evidence to support the need for legislation.

By focusing on children, the government was able to demonstrate that it was taking decisive action to protect online privacy, while also buying additional time for industry to get its act together. "Effective industry self-regulation . . . has not yet taken hold," the FTC report explained.[126] "Protecting children's privacy was the first and most urgent step," the commission said, promising that "further recommendations for measures to protect all Internet users" would be made later in the summer.[127] But as the hot Washington summer moved forward, the FTC issued no further calls for Congressional action. More government meetings were held, with U.S. officials holding the Damoclean sword of regulation over the heads of industry leaders. Online companies and trade groups continued their flurry of activities to come up with the long-promised self-regulatory regimes. And Ira Magaziner, joined by a team of U.S. officials, continued his forays into foreign capitals, intensifying the campaign to assure the EU that meaningful self-regulations were being put in place, and shoring up support from countries outside of Europe for the hands-off U.S. model.

On the eve of the Commerce Department's privacy summit in late June 1998, several industry groups announced their safeguard programs. Despite the online industry having had more than two years to develop the plans, it was clear that they were still works in progress, missing some of the most critical pieces. The newly created Online Privacy Alliance presented a framework of guidelines that called for disclosure of data collections and an opportunity for consumers to opt out. The group also announced that it was working on a set of guidelines for children's online privacy. But as the *Washington Post* reported, the organizers continued to "squabble over how to enforce the voluntary rules," promising to reach resolution by September.[128]

At this late date, the industry clearly was not in any way prepared to implement full-fledged self-regulatory programs. By July, the Better Business Bureau announced it was still working on its seal-of-approval program but explained that it needed much more time. Promising that its system would be in place by the end of 1998, the organization acknowledged that it would be at least another year before most companies would be using it.[129]

Privacy groups expressed strong skepticism about these moves. Marc Rotenberg, EPIC's executive director, called the seal-of-approval programs "do-it-yourself" privacy policies, charging the industry with launching PR efforts to stave off government regulation.[130] "It's time to move beyond public relations," he told the press, "and get on with the hard work of privacy protection," warning that "time is running out."[131]

Meanwhile, as the online industry continued to ask for extensions on the deadline to develop self-regulation, Congress already was moving on a new law to protect children's online privacy. The proposed legislation would authorize the FTC to develop guidelines requiring parental permission before commercial-Web-site operators could collect personally identifiable information from children.[132] While there were still bills in Congress that would have provided much broader protections for all Internet users, they did not appear to be going anywhere. "Congress and federal regulators appear to be leaning toward giving Internet marketers an additional six months to come up with effective self-regulation of privacy issues before resorting to new laws," observed *Advertising Age*.[133] But for the time being, the Congressional focus was on children.[134]

The Devil's in the Details

By the time the children's privacy legislation was introduced in Congress, the basic framework of the new policy already had been established. The legislative aides and FTC staff who drafted the language for the new law drew heavily from the CME/CFA proposal.[135] Under Senate bill 2326, introduced by Richard Bryan (D-NV) and cosponsored by John McCain (R-AZ), the FTC would be directed to develop a set of rules to govern children's privacy on the Internet. Commercial-Web-site operators would be required to: (1) provide clear notice of their information collection and use practices; (2) obtain verifiable parental consent prior to collecting personal information from children under thirteen; and (3) require the commercial-Web-site operator to ensure data quality and security. The legislation also allowed for "safe harbors" that would enable self-regulatory groups to develop their own guidelines, as long as the FTC approved them and found them to be in compliance with the federal regulations.[136]

One of its most controversial passages was the provision that would allow parents to find out what information their children age 16 and under had provided to commercial Web sites and to have that data removed. The American Library Association pounced on the proposal, warning that it "would prove damaging to the rights of teenagers and burdensome to their parents." Civil-liberty and education groups also opposed the provision that would require parental permission before teenagers could provide personal information to commercial Web sites, arguing that such a requirement would restrict teenagers' access to information online. "Should mature minors over 12 be treated as children for the purposes of privacy legislation?" the ALA asked in its Congressional testimony. "Teenagers have independent rights to free speech and privacy that would be severely compromised if parental notice were required each time they engaged in a transaction with a commercial Web site. If a teenager is able to sign up for a contest or a newsletter in a bookstore without parental supervision, what rationale is there for imposing parental notice requirements to do the same at Amazon.com?[137] Civil-liberty groups expressed concern that conservative families might not let adolescents have access to information about birth control, abortion, and other crucial issues. While the proposed law would not apply to noncommercial Web sites, there was concern that it might set a dangerous precedent that could be expanded beyond the marketing practices of commercial Web sites.

For the child-advocacy and consumer groups leading the campaign for children's online privacy, the age issue proved to be a vexing challenge.[138] The question of when a child is mature enough to navigate the Internet without the oversight of parents was at the heart of many public debates over the new digital media. In media law, there was legal precedent for safeguards that would apply to children 12 and under, including the FCC's guidelines on children's advertising.[139] The research that had led to those rules showed that young children would be particularly vulnerable to the appeals for personal data from online marketers.[140] But at what age were they old enough to supply this information without parental consent? In many legal matters that age would be 18, when a child reaches adulthood. If children were not considered old enough to sign a contract on their own, should they be allowed to give out personal information on the Internet without asking their parents first? There was no academic literature in this new area to provide guidance to policymakers.[141]

The Center for Media Education shared the concerns about teenagers' right to privacy and access to online information. CME and its coalition also knew that opposition from such powerful groups as the ALA and the National Education Association to this one provision of the legislation could undermine key support for the entire bill. They had to find a middle ground. Teenagers should not be left completely out of any new law to protect privacy on the Internet. Teens already were one of the most sought-after demographic groups online, a key target in the burgeoning e-commerce marketplace. And while adolescents clearly were more sophisticated than younger children, they still could be highly impressionable and less likely than adults to foresee the consequences of their actions. Teenagers do not suddenly become more privacy savvy once they turn 13. In testimony before the Senate, CME called for "fair information practices" on teen Web sites to "help ensure that these young people become thoughtful, responsible consumers." At the very least, the group argued, Web sites targeted at teens should display clear, understandable privacy policies, and the opportunity to choose whether or not to give any personal information (the "opt-in" model).[142]

But while CME, the Consumer Federation of America, and the National PTA fought to maintain some protections for teenagers, they were ultimately unable to do so. By the end of the summer, the Children's Online Privacy Protection Act was on a fast track in Congress. Sticky issues concerning age, as well as the mechanisms for securing parental permission, were worked out in a series of negotiations between industry and consumer groups. The FTC convened the stakeholder organizations together to hammer out the details of the final bill's language. The online industry had forged alliances with some of the most prominent organizations in the education and civil-liberty communities, creating a strong voice in the closed-door negotiations.[143] With the ACLU, NEA, and ALA opposed to privacy safeguards for teens, the language quickly was stricken from the bill. Industry also pushed to allow commercial Web sites more freedom to interact with children—through e-mail and online newsletters—with minimal or no involvement of parents.[144]

By the time the legislation reached the floor of Congress in October, its focus had shifted to ensure buy-in from key lawmakers on Capitol Hill. One outcome was a much stronger emphasis on child safety and less focus on marketing. In an internal memo to his colleagues in Congress, Senator Bryan laid out COPPA's goals:

> (1) to enhance parental involvement in a child's online activities in order to protect the privacy of children in the online environment; (2) to enhance parental involvement to help protect the safety of children in online *fora* such as chatrooms, home pages, and pen-pal services in which children may make public postings of identifying information; and (3) to maintain the security of personally identifiable information of children collected online.[145]

Even as the legislation headed for its final passage, some industry trade groups continued to argue for a self-regulatory approach. In the final Senate hearings in September, Elizabeth Lascoutx of CARU warned that "too much regulation could slow down the growth of the Web."[146] But the major companies already had decided to cut their losses and endorse the privacy legislation, particularly with the White House supporting the bill, the EU deadline imminent, and other bills affecting their business before the Congress. "Although we usually support self-regulation of electronic commerce, we believe it may be appropriate to consider targeted legislation in this area," Time Warner Vice President Arthur Sackler told a Senate panel.[147]

The online industry had much at stake in a flurry of legislative items that were being pushed through during the final days of the Congressional session. With midterm elections less than a month away, intense lobbying was underway on several key Internet bills that would benefit e-commerce businesses. The Digital Millennium Copyright Act was aimed at strengthening industry's ability to control copyright in new digital media, making it a crime to circumvent technologies designed to protect digital copies of software, music and videos, and literary works.[148] The Internet Tax Freedom Act, which was being pushed by the White House, would ban new Internet taxes, preventing state and local governments from imposing any levies on Internet businesses.[149] The children's privacy bill already had been attached to the Internet tax legislation in the Senate, along with a number of other amendments.[150]

On the overseas front, U.S. officials reported progress in their negotiations with European regulators. Secretary of Commerce Daley told a congressional hearing at the end of July that the European Union was likely to accept the privacy standards being created by U.S. industry. As he explained to lawmakers, U.S. representatives had made significant headway in their efforts to make EU officials "fully aware of the progress that U.S. industry is making in implementing privacy protections."[151] Though Daley did not mention the children's privacy legislation that was currently being decided in Congress, the Europeans knew that the pending law was part of the overall mix that the U.S. government had concocted to allay concerns that the United States had reached an "adequate" level of data protection. At the same time, the Europeans were not eager to disrupt business relations with the United States.[152]

To demonstrate its commitment to self regulation, the U.S. online industry launched a high-profile public-service-ad campaign on October 7, less than two weeks before the EU deadline. TRUSTe, in partnership with America Online, Yahoo!, and dozens of other companies, planned to flood the Web with ads aimed at educating the public about protecting their personal data and urging online companies to adopt privacy policies. Slogans for the month-long campaign included: "Concerned about online privacy? . . . (we are, too)," "Do you trust the Web? . . . should you?" and "Privacy is everybody's business." For privacy advocates, however, the slick PR campaign stood in sharp contrast to how the rest of the world protected consumer privacy. An EPIC report showed that "nearly all industrialized countries have either adopted or are in the process of adopting comprehensive privacy laws."[153]

A Congressional Field Day for e-Commerce

Having passed the House and Senate in early October, the Children's Online Privacy Protection Act (COPPA) was headed for the full Congress, bundled into a large, omnibus budget bill. Thoroughly vetted by all parties, COPPA's passage was practically assured. Sharing ridership with it on the budget bill by now, however, was a much more controversial piece of legislation with an acronym that appeared to parody that of the privacy bill. As industry lobbied for government favors, conservative lawmakers, joined by right wing groups, upped the ante on the Internet e-commerce interests by pushing through a revised version of the failed Communications Decency Act. The Child Online Protection Act (COPA), sponsored by Senator Dan Coats (R-IN) and Congressman Michael Oxley (R-OH), was fashioned more narrowly than its predecessor, in an effort to thwart constitutional challenges. The bill would "impose criminal penalties on commercial Web sites that allowed children to access material considered harmful to minors."[154] The choice of name appears to be no accident. The bill's authors clearly wanted to link the privacy and antipornography provisions together, not only in the legislative vehicle, but also in the public mind. In a statement released by key members of the House Commerce Committee, lawmakers promised that passage of the new law would help assure "a day when young people are safer on the World Wide Web." Congressman Thomas Bliley (R-VA), chairman of the House Commerce Committee, commented: "Parents do not want their children visually assaulted by pornography on the Web. And they do not want commercial Web sites collecting information from children without parental approval."[155]

As they had done during the debate over the Communications Decency Act, civil-liberty groups mobilized against COPA. But this time, neither the online industry nor the White House did much to prevent its passage. Some companies made efforts to protect their own financial interests, but with so much at stake in other measures before the Congress, the online industry did not campaign against it. As *Wired News* noted, "While few companies want to oppose the legislation in public, several major Internet players have tried to thwart or scale back the bill in behind-the-scenes lobbying. They argue the proposals could inadvertently subject them to liability." America Online and Walt Disney called for broad exemptions to protect companies that primarily were not engaged in selling pornography.[156] President Clinton and the Justice Department initially raised objections to the legislation's possible First Amendment violations, and threatened to oppose it, but later decided otherwise. "Instead," reported the trade publication *Newsbytes,* "the White House will let that act and the Internet Tax Freedom Act enjoy a relatively free ride on the omnibus spending bill that is readying itself for Congressional approval and the President's signature."[157]

With opposition to COPPA effectively silenced, the entire package of Internet bills quickly sailed through Congress, "altering the course of Internet policy as no U.S. Congress before it has done," Will Rodger noted in *Interactive Week:*

> In a matter of 48 hours, the White House and lawmakers agreed to put a moratorium on further taxation of the Internet; restrict children's access to online pornography; even as a constitutional challenge looms; protect internet users' privacy online; double the number of temporary visas granted to high tech workers; and require the federal government to post all its forms on the Internet, as well as accept electronic online signatures in lieu of the old fashioned pen and paper kind.[158]

Just days before, Congress had passed the Digital Millennium Copyright Act.[159] Commenting on the legislative measures, Ira Magaziner told the press, "I think we've gotten our major priorities. . . . We are feeling pretty good right now."[160]

EPIC, along with the ACLU, EFF, and other civil-liberty groups, immediately challenged COPA in court.[161] As for COPPA, EPIC's David Sobel told the press: "It's a start. But we think that everybody, regardless of age, needs privacy protection on the Internet." The group vowed to continue its campaign to extend government privacy safeguards for all Americans.[162]

The deadline for the EU Data Directive passed a few days after COPA became law. Despite the fact that the United States had failed to pass a comprehensive privacy law, there was no disruption of business. U.S. officials had succeeded in convincing the European regulators to agree to further discussions on a set of new "safe harbor" provisions to ensure data protections for consumers dealing with U.S. companies. The press noted that talks between the United States and the European Union were underway, "focused on introducing model contract clauses, as a middle ground between self-regulation and enforced legislation."[163] Secretary of Commerce Daley sent a letter to online-industry leaders a week after the EU deadline, informing them of the status of the negotiations:

> In an effort to find ways to bridge differences in our approaches to privacy, the U.S. Department of Commerce, on behalf of the U.S. Government, and Directorate General XV of the European Commission have been engaged in a dialogue on privacy for the past several months. We have discovered that, despite our differences in approach, there is a great deal of overlap between U.S. and EU views on privacy. Given that and to minimize the uncertainty that has arisen about the Directive's effect on transborder data transfers from the European Community to the United States, the Department of Commerce and the European Commission have discussed creating a safe harbor for U.S. companies that choose *voluntarily* to adhere to certain privacy principles.[164]

It would take another two years before a final agreement on the safe-harbor concept was reached, but ultimately the Clinton Administration succeeded in heading off passage of a comprehensive law to protect consumer privacy on the Internet. Continuing with its "sectoral" approach to privacy policy, the U.S. government passed several laws in the next few years to address privacy issues in the health-care and financial industries.[165] Privacy groups continued their advocacy efforts throughout the next decade to push for broader consumer protections, but in the wake of the terrorist attacks on September 11, 2001, these organizations were forced to shift some of their attention away from concerns raised by business to the growing privacy threats posed by new government homeland-security policies.[166]

Post-COPPA

COPPA took effect in 2000, at the same time that the dot-com market began its first downward move, prompting some industry complaints about the cost of complying with the law. "If you're a children's site and are not compliant, the damning part of that is the potential to lose ad revenue or partnerships or whatever is critical to your business," one industry analyst told the press.[167] While some children's Web-site companies went through the cumbersome task of offline parental permission before collecting information from children, many simply chose to stop the practice of collecting personal information from children altogether. At Barbie.com, Lego.com, and Nick.com, for example, children could register by using screen names, enabling them to interact with the site without disclosing any specific information about themselves.[168] A number of Web sites with extensive data-collection practices were forced to examine their lists of registered users and purge those who were under 13.[169] One online executive told the press that it would have cost his site $25,000 a year to screen for subscribers younger than 13 and obtain parental permission before collecting their personal information. "We kicked them all off rather than institute a credible verification," he explained.[170]

COPPA was by no means a perfect regulatory solution. Requiring Website operators to receive parental permission before collecting personal information from children was an extension of the new "parental empowerment" approach that the government had taken on online indecency and television violence. As such, it placed new responsibilities on parents of younger children, who were now expected to read carefully the privacy policies on children's Web sites before agreeing to let children participate in the data-collection efforts of these companies. For over-stressed parents who knew less than their children about the online marketplace, and even less about the business models of these digital enterprises, this was a daunting task. The law also contained some loopholes that made it easy to circumvent the requirement for prior parental permission.[171] Industry lawyers quickly plunged themselves into the nuances of the rules, advising clients on creative ways to comply without disrupting lucrative business practices. An online article titled "COPPA Loopholes, Demographics, Creative Samples" carried this piece of advice:

> If you're sending periodic email newsletters to children, all you have to do is notify the parent and that's enough. However, you can't do anything else with the child's data or email them anything but the newsletter. Announcements of sweeps or sales, then, would be out of the question—unless, of course, you send it under the guise of the newsletter.[172]

Nor did anything in the law prevent a child from simply lying about her age. On Web sites for teens and adults, for example, where a child's age was asked in order to enter the site, any smart kid could figure out pretty quickly what birth date to put down in order to gain entrance. Some industry insiders were quick to point this out. "If a 15 year-old wants to buy cigarettes, they will find a way to do it," commented a CEO of a popular teen site. "The Internet is no different. If someone wants to lie about their age, they will find a way to do it. The best protection is at the computer, and mom and dad need to supervise their kids. This law shouldn't be seen as a safety net. The only safety net is mom and dad."[173]

COPPA rules applied only to children under 13, offering no protections for teenagers. As a consequence, teens were fair game for marketers. Yet assumptions about adolescents' ability to make mature decisions in the online marketplace may have been off base. For example, a 2000 study by the Annenberg Public Policy Center found that, if offered enough money or gifts, not only children 10 and up but also teenagers as old as 17 were quite willing to give up their names and addresses, along with other personal information, to commercial Web sites. As one commenter noted: "The FTC rules may consider youngsters 13 and over to be 'adults' when it comes to disclosure of information on the Web, but 60 percent of parents surveyed say they're more worried about what information a teenager would give away to a Web site than a younger child." In fact, public opinion appeared to be in sharp conflict with public policy, with almost all the parents in the survey (96 percent) believing that teenagers should be required to get their parents' consent before giving out personal information online.[174]

But despite these flaws, the new children's privacy rules succeeded in curtailing some of the more egregious practices that might have become commonplace without some government oversight. A CME study a year after COPPA took effect found that the number of children's Web sites collecting street addresses from children had dropped significantly, from 49 percent in 1998 to less than 20 percent in 2001. Only 10.7 percent of the sites were collecting telephone numbers, compared with nearly 25 percent three years before.[175] The COPPA rules, applying across the board to all online commercial ventures targeted at children, placed a damper on the explosion of database business models that were so swiftly moving into place on the Web. At least where younger children were concerned, it would be harder to set up Web sites that were built on massive amounts of data collection. The law stipulated the data collection had to be minimized and required that companies clearly disclose what they were collecting from children and what they were doing with the information. This kind of transparency put these companies into the spotlight, where advocates as well as parents could scrutinize their practices more closely, and could publicize them to the press. Marketers knew this and were reluctant to take that risk during this period of heightened awareness.

In 2001, the FTC imposed fines on several sites found in violation of COPPA, charging the companies operating Girlslife.com, Bigmailbox.com, and Insidetheweb.com with collecting full names, phone numbers, home addresses, and e-mail addresses without first obtaining parental permission.[176] Yet when EPIC filed a complaint against Amazon.com two years later, the commission denied it. EPIC charged that the Toy Store section of the Amazon.com site was aimed at children and had unlawfully collected personal information from them. The site encouraged children under thirteen to write reviews of products and sometimes the children provided their own names and other personal information. But the FTC ruled that Amazon.com did not know that children under thirteen were supplying such information.[177]

COPPA contained its own safe harbor provision designed to enable industry self-regulation to work hand-in-hand with government oversight. Self-regulatory guidelines had to be vetted first with the FTC, with input from consumer groups, before they would be approved. As a consequence CARU, TrustE, BBBOnline, and other self-regulation organizations that had been actively seeking to prevent passage of privacy legislation found themselves in the position of implementing the new law. Their role in educating the industry and monitoring compliance with the law helped encourage widespread adoption of the safeguards.[178]

But while COPPA forced online marketers to change some of their operations, many managed to shift their strategies in order to continue reaching their target demographic group of digital kids. Five years after the law took effect, industry observers looked back on the development of the children's Internet marketplace. "Part of the reason kids advertisers didn't originally flock online is that it is one of the most restricted media venues out there," noted *Mediaweek*. "The basic gist of COPPA and CARU's rules is don't sell to kids directly, don't deceive them and don't take their personal information without parental permission. Therefore, Web sites tend to err on the side of caution." But "there is more than restraint at work here," the article noted. As one industry insider put it, "That stuff doesn't work against the target." Marketing to kids—"who have been online since they were old enough to click a mouse"—required "a far more interactive marketing approach," the article explained. Therefore online advertising to kids began to move toward "immersive" games and contests (or what had become known in the industry as "advergaming"). Among the businesses engaged in these online-marketing strategies were some of the very companies CME criticized in its 1996 report—including Post's Postopia.com, Kraft's Candystand.com and Frito-Lay's INNW.com.[179]

Long after its high-profile role in the controversy over children's privacy, KidsCom remained on the Web, offering a colorful mix of "cool stuff," "chat & buzz," "adventure," and interactive games. Children who registered on the site still could win KidsKash points to redeem in the Lootlocker. But the site's privacy policy had been revamped carefully to comply with the new privacy safeguards:

> KidsCom.com follows the FTC guidelines for compliance with the Children's Online Privacy Protection Act and Rule. Verifiable parental permission is required prior to the collection of any personal information. No personal information provided to KidsCom.com by its registered KidsCom.com Club users is sold or rented to any third parties.[180]

KidsCom founder Jorian Clarke, whose company was renamed Circle 1 Network, was frequently on the conference and press circuit, billed as a seasoned expert, "providing advice on kids, tweens, and teens on the Internet and online safety since 1995." Listed among the highlights on her bio was the fact that Ms. Clarke "has been involved in setting direction for industry through talks (by invitation) for the U.S. Commerce Department and the Federal Trade Commission on Internet safety and privacy."[181]

5 Born to Be Wired

"I have erratic mood swings, and can go from mild-mannered bookworm to psychotic mad-girl on a rampage in less than five seconds."[1] These angst-ridden words, written by fourteen-year-old "Katie," expressed the common frustrations and emotional fluctuations of many teenagers. In an earlier time, Katie might have handwritten them in a leather-bound diary, snapping the tiny lock shut before stashing her private broodings away in a drawer or closet. But in the digital era, she posted the words on the Internet for millions of people to read. This candid entry was part of the Me page of her personal home Web site, presented with an array of other very personal information carefully selected to show the world who she was. Her musings were documented in Professor Susannah Stern's study of girls' home pages.[2] On the Web, Stern explained, these girls appeared "eager to disclose thoughts and ideas they feel unable to utter in 'real' life." They often used their home pages as "venting stations, to get off their chests the dark and unlady-like feelings that engulf them." Their home pages functioned as public sounding boards, places where they could "receive feedback from faceless comrades and critics regarding their creativity and insight."[3]

In many ways, teens were the *defining users* of this digital media culture. With nearly three-quarters of twelve- to seventeen-year-olds online by the end of the twentieth century, teenagers embraced this new online world with great enthusiasm, responding eagerly to its invitation to share ideas, contribute content, and otherwise place their stamp on a media system that they themselves could create and manage. The Internet played a pivotal role in their lives, influencing their family and social relationships. Teens far exceeded adults in their use of instant messaging, visits to chat rooms, and playing or downloading games.[4] At the forefront of technological innovation, they eagerly adopted cell phones, pagers, and other new electronic gadgets that offered unprecedented levels of communications and information retrieval at their fingertips.[5] The properties of the interactive media were uniquely suited to their developmental needs. Online communication tools enabled instantaneous and constant contact with peers; personal Web pages offered compelling opportunities for self-expression and identity exploration; and ubiquitous portable devices facilitated mobility and independence. The Internet's extensive reach and its promise of anonymity created an environment that encouraged information-seeking on a wide variety of topics inaccessible or taboo for earlier generations. Anonymous chat rooms and forums allowed teens to engage in discussion and debate without fear of exposure. The digital media were challenging many of the conventions and institutions of the past, blurring or obliterating the boundaries between public and private, commercial and noncommercial, school and home, local and global.

Teens were also playing a central role in the new digital marketplace. Long before the advent of the Internet, market researchers engaged in an ongoing campaign to track the valuable—but elusive and quixotic—teen demographic, frequently enlisting teenagers in a variety of street marketing and "coolhunting" schemes to identify the latest fashion and music trends.[6] The rapid growth of interactive technologies enabled marketers to gain unprecedented direct access to teenagers. "Cyberteens" were key pathfinders in the unfolding digital landscape. "The current generation of teens is more tech savvy than any segment of the population," noted a report by Cheskin Research. "As this generation matures, their attitudes and experiences on the Internet will vastly influence its future evolution. By investigating teen behaviors and attitudes today, companies can not only address the needs of this market, but also predict future trends in technology."[7]

Unrestrained by government regulation and largely under the radar of parents and policymakers, advertisers were able to develop a full array of new digital-marketing strategies, especially tailored to the developmental needs and interests of adolescents. Enlisting the expertise of psychologists and anthropologists, marketers continued to probe the inner workings of the teen psyche, designing marketing strategies that would cater to them. The interactive media ushered in an entirely new set of relationships, breaking down traditional barriers between content and commerce, and creating unprecedented intimacies between teens and marketers. Marketers were able to shadow teens' every move through the expanding media environment, identifying new targeting opportunities along the way. Brands and corporate logos tagged along with teens in their journeys through cyberspace, inserting themselves into instant messages and chat rooms, popping into e-mail in-boxes, and beckoning from cell-phone screens.

As the features and functions of this new digital online marketplace were being tested and refined, teens served as subjects for continuous and pervasive market research, connected through electronic umbilical cords to a phalanx of companies that monitored their every online move. As a consequence, a unique symbiosis developed between the corporate creators of this digital culture and its most avid users; many teens were eager and willing partners in the design and implementation of an array of cultural products and practices that would help shape the electronic media system in the twenty-first century.

The Quest for Identity

As children make the critical transition through the teenage years, they undergo a number of biological, cognitive, and emotional changes. One of the most fundamental is a quest for their own unique identities.[8] Media always have played a role in this process, exposing teens to a wide variety of cultural role models and enabling them to explore their "possible selves."[9] Adolescents "use media and the cultural insights provided by them to see both who they might be and how others have constructed or reconstructed themselves," observed communications scholar Jane Brown.[10]

In their study of "adolescent room culture," Brown and her colleague Jeanne Steele discovered that a teen's bedroom serves as a key site for identity formation. Bedrooms are "an important haven for most teenagers, a private, personal space often decorated to reflect teens' emerging sense of themselves and where they fit in the larger culture." In these hallowed and private spaces, they observed, "teens listen to music, read magazines, watch television, do homework, and consider the events of the day," where they "appropriate and transform media messages and images to help them make sense of their lives."[11] Media scholars also learned that teens are "active users" of media, choosing from a wide range of music, magazines, movies, and television programs to create their own personalized cultural collages. Teenagers often selectively use media when they are alone to cultivate a "private self."[12] For example, music not only helps teenagers grapple with stress, loneliness, and depression, but also conveys important messages that can help them differentiate themselves from their families and other people.[13]

With the advent of the Web, teens had new tools to help them with this essential psychological process.[14] The unique nature of the new digital environment presented an unprecedented array of possibilities for communication, interaction, and expression never before available.[15]

Home (and Abroad) on the Web

Home pages served as the virtual equivalent of a bedroom wall, where teens could select, arrange, and display the images from popular culture, as well as their personal writings. In "real life," adolescents could not easily change friends or clothes on a daily basis. On the Internet, however, a teenager could try out a new self (or selves) literally every day. The teen girls that Susannah Stern studied used their personal home pages for "self-clarification," employing a variety of techniques and tools to "explore their beliefs, values, and self-perceptions" to help them answer the fundamental questions: "who am I, and who will I become?"[16] As Stern explained:

> Not only did the girls provide elaborate descriptions of themselves as they are now, but many chronicled their pasts and speculated about their futures. In addition to detailed descriptions, most also passed judgement on themselves and their histories, and indicated distinctions between who they considered and wished themselves to be. Images and links assisted the girls in clarifying themselves, because they suggested who and what girls identified with from among the vast array of possibilities offered by their culture. Finally, girls articulated their thoughts on a number of substantive issues, especially religion and sex, differentiating their personal perspectives from those assumed to be more widely accepted.[17]

Online companies helped facilitate the creation of home pages by offering hosting services and access to user-friendly tools. Many Web sites provided special areas that encouraged young people to create their own personal Home pages. Teens could assemble entire collages of media products gathered both on and off the Web with great ease. Merely by clicking on an image on another Web page and saving it to a file, they could accumulate pictures of events, places, and personalities to paste onto their own pages. They could feed video and audio into their computers so visitors to their sites could experience their musical and visual tastes, and they could scan in images they had clipped from magazines or other printed media. Online companies streamlined the incorporation of pop-culture images into these new virtual bedroom walls. For example, Angelfire, a division of Lycos and one of the more popular personal-Web-site hosting services, offered a guide to the construction of celebrity fan sites.[18] These new tools for self-expression could be very empowering. "Being able to produce Web pages," wrote scholars Daniel Chandler and Dilwyn Roberts-Young, "is like owning your own printing press."[19] By 2000, 24 percent of all online teens had created their own Web pages.[20]

More than half the girls in Stern's study had taken advantage of the free space provided by commercial online-hosting services. As she pointed out, however, there was a quid quo pro: "in exchange for sponsoring their home pages, such companies retained the right to advertise on their members' home pages." This arrangement bothered some of the girls, who saw the advertising as an intrusion. Stern quoted a complaint posted by one girl:

> Ok, first I'm gonna bitch about these freakin Angelfire banners they foreced us to put on the sight [sic]. . . . Yes, they've made us put adds on our pages, yes it's a pain in the ass, yes I'm NOT moving my page. I'm just not. I don't have the time or see the point in moving my page when possibly where ever I relocate to they'll make me eventually put up adds [sic] there.[21]

But for most teenagers, carrying a banner ad was a small price to pay for the opportunity to own one's small piece of the Web.

With cyberspace, there also was a limitless and anonymous audience. "On the Web," noted Daniel Chandler, "the *personal* function of 'discovering' (or at least clarifying) one's thoughts, feelings, and identity is fused with the *public* function of publishing these to a larger audience than traditional media have ever offered."[22] The teen girls in Stern's study created their personal Web sites with the clear intention of communicating with the rest of the world. "With diaries," she observed, "there is always the risk that someone will find it; with home pages, the risk becomes an expectation." But the distinctions between public and private were not always that clear. "Perhaps the sense of one's homepages as being part of oneself," Chandler and Roberts-Young point out, "leads some authors to slip into feeling that it is a purely private dream space."[23]

While unlimited opportunities for expression were liberating for many teens, they sometimes could create problems. At a middle school in Massachusetts, three eighth graders were suspended after making unkind comments on the Web about their classmates.[24] When a high school senior in Belleview, Washington, created a Web site that was a parody of the school newspaper, school officials retaliated by withdrawing his recommendation for a National Merit Scholarship. The ACLU came to the student's aid and managed to win a "very favorable settlement" from the school district.[25] Such incidents sparked debates about the boundaries between school and home, the roles of traditional authorities, and the level of First Amendment freedoms for young people in the Digital Age. School administrators, already grappling with "acceptable use" policies necessitated by fears of online safety, indecent Internet content, and filtering software, were forced to come up with more rules to govern what students could say in the privacy of their own homes.[26]

Chat-Room Culture

If personal Web pages created a new virtual forum for identity exploration, chat rooms offered another set of possibilities, allowing teens to create personas quite distinct from their everyday identities, and to act them out in cyberspace. A 2001 study of media use by the Kaiser Family Foundation found that nearly three-quarters of teens were participating in chat rooms.[27] Another survey by the Pew Internet & American Life Project reported that more than half of all online teens had visited a chat room.[28] A quarter of the teens in the study told researchers they had pretended to be a different person when they were communicating online, either in a chat room, instant messaging, or sending e-mail. Because they cannot be judged by their appearances when they are online, many teens felt they could be their "true selves." Teenagers also were learning that deception was very easy in the online environment, and chat rooms were becoming a place where such practices were commonplace.[29]

Teen chat-room visitors created entirely new languages and codes for describing themselves to others online. On one level, these codes were necessary tools to provide information (such as age, sex, and location) that are taken for granted in face-to-face interactions, but missing in online communication. But as a team of UCLA researchers found, the names that teens created for themselves often were designed to display characteristics that would "proclaim identity, fit in with the peer group, and attract potential partners." Playful online nicknames ("nicks")—like PinkBabyAngel1542, MizRose76, Rollerbabe904590, Sportyman04, and DustinKnowsAll—were the virtual equivalent of the teen fashion statement, broadcasting to others a particular image of how teens wanted to see themselves and have others see them. As the authors explained, these "chat codes" helped teens "address important developmental issues, such as a concern with their sexually developing bodies, romantic partner selection, and gender identity in an environment where there is no physical embodiment of physical identity." Teenagers quickly assimilated into a new "common peer culture" on the Internet, with unique abilities to tap into some of the "essential concerns of adolescence."[30]

Teens also began to use chat rooms to explore their developing sexual selves. Research by the UCLA team found that the online medium provided "a relatively safe place to 'practice' new kinds of relationships such as dating, that can be risky in the real world." Teens entered chat rooms to talk about sex, learn about sex, and engage in what experts referred to as "pairing off" with potential sexual partners. But while in the offline world, these encounters were taking place in the high school cafeteria, the movie theater, or other social spaces, the online environment required its own set of virtual meeting places. For example, participants seeking sexual partners could engage in the "cyber pickup," where two individuals in the chat room decide to move away from the larger group into the more private, one-on-one space of instant messaging. "Internet dating allows teenagers to maintain their anonymity and may lessen the emotional pain often associated with face-to-face dating," the researchers concluded. "It may also allow girls to assume more authority in their interactions since obtaining and maintaining relationships is based on verbal skill, generally accepted as a female strong point."[31]

Not all of the encounters in chat rooms were completely innocent, however. UCLA professor Patricia Greenfield entered chat rooms herself to experience what it was like for teens and found that teens routinely were exposed to pornography and sexual advances from strangers. "The sexuality expressed in a teen chat room was public, linked to strangers and had nothing to do with relationships," she explained in a press interview. "It was very explicit and focused on physical acts, and often associated with the degradation of women."[32]

But despite these troubling incidents, research found that most online interaction by teens was with friends and family.[33] While many parents remained fearful of teenage forays among strangers in the wild frontiers of cyberspace, the Internet was evolving into a social networking and communications medium, an increasingly vital tool for peer-group interaction.[34] Far from an isolating force, the Internet could help young people forge bonds with others and manage their social networks.[35] An AOL study found that "The Internet has become the primary communication tool for teens, surpassing even the telephone among some groups."[36]

The Instant-Message Generation

By 2000, nearly 13 million teenagers were instant messaging (IMing), almost double the number of adults engaged in the practice. "Talking to buddies online," explained a study by the Pew Internet & American Life Project, "has become the information-age way for teens to hang out and beat back boredom."[37] Instant messaging was so widespread among teens that the Pew Project tagged them the "Instant Message Generation."[38] Most of the IM services were provided free, and the most popular were AOL's Instant Messenger (AIM), ICQ, MSN Messenger (MSN) and Yahoo! Messenger.[39] AIM dwarfed all other competitors in the field, and its feature set, which included a collection of graphic "emoticons" for the shorthand expression of a range of feelings, proved enormously popular with young people.[40]

No other feature of Internet communication fit so well with the needs and desires of the American teenager. Just at the point in their lives when adolescents were turning away from their families to their network of social relationships, the Internet provided a perfect vehicle for facilitating that process. IM was particularly suited to the hectic lives of the so-called "hurried child."[41] School often left too little time to be with friends, so the first thing teenagers would do is to "log on and open the IM channel to meet them again."[42]

Instant messaging helped young people conduct some of the difficult aspects of personal relationships that plague teenage years. Many of the teens in the Pew study reported that they were using IM as a key tool for dating, finding it a less daunting way than the telephone to ask someone out for a first date. Although teens reported that lack of visual and aural cues in instant messages sometimes led to misunderstandings, this feature was also viewed as an attribute. For example, many teenagers used IM to say something they would not have been able to say to someone's face. As one seventeen-year-old girl reported: "Online you can think things over, and erase them before you look stupid, rather than to their face, where you can't always take things back." Some also said they found IM an easier way to break up with a boyfriend or girlfriend and 13 percent of the IM users said they had used it for this purpose. Others noted that "sometimes it is easier to say what is in your heart online. . . . You can type the words and hit send instead of freezing up in person."[43]

Like Home pages, IM often confused the boundaries of what is public and what is private. For example, unlike e-mail, where a record of the communication is kept automatically, instant messages are ephemeral, vanishing as soon as the exchange takes place. But they also can be logged or copied and pasted into e-mail messages and sent to others. Thus, an intimate conversation between two friends could be broadcast to others, betraying an expected confidentiality. These new, sophisticated tools enabled a host of complex online relationships, but they required special skills to manage not only teens' social relations, but also the technology itself. Teens quickly became adept at carrying on simultaneous IM conversations with their friends. As one study found, teenagers could participate in a group discussion and at the same time conduct multiple one-on-one conversations on the side, "often with some of the same people involved in the group conversation."[44]

Just as teenagers regarded their bedrooms as a private haven, many also used IM to "carve out a private world within the public space of the home." As researchers Rebecca Grinter and Leysia Palen found, some teens often "underplayed" their involvement with IM "as a means to help create this separate world, keeping it below the horizon of notice." This was particularly important for younger teens who had less independence in the family, and "who appear to benefit most from this by creating private spaces over which they can exert newly found control." While many parents might not want their teenagers to talk on the phone or have friends over during certain times of the day, they often did not even notice that their children were IMing. And teens were not going out of the way to let their parents know. As the researchers noted: "In domestic ecologies, IM operates 'below the radar': it is a quiet technology that is easily integrated into the conduct of other activities. . . . Use can be unobtrusive, go unnoticed, or even be covert."[45]

Though IM still lacked a workable business model for "monetizing" its services, its value was unquestionable, particularly in the teen online market. "Instant messaging might seem like nothing more than a touchy-feely service for teen-agers now," wrote Louise Rosen in *UpsideToday*, "but it's one of the stickiest applications on the Web, and in the future there are potential revenue streams that could yield billions of dollars. IM can drive up a site's traffic and brand awareness. It will be an important feature of interactive television; it has the potential to be used as a killer direct marketing tool, and can add real-time customer services to a site."[46] "Stickiness" became a core strategy in the digital marketplace. Sticky applications were designed to keep consumers tethered to the Web, engaged for long periods of time in online activities, and eager to return.[47]

Commodifying Identity in the Dot-com Era

Beginning in the mid 1990s and continuing through the dot-com boom, the Internet spawned hundreds of Web sites, portals, and other content networks created especially for teens. The creators of this online commercial teen culture included both the familiar and established media conglomerates as well as new Internet start-up companies.[48] Stressing interactivity, communications, and creative new marketing practices, these sites offered a preview of cultural trends that soon would become part of mainstream media. Much of the content on these teen Web sites was an extension of the popular culture aimed at teens in such "offline media" as music, games, movies, and television. But the unique nature of the Web enabled it to provide a much richer experience than was available in conventional media. Teens also could find an abundance of valuable information online about a broad range of topics that were directly relevant to their lives. The Internet allowed teenagers to form communities with their peers, express themselves through writing and art, and engage in political activism. At the same time, the myriad activities that defined much of this Web culture for teens were deeply rooted in the business imperatives of the new digital economy.

Teens were at the center of the fundamental reconfigurations in the way that music was distributed and consumed in the Digital Age. Music videos had been available for online viewing for a few years, on Web sites such as MTV.com and individual record label and artists' sites. Fans could go to these sites to learn more about their favorite performers and (if they had a sufficiently fast connection) to view entire music videos. Some of these online music videos began taking advantage of the Internet's capacity for interactivity. MTV.com was experimenting with a type of music video called "Webeos," videos designed for Web broadcast that allowed viewers to control some of the sounds and images.[49] The Internet made it possible to bypass the music store and download the music directly to one's own computer. With the 1999 introduction of Napster—a commercial peer-to-peer music downloading service that quickly attracted more then 50 million users—this practice became one of the most popular uses of the Internet among teens.[50] Music file-swapping became particularly prevalent among teen Internet users. High school faculty reported that downloading music was one of the most popular uses of their schools' high-speed computers. "Buying a CD just isn't very useful anymore," commented one high school sophomore. "I can get any CD I want off the Internet."[51]

Brands tailored their Internet marketing strategies to teenagers' fundamental needs for identity development, self-expression, and peer group communication. For example, the prevalence of surveys and questionnaires on these Web sites served dual functions. For teens, they were testing grounds for identity exploration, helping them work through how they perceived themselves and how they wanted to be seen in the world. For companies, they were effective tools for ongoing market research. "As teens' life-stage task is to sort through all kinds of identity issues," explained Julie Halpin, CEO and cofounder of the Geppetto Group, a youth-marketing firm, "the money they are given or earn all goes to fuel that drive. 'How can what I buy help me define who I am, to myself or the people I care about?' "[52]

The interactive nature of the medium created a perfect vehicle for a proliferating arsenal of "data mining" tools that had become a dominant feature in Web sites for teens. Marketers also routinely monitored the chat rooms, bulletin boards, and discussion groups where young people spent so much of their time, gathering valuable insights about the latest trends, hottest products, and most compelling obsessions of the teen world. As one industry publication stated, "Web sites designed to get kids talking are big winners. They're not just a draw for kids, either—they provide marketers with powerful insights on kids' lives."[53] In this way, the Web became a potent surveillance tool that enabled constant and unobtrusive monitoring of teen subcultures. As teens freely shared their views about a wide range of topics, either in direct response to the site itself or in their online discussions with peers, marketers were taking the collective pulse, moment by moment, of this lucrative demographic group. Chat rooms, discussion groups, and bulletin boards became massive electronic focus groups with a constant influx of participants eager to share their personal and public lives.

Market research became such an integral part of the online teen culture that it not only shaped the content, in many cases *it was* the content. Companies for a variety of teen-oriented products—from cosmetics to fashion to music—easily could set up a Web site for the primary purpose of gathering market data. David Conn, vice president of marketing for the teen apparel maker Candie's, explained the ethos behind the design of Candie's Web site: "We felt that if we could build a site that became a prime gathering point for our customers, that gave them a platform and entertained them, we could data-mine and learn a lot about our market, which is really important when marketing to fickle teenage girls."[54]

As teen Web companies experimented with various ways to sustain their online operations, market research was proving to be one of the more promising business models, in some cases surpassing advertising and sponsorship as the key source of revenue. Market research became increasingly woven into the very fabric of the new teen Web culture, creating a rich and colorful tapestry of interactive features designed not only to engage and entertain, but also to gather valuable intelligence about this profitable market segment. The Web itself was becoming an extension of market-research practice, creating a constant feedback loop that monitored not only the interests and tastes of teens, but also some of their most intimate communications and patterns of online behavior.

While the 2001 dot-com crash knocked many online start-ups out of business, the basic elements of the emerging digital marketplace remained in place. Even as online companies struggled to find workable revenue models in an uncertain economy, marketing and advertising already were shaping twenty-first-century media culture fundamentally, creating new hybrid forms that blended communications, content, and commerce.

"It's Not about Teens, It Is Teens"

With offices in a large space in downtown Manhattan's Silicon Alley, Bolt.com was emblematic of the many teen-oriented Web sites that began dotting the digital landscape in the mid-1990s. Its cofounders, Jane Mount and Dan Pelson, were among the new generation of young "netpreneurs" who viewed teens as central players in the emerging "e-economy" of the Internet. As Pelson told the press, "When we started the company, we were looking at massive growth in this marketplace combined with social trends that were making this audience even more important. Twenty years ago, even if the audience was as big as they are today, they didn't have the economic power that they now have." Like many of his Internet-industry cohorts, Pelson spoke enthusiastically of the Internet as an "empowerment medium" for young people. "Teens are the epitome of a disenfranchised community," he explained. They are also the demographic group that is "driving the success of the AOLs, the Yahoo!s, and the other community sites available online. . . . Take all those factors, and we saw a very clear void, which was that it would be great if we could solve the needs of the consumer and the marketers at the same time."[55]

Launched in 1996, Bolt quickly became a highly popular teen site on the Internet. By blending the Web's compelling interactive features with its powerful data-collection capabilities, the company forged a particularly lucrative business model that was perfectly suited to the teen market. It also served as a prototype for the new digital media culture.

The original idea for Bolt was to create an e-zine (online magazine) for teens, recalled cofounder Jane Mount. But she and her colleagues soon found out that "the most compelling content for teen consumers" was "the kind they develop and create themselves." The main attraction for teens online, she said, was the ability to communicate with each other. So the site developers decided to "let teens own the site." What Bolt was doing, she explained, was "creating a context, a framework, a common platform," that enabled teens to find other people like themselves.[56] This strategy was much cheaper than an e-zine. With teens supplying 95 percent of the content on the site, "we never had to have 50 writers trying to crank up funky, cool content for 18 year olds."[57]

For teens, Bolt.com was a fun-filled online "destination," complete with many of the sticky features that were becoming state-of-the-art on the Web. The site offered a full range of communication, personal-management, publication, and shopping services, including e-mail/voice mail, instant messaging, chat, personal calendars, message boards, Web site publishing, polls and surveys, and an online store.[58] First-time visitors found a highly charged, vibrant, pulsating menu offering a collage of activities available at the click of a mouse. Teens could vote in the day's survey panel, responding to such questions as "Do you believe in ghosts?" or "How much TV do you watch every day?" Visitors could find like-minded teens interested in a variety of subjects, with over 7,000 online clubs that ran the gamut of teen interests. Like other online communities, visitors needed to become members in order to fully participate on the site. With literally thousands of clubs, discussion forums, and other online presentations (e.g., photo albums, profiles, and polls), the site explained, "Bolt members could express themselves and speak their minds on whatever they want. From dating to current events, final exams to MP3s—the discussions on Bolt are driven by our members."[59]

For companies seeking to reach the lucrative teen market, Bolt offered another set of new high-tech features. In its press materials, the company boasted of its unique ability to conduct extensive, ongoing data collection and profiling, using interactive features that were built into the structure of the Web site. The site was described as "a communications platform . . . which serves as the gateway for marketers looking to establish a dialogue with 15- to 24-year-olds . . . a true grassroots phenomenon [that] . . . has grown through word of mouth and through strategic partnerships with some of the largest brands on the Web."[60] In the words of Bolt CEO Pelson: "Bolt is a 24 hour a day, 7 days a week focus group of hundreds of thousands of individuals on a daily basis, and millions of individuals over the course of a month from across the globe. They are saying what is important, and what isn't important to them. The information is all there."[61]

The Bolt formula illustrated the key features of a new digital-marketing paradigm, deeply rooted in the special capacities of Internet technology.

By harnessing the interactive capabilities of the Web, the company was able to place itself at the hub of teens' daily lives, tapping into their developmental needs for autonomy, social connection, and identity exploration. Its popular Web site played multiple roles for teenagers: an online community, a communications tool, and a self-expression platform. But Bolt's fundamental role was to serve as an elaborate data-mining operation. It traded on its understanding and trust of its teen "members," marketing itself as a kind of digital soothsayer, advising corporate clients on the exotic, elusive, and little-understood workings of the teenage subculture and where it was headed. As an ad in the *Wall Street Journal* put it: "Bolt. It's not about teens. It *is* teens."[62]

Following the principles of customer relationship management (CRM), Bolt used surveys and clickstream analysis to compile detailed profiles of each member, generating about "200 data points on the average" per individual. According to company executives, the Web site generally did not collect full names, addresses, or phone numbers, except when doing a "sampling campaign."[63] However, with the sophisticated data-mining software and interactivity of the Internet, these old-fashioned kinds of personal information no longer were necessary. As Pelson told the press, the site "doesn't need to identify its members for targeting—it can do that from data points like the bands they enjoy or their hair color."[64] David Titus, Bolt's head of business intelligence, added: "Our information will be very accurate because it's behavioral data. I can watch what people do, where on the site they go."[65] Using this information, marketing partners then could send highly targeted e-mail messages to individual Bolt members.[66]

In addition to creating a "dialogue" between teens and corporations, Bolt offered a full panoply of other services to marketers, including: "teen consumer insight and intellectual capital on all things teen," "expertise on marketing message, creative strategy and execution," "customized rich media ad units and integrated content sponsorships," "targeted banner advertising (by age, gender, and geography)," "email and database marketing," "customized market research," and "customized events and promotions online and offline."[67]

Bolt's intimate knowledge of online behaviors enabled it to engage in "viral marketing" campaigns, the Web-facilitated version of word-of-mouth promotion. Viral marketing quietly promoted products or services by including product information, such as a clickable URL, with every communication sent from one user to another. As one observer noted in an enthusiastic account of such promotional techniques in *Fortune* magazine, "marketing messages spread like the flu, passed by word of mouth from one friend to another to five more, until there's a full-blown epidemic and products are flying off the shelves."[68] The practice soon became one of the core strategies for marketing and advertising to teens, tapping into their need for social connection and peer influence, and taking full advantage of their affinity for IM, e-mail, and other forms of peer-to-peer digital communication.[69] Viral marketers sought out "connectors," those teens with strong social connections, long IM buddy lists, and the ability to promote brands to their friends. For Bolt, this task was easy. As Pelson explained:

> Because we have all this data on our 7 million members, we are able to identify who out of those members the "connectors" are. We can identify who the highly influential teens in the real world are, based on their activity on the site. For example, it is great to see how many people they have on their friends list, but it is also important how many people have put them on their friends list. It is important how many times they have posted [messages], but it is more important how many people responded to their posts. Using this kind of data, we can score people in terms of their "connectivity." We then work with say a soda company, and then send coupons to 50,000 highly influential teens around the country, with the knowledge that they will influence 10 more people.[70]

Like many other Internet companies, Bolt was forced to cut its staff in the wake of the dot-com crash.[71] But the company survived by "operating on a very lean structure, and consistently marketing itself to advertisers as a digital platform through which to reach the fickle teen consumers."[72] Ultimately the enterprise became highly successful, expanding into global markets, and spreading its popular brand into the United Kingdom, Canada, and Western European countries. The Web site platform also moved onto cell phones, through the company's deals with AT&T Wireless, Sprint PCS, Verizon Wireless, and other wireless companies.[73]

One-Stop-Shopping in the Digital Marketplace

Launched in 1996, Alloy.com was another online venture created to cash in on the lucrative teen demographic. Aimed at girls age thirteen to nineteen, the site offered a glitzy package of fashion, entertainment, and shopping.[74] Alloy quickly made its mark in the early dot-com era, registering more than 3.7 million users by 2001 and placing itself among the few commercial teen Web sites to earn a profit.[75] Unlike Bolt, which was primarily a virtual operation, Alloy maintained a combination of online and offline properties, including a successful fashion and accessory catalog business. So even as other Web ventures succumbed to the economic downturn in 2001, Alloy was able to survive by generating revenue from a variety of sources. As chairman and CEO Matt Diamond explained in a press interview, teens were less affected than adults by economic crises. "They don't have mortgages," he pointed out. "They're still spending a lot of money."[76] In 2003, Alloy acquired the retail chain dELiA*s, which included a popular teen Web site and several dozen stores in malls around the country.[77] Within a few years, the company expanded its holdings to include a full range of direct marketing, retail, market-research, and ad-agency services.[78]

Today, Alloy's empire of youth properties embodies the integrated nature of contemporary commercial culture, in which content, advertising, data collection, and direct sales are combined into a seamless web of influence. The company has outlets in thousands of elementary schools, high schools, and colleges across the country. For example, 360 Voice, which is aimed at kindergarten through eighth grade, features large, prominently placed posters where advertising messages are placed alongside "engaging, fun and easy to consume educational content."[79] Alloy's marketing arm, 360 Youth, serves more than 1,500 clients every year, promising marketers a "powerful and efficient one-stop-shopping resource" and access to more than 31 million teens, tweens, and college students.[80] Its arsenal of advertising and marketing weapons includes "e-mail marketing strategy and implementation," "viral applications," "interactive and multi-player games," and "quizzes and polls."[81]

The 360 Youth network operates a stable of Web sites designed for specific demographic segments of the teen market. According to the company's Web site, Alloy.com and dELiA*s.com are:

> the largest and most dynamic destinations for girls, providing the coolest content, hippest fashions, and all that is happening in entertainment. Designed as a virtual mall with all of the social trappings of teen-hood, Alloy.com and dELiA*s.com invite girls to interact, browse, shop, and connect.

At ELLEgirl.com, teens around the world can find out about "the hottest fashion trends, score some beauty advice, browse fun content, or win prizes." For boys, CCS.com and Danscomp.com offer "bold sites packed with content, community, and commerce, reaching millions of action sports enthusiasts and providing an ideal platform to reach teen boys."[82] Careersandcolleges.com, eStudentLoan.com, and AbsolutelyScholarships

.com target college-bound high school students seeking information about colleges and universities, student loans, and scholarship programs.[83]

The heart of Alloy's business strategy is its ability to amass enormous amounts of data about its teen consumers. As the company told the Securities and Exchange Commission in its annual report, "We believe we are the only Generation Y–focused media company that combines significant marketing reach with a comprehensive consumer database, providing us with a deep understanding of the youth market."[84] While Alloy uses a variety of methods to build and mine its database, the company's growing network of teen Web sites is one of its most powerful tools. Each site is packed with pages and pages of interactive activities, including contests, sweepstakes, and surveys tied to the icons of teen popular culture. Along with collecting voluntary information from visitors, the sites use cookies and other online tracking tools to monitor online behavior.[85] Teenagers are offered a range of incentives designed to lure them into signing up for e-mail newsletters from Alloy and its marketing partners, enabling the company to collect a wealth of personal data that is both deep and wide. "In addition to names and addresses," the company informed the SEC, "our database includes a variety of information that may include age, purchasing history, stated interests, online behavior, educational level and socioeconomic factors." In 2005, the company reported that it was "interacting with our registered online user base of more than 4 million customers."[86]

Alloy's social-networking sites illustrate how companies can use the Web to insert their data-collection and marketing operations into the most intimate aspects of teenagers' daily lives. For those struggling with the anxieties and frustrations of dating, for example, eCRUSH is on hand to help them out, promising to ameliorate much of the pain and risk involved. Described as "the original 'crush' site—a way to find out anonymously if someone you like feels the same about you, with no chance of rejection," eCRUSH was launched on Valentine's Day 1999 and by 2005 claimed to have more than 2.5 million registered users.[87] To sign up, teens supply their names and e-mail addresses along with those of their "crushes." The Web site then sends e-mail messages to the crushes, asking *them* who *their* crushes are. "If you match we tell you both at the same time. If you don't, they never know it was you."[88] Truthquiz, another Alloy Web site, allows teens to explore their own individual identities and test out how others perceive them. "Find out what people *really* think about you," the site teases. All a teen has to do is pick a theme (e.g., Am I a Punk? Am I a Prep? Could I Make It in Hollywood?), and then use the ready-made questions to create a personalized quiz, send it to friends or "post on your blog, bathroom wall, whatever . . . We'll show you the results—though you will never know who said what." Teens who sign up for the service also are offered the chance to take surveys in order to "influence cool brands like Abercrombie, Steve Madden, Dreamworks & more."[89]

eSPIN-the-Bottle offers "the premier way for gen-y to connect, to flirt or to make new friends online." Described as "the old junior-high game with a modern twist," the site promises teens the chance to meet new people online without having to reveal their true identities. To play, a member creates an online profile including interests, hobbies, favorite stars, etc., as well as a personal photo. Then she describes the person she wants to meet. The "magic bottle" spins around and searches through other profiles to find a match. If she likes the profiles it offers her, she can add them to her list. If they also choose to add her to their lists, the site lets her know. Then she can contact them through the proprietary e-mail system Match Mail. The site's IM system, Note Passer, allows members to "pass notes" to people they are interested in. Launched in 2001, eSPIN-the-Bottle claimed more than 600,000 active users by 2004, proudly announcing that "the bottle's been spun over 50,000,000 times!"[90]

Given the sensitive privacy issues related to teen online dating, Alloy is very careful to explain that eSPIN-the-Bottle is not engaged in passing personal information about teenagers along to strangers, and to caution participants on how to behave safely online. The site offers advice to teens on how to protect themselves, and provides an online hotline to report any inappropriate advances. "If you're under age, check with a parent or guardian before using the eCRUSH network," the Web site advises. "Watch what you say! It's fun to flirt, but take it easy on intimate innuendos or way-too-personal information."[91]

But while the site takes precautions to protect teens from danger, it encourages its marketing partners to gain easy access to them. By clicking on the Privacy Policy link (in tiny letters at the bottom of the Home page), a visitor can learn that it is possible to "opt out" of the process.[92] A careful reading of that policy also reveals, however, that the default on this site (as in Alloy's entire network of teen sites) is data collection and personalized advertising and promotion. "By not opting out, the user is agreeing to receive current and future offers. For our cobranded offers at registration we obtain your consent to sell or rent the email and/or postal addresses of registered users to bring you great offers directly from our sponsors, partners or advertisers."[93] This kind of language has become boilerplate in the privacy policies of thousands of commercial Web sites, for adults as well as teenagers.[94]

Commercializing Culture

As marketers closely watch how youth are interacting with digital media, they continue to probe the psychological links between technology and adolescent development. "The Internet has evolved to become the 'hub'—or primary medium" for young people, explained the market research report, *Born to Be Wired*. "Teens and young adults are searching for independence and control, and the Internet gives it to them like no other media can." To develop the most effective strategies, Greg Livingston of the Wonder Group advised, brands need to take into account the "basic motivators" for this developmental stage—"freedom, independence, power, and belonging." But companies also must understand fully how members of the Millennial Generation are engaging with digital media. With this knowledge, marketers can "be part of the communication structure of teens."[95]

Exploring the subcultural terrain of the Digital Generation, market researchers divide teen consumers into clearly definable market segments, based on their use of new media. "Chic Geeks" are "early adopters of technology and heavy users of gadgetry . . . with cell phones as constant companions." They have "wide social networks that they actively cultivate," and they are "conspicuous consumers . . . looking to brands to get them noticed" and desiring "new news in their messaging . . . to be the first to hear it." Therefore advertisers should "give them a sense of exclusivity with the information you provide to them," remembering that "image is important to them." The "Now Crowd" also are heavy media users with large social networks. A "young segment, largely suburban," with "a relatively high proportion of African-Americans," the Now Crowd are "wannabe" leaders. To best appeal to this group, marketers are advised to offer them "social currency—information that will raise their standing and credibility" among their peers. "The role of brands to the Now Crowd is to help 'brand' them as being the 'Now' Crowd." But companies are warned not to let members of this subgroup down because "they are influential and can spread negative word-of-mouth quickly."[96]

In their efforts to become part of the communication structure of young people, marketers have developed a set of practices specifically designed to target teenagers, tweens, and young adults. These practices, forged in the formative period of Internet ecommerce, have evolved and migrated to mesh with a growing number pastimes and "platforms," as companies seek to ensure their brands become a pervasive presence throughout the digital media culture.

Mobilizing the Tremor Nation

Procter & Gamble (P&G) has been at the forefront of innovative teen-marketing strategies, launching its peer-to-peer network Tremor in 2002.[97] The venture is designed to identify those teen "connectors" who are socially active and can influence their peers' buying decisions.[98] This select group of young people—largely female (60 percent)—is then given inside information about new products and previews and programs and encouraged to tell their friends about them.[99] Not any tween or teen can become a "Tremorite." P&G carefully chooses its members based on "Internet usage, social networks, and willingness to advocate to peers" and then gives them coupons, new product samples, and advance copies of TV-program scripts to encourage their participation in research and promotion. Tremor claims great successes for its clients, which have included entertainment companies such as WB, AOL, and Time Warner, as well as numerous product brands. Savvy Tremorites are credited with helping Coca Cola choose "Nothing Else Like It" as a billboard slogan.[100] And when Tremor sent teens in one city an advance script for an episode of *Dawson's Creek,* the ratings for the popular show tripled in that market.[101]

Within two years of Tremor's launch, P&G boasted "a sales force of 280,000 teens . . . part of a massive focus group and word-of-mouth marketing drive to counter the eroding influence of conventional advertising." The company predicted that its Tremor Nation was only the "first of what's expected to be a mass movement as advertisers seek new ways to attract audiences and build loyalty." Through its partnership with Blue Dingo, a New York ad agency specializing in "building Web-based interactive marketing programs and online social networks," P&G created a high-powered new Web site for its Tremorites, billed as "the virtual equivalent of a gated community, a members-only Web site replete with word-of-mouth and community building features." Underscoring this exclusive insider status, the site spells out the various perks of membership:

> Tremorites can quickly view Tremor programs currently available to them, search for and hook up with other members with similar interests whom they can Instant Message, send personal messages to one another and receive notes from the Tremor Team, see what's hot and happening in their own hometown and throughout the

U.S., access unreleased music and get special offers negotiated exclusively for Tremorites.[102]

Peer-to-peer marketing (sometimes called "buzz" or "viral marketing") has become a staple among youth advertisers.[103] The practice spawned its own Word-of-Mouth Marketing Association (WOMMA). Its Web site features mini–case studies of successful campaigns, offering a rare glimpse into the inner workings of online marketing. When Elizabeth Arden launched its new perfume, Britney Spears's Curious, its debut took place with almost no publicity. Yet within only four months, it had become the leading fragrance launch for 2004, generating sales of $36 million. "How did a mature, fit-for-the-elderly brand like Elizabeth Arden manage to attract teenage girls in such a discreet and succinct manner?" the Web site asked. "The answer: Unconventional/Viral marketing." In the weeks before the perfume was launched, "a discreet banner with a photo of Britney" was placed on five Web sites popular with teens, including Bolt.com and Alloy.com. The ads asked girls to provide their cell phone numbers and zip codes. Almost 30,000 girls responded in order to receive a 45-second recorded voice mail from Britney herself, in which the pop star thanked her fans and shared her excitement about the new fragrance she was working on. "This marketing campaign worked because it was done in non-marketing speak," explained the Web site. "Since these girls felt like they knew a secret from Britney about Britney, they forwarded the voice mail to their friends, even posting the text on Britney-related online forums." To keep the buzz going about the yet-to-be-released product, marketers followed up with a series of text messages from Britney, including a list of nearby stores where the girls would be able to buy Curious. "In the end, with its viral effect, the campaign reached about 300,000 girls. The strategy was a first for the fragrance industry."[104]

Bot Buddies and Other Virtual Friends

Given the popularity of instant messaging among teens, marketers devised strategies for penetrating these online-communication spaces. One innovation was the "ELLEgirl Buddy." ELLEgirl magazine partnered with a company called ActiveBuddy to launch a "bot" on AOL's AIM platform, designed to promote the magazine and drive traffic to the ELLEgirl.com Web site. Bots (digital robots) are software applications that can be programmed to engage in online conversations with people and search databases for information. Bot buddies are promoted through word-of-mouth, with IM users then choosing to add the virtual characters to their own buddy lists. When someone clicks on the buddy's screen name, a computer server uses key words to analyze the message and formulate an appropriate answer. The buddies also can serve as personal research tools. If people do not want to search the Web themselves for information, they can send the buddy an IM and it will go out into cyberspace, find the answer, and bring it back. By 2002, at least a dozen companies had begun using bots as online marketing tools. "Hooking up with human pals through instant message services," explained Christine Frey in the *Ottawa Citizen,* bots "urge people to buy Ford trucks, check out the eBay auction site and take in *The Lord of the Rings.*"[105]

The ELLEgirl buddy was targeted at teen girls. "Teenagers are moving targets and it is really difficult to keep updated lists" an executive of ELLEgirl's parent company told the press. "With the IM buddy, we think we can connect with teenagers in an environment where they are already comfortable."[106] The bot was given an entire profile, complete with details of her personal life, home, family, and hobbies. The fictitious "ELLEgirl-Buddy" lived in San Francisco with her mother, father and an older brother.

> Her favorite book is *Catcher in the Rye*. Her favorite television show is *Buffy the Vampire Slayer*. Her favorite band is No Doubt. When she grows up, she wants to design handbags, own a bookstore café and work overseas as a foreign correspondent.[107]

The bot IM conversations, designed to mimic those of a 16-year-old girl, gushed with teenage enthusiasm, as the virtual friend offered advice on fashion, beauty, and horoscopes. "I (lower case) looove making my own clothes. I use gap tees a lot. You just shrink em and add ribbons. Insta-chic. . . ."[108] The personality of the ELLEgirl bot was so developed that some girls saw it as a "cyber confidant, writing to and about bad haircuts and image problems. . . . While gabbing about lip gloss and prom gowns, it interjects occasional promos for the magazine, urging girls to click on a link and 'give the gift of beauty—give a gift subscription to ELLEgirl magazine, get billed for it later!' "[109]

As the trade publication *Marketing News* explained, bots can "capture imaginations, reinforce brand values and catch consumers at the spending point."[110] New Line Cinema created a bot to promote its 2002 film *Austin Powers in Gold-Member.* Ads on AIM invited users to put the Powers interactive agent on their buddy lists, where he dispensed with "groovy" horoscopes, "secret spy names," and information about the upcoming movie. According to industry executives, the bot campaign resulted in a 75 percent click-through rate to the movie's Web page.[111] When Johnson & Johnson wanted to introduce a new line of Acuvue contact lenses, the company launched a "personalized" marketing strategy that engaged young customers in a series of "customized conversations" with bots on a variety of digital platforms, including Web sites, cell-phone text messaging, e-mail, and IM.[112]

In 2005, AOL launched an upgraded version of AIM. "The AIM service has become the front door to more than 43 million Americans' social networks," said Bill Schreiner, vice president of programming for AOL, in the company's press announcement. "The New AIM Today is designed to help them integrate music, entertainment, games and social networking features like AIM® Fight into that setting, and easily share them with friends, family members and colleagues."[113] Among the new features was a souped-up flock of "bots" that "allow users to interact, play and get essential information updates in real time."[114] The new IM service is also designed to better service advertisers and marketers, the press release explained: "The New AIM.com and AIM Today present marketers and advertisers with targeted and 'day-parted' reach into AIM's dynamic user base, half of which is comprised of attractive yet hard to reach teen and young audiences."[115]

At British Web site HabboHotel, teens can find an entire community of virtual friends. The colorful site combines the appeal of such popular simulation games as *The Sims* with the real-time communication functions of a chat room. Teens (and younger children) can create their own avatars to interact with other online characters representing the game's hundreds of thousands of participants around the world.[116] Inhabitants of the Habbo Hotel also can use real money to buy their own virtual furniture through an online credit-card purchase. Marketers seeking to reach young consumers have found this highly popular online community rich with opportunities. For example, new movies often are featured on the site's virtual movie theaters, and chat room participants hawk Coke and other products. Capri Sun's Burp Campaign, which was aimed at the 7–12 age group, placed ads on Habbo Hotel and FoxKidsUK to drive kids to the PlanetJuice.com site. Once there, they were invited to play the burping game, which involved drinking Capri Sun as quickly as possible, and using special digital devices on Capri Sun packages to record their "weird and wonderful burp sounds."[117]

In 2005, Alloy.com forged a strategic partnership with the Sulake Corporation, the company that operates Habbo Hotel, to launch a "fully integrated marketing campaign" for promoting the highly popular Web site.[118] As the "exclusive advertising representative for Habbo Hotel," Alloy is "handling all advertising sales and promotional opportunities with marketers and agencies seeking to reach teen consumers through non-intrusive online immersive branding." The company also announced plans to promote the Web site through its "proprietary in-school media network, reaching more than half of the nation's high school students."[119]

Advergaming, Game-vertising, and Immersive Advertising

The growing popularity of video games among young people has created an opportunity that advertisers simply cannot pass up. By 2005, 81 percent of the online teens—or two-thirds of the teenage population—were playing video games, a 50 percent increase from four years earlier.[120] Video games have become the fastest-growing form of entertainment, with teens and young adults at the forefront. Gaming takes place across many digital platforms in addition to the Internet—from consoles, to hand-held devices such as Nintendo, to cell phones.[121] In industry surveys, youth consumers consistently rank the Internet and videogames higher than television in importance.[122] More than 70 percent of young men have become avid gamers, spending as much time or more playing games as they did watching television.[123] And games are not the exclusive domain of young males; one third of gamers are women. Video games in the United States have grown into a $9.4 billion business, surpassing even the movie box office in earnings. "Video games are part of the DNA of the youth market—they're in the bloodstream," Howard Handler, Virgin Mobile's chief marketing officer, told the press, "The contemporary youth lifestyle is music, action sports and video gaming all blurred together."[124]

Games also provide the perfect "immersive environment" for incorporating brand messages seamlessly into the content. The quintessential sticky content, video games can engage players for hours on end, inviting them to come back again and again to play. "Because players spend so much time with a single game—on average, between two and four hours at a time—brands want to be part of that hyper-reality," reported *Advertising Age*.[125] "Advergaming" is considered particularly effective with young audiences who tend to get annoyed with Internet banners and pop-up ads.[126]

Not surprisingly, marketers have been flocking to videogames.[127] Some of the biggest brands in the youth market have been woven into the content of popular games, including Levis, Coke, McDonalds, Kraft, Nike, AT&T Wireless, and Nokia.[128] In-game advertising is considered particularly ideal for "reaching that elusive male youth demo," explained one trade publication. Massive, a New York-based advertising network, offers opportunities for "dynamic product placement (which allows advertisers to update ads) in video games like Ubisoft's Splinter Cell: Chaos Theory and Funcom's Anarchy Online." The company signed deals with a dozen U.S. advertisers, including Paramount Pictures, Nestle, Honda, and Dunkin' Donuts.[129]

Internet games create unique opportunities to microtarget individual game players. In addition to integrating brands into the storyline of a game, advertisers can respond to a player's actions in real time, changing, adding, or updating advertising messages to tailor their appeal to that particular individual. "The silver bullet is really doing custom solutions for clients," explained one industry executive.[130] The technology also enables precise measurements of customer reactions to the online advertising.[131] Digital marketers have perfected software for tracking consumer behavior in the interactive market. DoubleClick's Motif, for example, offers a full spectrum of sophisticated tools in its Audience Interaction Metrics Package. "You'll automatically get metrics on how long each ad was displayed or how the viewer interacted with the ad . . . so you know what works best. . . . You can track audience video plays, completions, pauses, stops, restarts, mutes, average view time, and custom video interaction metrics." Among the hundreds of Web sites subscribing to DoubleClick's Motif service are some of the top destinations for children and teens, including: Alloy.com, DELiA*s.com, AIM, KidsWB.com, Nick and Nickjr.com, TeenPeople.com, and Timeforkids.com.[132]

Online games are a particularly good vehicle for advertising snacks and other impulse food products. According to Nielsen//Net Ratings, buyers of Tombstone Pizza were among the most avid players of such online games as cheatplanet.com, addictinggames.com, and Lycos Network entertainment.[133] In 2005, Sony partnered with Pizza Hut to take "dynamic product placement" a step further, building into its Everquest II videogame the ability to order pizza. "All the player has to do is type in the command "pizza," and voila—Pizza Hut's online order page pops up," explained the trade publication *Strategy*. "While it's just pizza now, the in-game purchasing potential is wide open."[134] One expert predicted that advertising and product placement will soon be as integral to video games as story lines and action.[135]

When Nickelodeon bought the highly popular online game site Neopets in 2005, to become part of the new TurboNick Web site, one of its goals was to "monetize" the huge amount of traffic the site enjoyed by inserting more brands. Already, McDonalds and other familiar icons can be found among the businesses in the site's Shop section. In a game where the object is to keep your neopet alive by feeding her regularly (ensuring your repeated visits to the site), executives envision a future scenario in which game players "will be feeding their pets with food products from major brands."[136]

The Mobiles

Offering freedom, autonomy, and constant connection, cell phones are one of the fastest-growing digital products among youth. In 2005, nearly half of all teenagers had their own cell phones, with girls more likely to own one than boys. Teenagers were getting cell phones as early as middle school, with two-thirds of high school juniors and seniors already avid users.[137] "The cell phone has become a primary mode of socializing for teens," explained Robbie Blinkoff in an interview with *Wired* magazine.[138] An anthropologist by training, Blinkoff has turned his academic expertise into a marketable commodity, probing the cultural mores of technology users for an impressive list of clients, including American Express, Bristol-Myers Squibb, and Procter & Gamble.[139] Blinkoff enlisted the involvement of 3,500 anthropologists around the world to study the behavior of 144 cell phone users in more than a half-dozen countries. The report, "The Mobiles: Social Evolution in a Wireless Society," concluded that wireless technology already was producing significant changes in society. At the forefront of these changes are "the mobiles," those "people who have seamlessly integrated wireless products into their daily existence."[140] Teens are quintessential "mobiles." When researchers took cell phones away from teens, the "backlash was horrible."[141] Cell phones quickly are becoming an essential prerequisite to acceptance by teen peers, according to researchers, and those without phones are considered outsiders. "Next time a teenager says, 'Mom if I don't have a phone,' or 'Dad, if I don't have a phone, I'm going to be a nobody,'" Blinkoff explained, "they are being serious."[142]

Marketers have seized upon the cell-phone phenomenon to introduce "mobile branding" to teens.[143] "Mobile phones are the nexus of communication for today's youth," one marketer told the press, "combining fashion, sports, entertainment, with the technological proficiency of today's youth to take advantage of this medium."[144] One of the companies on the cutting edge of the market is Virgin Mobile, a joint venture with Sprint widely promoted through a variety of retail outlets, music stores, and cable-television channels. In partnership with MTV, Virgin Mobile offers an enticing blend of content and services tied to the cable youth network's popular programming. Wireless users are given "customized sneak peek" updates on their favorite TV shows, breaking stories from MTV News, power to choose new music videos for the channel through "video voting," and audio postcards from music and TV celebrities to send to their friends.[145]

Virgin Mobile offers an array of customized features tailored to youth lifestyles and designed to become indispensable daily tools. "Style out your phone," the Web site urges, encouraging the use of ring tones based on the latest popular songs and downloads of a variety of games and other entertainment. "A download a day keeps boredom away."[146] To avoid being "stuck in a nightmare blind date, meeting, class, or conversation with no way out," users can simply "set up a 'Rescue Ring' for a predetermined time and your phone will ring. Pretend to have a conversation and take it from there."[147] Those who do not want to set their alarms can arrange for personalized "Wake-up Calls" and be greeted by a favorite rock star or TV personality.[148] If they need help managing the complexities of contemporary life, Virgin Mobile Gopher provides "real live operators who are ready to help you 24/7. Get driving directions, cross street info, movie times, concert info, sports scores and more."[149]

Virgin Mobile has teamed up with AOL Instant Messenger to offer IM text messaging.[150] Already a popular pastime among teens in Europe and many other countries, text messaging (called "short messaging service" or SMS on phones) has been aggressively marketed by U.S. wireless providers. "Now you can chat with your buddies right from your Virgin Mobile phone—whether they're logged-in at their computers or on their own phones," the Web site promises.[151] These marketing efforts have been successful. A third of American teens were sending text messages by 2005. Teen girls have found the practice particularly engaging, with more than half of online girls aged 15–17 having sent or received them.[152]

Cell phones offer multiple ways to generate revenues from the teen market. Instead of expensive contracts, for example, Virgin Mobile offers a pay-as-you-go package, enabling users to pay on a "minute2minute," "day2day," or "month2month" basis.[153] Downloadable content, such as games and ring tones, provide an additional source of revenue, and teens are among the most avid consumers of these optional services.[154] Mobile phones also are effective vehicles for database marketing. Though privacy laws on cellular telephone service require companies to get permission (opt in) before sending marketing messages to subscribers, most teenagers and young adults are eager to sign up, and in most cases have to do so in order to use the many add-on features offered.[155] Thus, cell phones can be ideal platforms for sending personalized messages to individual users.

In addition to cell phones, iPods and other mobile music devices have become attractive to marketers. A 2005 Pew survey found that use of paid music services had doubled from the previous year.[156] With teens among the most eager iPod users, companies such as Pepsi and Gap began offering free iTunes downloads as part of their promotions. As one marketer put it, "If you want to be hip and stylish and in-line with Generation Y, get yourself attached to the iPod or iTunes in any way you can." The arrangement also has been beneficial for the music-downloading service. While iTunes's share of digital downloads was 67 percent in 2004, according to one trade group, during the brand-promotion campaigns it spiked to 75 percent.[157]

The Social Network Sell

Social networking sites like MySpace.com, Friendster.com, and Facebook.com enable users to create personal Web sites, where they can post profiles, photos, videos, and music. But unlike standard Web-hosting services, these sites are designed to serve as powerful word-of-mouth engines, where people can find new friends, connect with old ones, and manage their social lives. Like IM, social-networking sites have transformed the informal social relationships that are part of everyday life into digitized personal networks.[158] In a very short time, these sites exploded in size and influence, becoming particularly popular with teen girls. "Teenagers say they get hooked on building and browsing the social networks, playing online games and music posted by their friends, answering lifestyle quizzes, and exchanging comments online into the wee hours," writes Andrew Trotter in *Education Week*.[159] Such "super hot sites," noted *Advertising Age*, offer "a potential eyeball windfall to advertisers."[160]

MySpace began as a way for musicians to promote themselves and their music, enabling fans to gather together and talk about their favorite celebrities.[161] Within a year, more than 20,000 established and up-and-coming artists had taken up online residence there, attracting a huge fan base, with more than 40 million members by 2005. The small upstart had become such a hot property that Rupert Murdoch's NewsCorp (owner of the Fox entertainment empire) quickly bought it up. MySpace also attracted hundreds of advertisers eager to tap its rich demographic mother lode. Procter & Gamble chose MySpace as the place to introduce teen girls to its new brand of deodorant, Secret Sparkle. By placing the product logo on the home page for singer Hillary Duff, the company encouraged teens to promote the brand to each other through their social networks. The campaign also featured an iPod sweepstakes. "We have to be where they are in this online world," a company spokesperson told the press.[162]

As companies insert themselves into these new social networks, they are intentionally seeking to blur the lines between advertising and content. Social networks are "breaking down that wall between what is marketing and what isn't," commented one youth marketing expert. "[S]ometimes the marketing is so embedded in the social network sphere that it draws users to interact with the brand as if they were e-mailing friends," explained an article in *Marketing*. For example, to promote its film *The Ringer*, Twentieth Century Fox created special Web pages on MySpace for the fictional characters, as if they were real users. The tactic proved successful. The character Steve "has more than 11,000 friends," observed the article, "in other words, more than 11,000 consumers who visited his page and requested to become a part of it. They respond to Steve's fictional blog entries and become involved in the story."[163]

The social-networking phenomenon is likely to grow into an even more successful venue for advertising and marketing. In October 2005, MySpace.com alone accounted for 10 percent of all ads viewed online.[164] Marketers are particularly eager to take advantage of the large, highly detailed user profiles and expanding lists of "friends" on these sites. "The targeting we can do is phenomenal," one industry executive told the press.[165]

Blogging for Dollars

"Blogs" (short for Web logs) are online journals where users can chronicle their lives, spout off about their political views, and engage in other forms of personal expression.[166] In many ways, they are an updated version of personal Home pages. But while creating a personal Web site requires a certain level of knowledge, skill, and time, blogs are an instant online outlet. With user-friendly software and free sites to host bloggers, the number of people reading blogs, contributing to the blogs of others, and creating their own blog has grown phenomenally. "By the end of 2004," Pew reported, "blogs had established themselves as a key part of online culture."[167] The following year, the expanding "blogosphere" reached huge proportions. In July 2005, Nielsen/NetRatings reported that "20 percent of active Web users, or 29.3 million people, accessed blogging or blog-related Web sites, growing 31 percent since the beginning of the year."[168]

Like social-networking sites, blogging has become a popular teen pastime—especially with girls, who flock to such popular youth-oriented sites as LiveJournal and Xanga, where they can post music, pictures, videos, and other material along with their personal musings.[169] Of the 2.7 million users listing their ages on the LiveJournal site, for example, 47 percent said they were 13–18.[170] David Huffaker, who studies blog use among teens, identified several key features that make them not only appealing to teens, but also a useful tool for adolescent development. "Constructing identity can be a continual process for adolescence, and one they can refer back to," he explained. Blogs offer easy ways to archive information and knowledge, opportunities for others to comment or provide feedback, as well as links to online blog communities. Like instant messaging, blog software on such popular Web sites as LiveJournal and T-Blogs also come with tools for expressing one's mood. "For each blog post, the author can write a creative text to represent a current mode: for example frustrated, bored, ecstatic, tired, excited, or cranky and still add an emoticon for additional emphasis."[171]

As blogs zoomed in popularity, marketers sought ways to tap into them, not always with the best results. When Nike tried to launch a blog-based campaign, a number of bloggers were less than enthusiastic about the idea. As Patrick Phillips, the publisher of the Web site IWantMedia.com, told the *New York Times*, "I'm skeptical that a lot of online readers would be interested in reading an advertorial blog. . . . If you go to a site and you know you're being sold something, I don't know that there's going to be a lot of interest."[172] Months later, Dr. Pepper/7Up tried to market Raging Cow, its new sweetened-milk drink, to several hundred bloggers with large readerships among teenagers and young adults. The plan backfired, though, with at least one blogger calling for a boycott against the product.[173] But despite these early rough spots, advertisers have found blogs increasingly receptive to their messages, and the practice quickly became institutionalized.[174]

Marketers also discovered that blogs can be the "ultimate focus group."[175] Cellular Wireless, working with youth-focused ad agency G Whiz, monitored blogs to find out what their potential college-aged customers were saying about their cell phones. Operating as a self-described "fly on the wall," the ad agency eavesdropped on blog conversations and used linguistic analysis to identify members of U.S. Cellular's target market and to collect postings about cell-phone use. The process revealed a wealth of useful insight. For example, bloggers "complained about unwanted calls that drag on and on, eating into their minutes and forcing them to turn off their phones toward the end of their billing cycles, leaving them feeling isolated from their friends." G Whiz quickly created a series of television spots playing off the same theme, and promoting "unlimited call-me minutes." As *Ad Week* observed, U.S. Cellular's experience showed the potential for blogs to capture "unvarnished opinions that can affect product choices and marketing strategies."[176]

Other youth-oriented companies have seized upon blogs as a valuable market research tool. "We find Web logs are a very rich source [of market intelligence]," explained one executive, "because people don't just go to talk about a recent movie, they go to talk about their lives."[177] Another added: "Blog feedback can be cheaper and quicker to obtain than traditional research, while also being free of biases inherent in paying people to be in focus groups. . . . As blogs mature, advertising in them will not be a big deal. Using them to gain insights will provide lasting value."[178]

At Look-Look, a Hollywood-based youth-market research firm, blogs are only the latest in a long line of innovative methods for studying the youth demographic. Serving such clients as Pepsi, Calvin Klein, Nike, and Coca Cola, the firm operates a 10,000 member Team Look-Look, enlisting young people to serve as on-the-ground "photo journalists," documenting the behaviors, lifestyles, and fashions of their peers around the United States and abroad and then uploading their findings on the Web. "From the back alleys of Los Angeles to the streets of Saigon," explains the Look-Look Web site, "up-to-the-second discoveries travel from our Team Look-Look correspondents back to the home office, where it is compiled by a group of Youth Information Specialists."[179]

Always on the cutting edge, Look-Look has been "leveraging blogging in youth culture research." As company representatives told fellow marketers, "We are interested in providing youth with methodologies that work within their existing lifestyle, and . . . youth feel comfortable with the technology and format" of blogs. Youth bloggers are encouraged to embellish their postings with photos, movies, and drawings. Advertisers also can be active participants in the blogging process, part of Look-Look's effort to provide a "two-way dialogue between our clients and these subcultures."[180] A PowerPoint presentation by one of Look-Look's executives displayed the following posting from a young blogger who fantasized about his ideal car:

> I would want a fast, sexy car. The body design would be relatively conventional, proably [sic] a hatchback, but it would be lowered with low profile tires and rims to match the black paint job. Ideally it would have just a straight up desktop PC integrated into the dashboard, with navigation, video games, Internet access, Music player, etc. Leather seats are nice to have. A built-in breathalizer would be key so people could check if they are ok to drive or not.[181]

Some advertisers have jumped on the blog bandwagon by offering their own "branded blogs" as a new tool for viral marketing. Members of mycoke.com found an announcement in their e-mail boxes during the summer of 2005. "Come check it out," the e-mail beckoned, "new features now and more to come later. The big news is BLOGGING! Post your thoughts on a favorite subject, upload pictures, allow others to comment, and share with everyone. Wonder how to get traffic to your blog? Send your blog's link to your friends!"[182]

The My Media Generation

In July 2005, Pew released its second survey of teen Internet use. Updating its research from five years earlier, the report found that the vast majority of U.S. youth between the ages of 12 and 17 (87 percent) go online, a 24 percent increase since 2000.[183] The report also found that "the scope of teens' online lives had broadened." Half of online teenagers accessed the Internet through a broadband connection, enabling a richer and faster set of interactive experiences in cyberspace. "Wired teens," the study noted, "are now more likely to play games online, make purchases, get news, and seek health information." An overwhelming majority of teens (84 percent) reported owning at least one personal digital device—either a laptop, cell phone, or personal digital assistant (PDA)—and close to half (44 percent) owned two or more.[184]

Teens also had taken ownership of these new tools, putting their own individual imprints on cell phones, instant messaging, and online journals. Some created their own photos, music, and artwork. Others chose elements from an endless supply of manufactured items, from cell-phone "skins," to celebrity pictures, to IM bots and buddies. Some IM programs enabled teens to create cartoon avatars of themselves, supplying them with a full palette of choices—including gender, hair, eye and skin color, clothing, and setting—in order to act out their identity fantasies.[185]

Technology did not completely subsume teen life, according to the report. Teens still spent more time doing things socially with their friends outside of school than they spent interacting with them digitally. Nonetheless, "American teens live in a world enveloped by communications technologies," the researchers concluded. These digital tools had "become a central force that fuels the rhythm of daily life." Teens were, in fact, "leading the transition to a fully wired and mobile nation."[186] The report also noted that a significant number of teenagers remained disconnected from the digital universe (13 percent), and not out of choice. Despite efforts throughout the 1990s to bridge the digital divide, the gap was still there, affecting African-Americans in disproportionate numbers.[187] As the report explained, "those who remain offline are clearly defined by lower levels of income and limited access to technology." In the Digital Age, being "offline" meant being disconnected from society. In only a dozen years since the release of the World Wide Web, Internet access had become the new normal.

Yahoo! conducted its own follow-up study of the Internet and youth in 2005, this time with OMD Worldwide, a global market-research and advertising firm. Young people were no longer "the New Millennials." Now they were the "My Media Generation." What makes this generation unique, according to the study, is their ability to "customize and personalize everything in their world and daily experiences in ways previous generations never could." Technology companies have created a host of new liberating media tools, all designed to address the three core needs of young people: community, self-expression, and personalization. And youth have quickly integrated these new tools into their lifestyles. "We've moved from broadcasting to podcasting in just a short period of time," Yahoo!'s chief sales officer, Wenda Harris Millard, declared: "The My Media Generation increasingly filters the flow of advertising messages—letting in only those that are relevant, entertaining, or delivering value. While that raises serious challenges for marketers, it also brings the promise of new, more powerful channels for reaching youth and having them willingly and enthusiastically engage with brands."[188]

6 Social Marketing in the New Millennium

When it premiered on MTV in 1992, *The Real World* generated mixed reviews from TV critics. A cross between a soap opera and a documentary, the show recruited seven people in their twenties and set them up in a plush apartment, where their lives were videotaped, day and night, for the next four months. The edited footage, with its steamy sex and raunchy language, was then set to a rock soundtrack featuring the latest hot music groups. "This is not reality as we know it," wrote Ginny Holbert in the *Chicago Sun-Times,* "but a highly artificial setup designed to turn ordinary life into an extended music video."[1] Despite its obviously contrived nature, the show's success helped usher in a new genre of "reality programming" that by the end of the decade had begun to dominate programming schedules. With no union writers to pay, these shows—*Big Brother, Survivor, American Idol,* and the like—proved cheap to produce as well as highly popular with younger demographics.[2]

By 2002, *The Real World* had become one of MTV's staple shows, selecting a fresh new cast of young people each year to act out their lives before millions of television viewers. Among the seven strangers thrown together that season—this time in a posh Las Vegas apartment—was Trishelle, a young woman from the rural town of Cutoff, Louisiana. Within the first few episodes, she succumbed to several serious temptations. She fell for a married man, got drunk, and had sex, all against her better judgment. She even kissed another girl, even though she was not gay. In angst, she called home to confide in her stepsister, only to have her father lambaste her for what he saw as sinfulness. Then, on top of everything else, her period was late.[3]

It took Trishelle three episodes to get up the courage to take a pregnancy test. To her great relief, it turned out negative, but before the series ended that year, MTV turned the pregnancy scare into an educational opportunity.[4] "Have you ever been late?" the show's Web site asked MTV's young viewers. "If so, you're not alone. To find out how to prevent being late or what to do if you are, click here." The next link provided a toll-free number for Planned Parenthood, advice on birth-control methods, and a hotline for access to emergency contraception. Viewers also were invited to take part in an online talk show with Trishelle, other cast members, and a representative of Planned Parenthood, who was on hand to answer questions about pregnancy prevention, sexually transmitted diseases (STDs), and communication strategies for dealing with sex partners.[5]

These efforts were part of the multimillion dollar public-education initiative Fight for Your Rights: Protect Yourself. A partnership between MTV and the Kaiser Family Foundation, the project's goal was to educate young people about sexual health and to promote safe sex. Because the campaign targeted teens and young adults through their own media, it was able to address sensitive sexual themes and controversial topics that might have sparked an outcry if they were more visible to the public.

For decades, youth have been the target of hundreds of "social marketing" campaigns, aimed at such vexing problems as teen pregnancy, drug abuse, and smoking.[6] Over the years, nonprofit groups and government agencies have produced public-service advertising messages aimed at teens and young adults, often enlisting popular celebrities to promote pro-social messages in television spots, billboards, and magazine ads.[7] In the new fragmented digital-media universe, these public-education strategies have evolved into highly sophisticated interactive campaigns, often employing the same tools of the trade that advertisers use to promote brands to young people. Just as marketers follow youth through the new digital landscape, closely tracking their every move, and inserting sales pitches into every possible venue, social marketers have crafted their campaigns to mesh with the media habits and journeys of teens and young adults. With the growth of the Internet and the proliferation of teen TV, public-health and social-issue organizations are able to incorporate their messages throughout the youth media culture, reaching their demographic targets with precision. Interactive media also make it possible to engage young people as never before, enlisting them in the campaigns, and encouraging them to spread the word among their peers. Sometimes these efforts generate controversy over their unorthodox methods and edgy messages. But much of the time they operate freely within the confines of a youth media world where adults seldom venture, addressing topics in a frank and direct manner that would be taboo in the mainstream media.

Sex Ed in the Digital Age

In August 1996, less than a month before the Democratic National Convention, President Clinton signed the Personal Responsibility and Work Opportunity Reconciliation Act of 1996 (commonly called the Welfare Reform Act), fulfilling a campaign promise he had made in 1992 to "end welfare as we know it."[8] The landmark law eliminated more than sixty years of federally guaranteed assistance to the poor and enabled states to develop their own programs to move people from welfare to work. The legislation triggered a fiery debate in the midst of this election year. Clinton's support for the law prompted outcries from liberal groups, including the Children's Defense Fund and the National Organization for Women, who charged the White House with abandoning the Democratic Party's long-standing commitment to women, children, and the poor.[9]

Among the law's new federal mandates was a little-known provision that had been pushed by conservative groups and slipped into the legislation at the last minute, requiring that $50 million a year be made available to the states to fund "abstinence-only" sex education. The new program, to be administered through schools, public agencies, and community organizations, was designed to deliver a clear and consistent message to young people: abstaining from sexual activity was the "only certain way" to avoid pregnancy and sexually transmitted diseases; the "expected standard" for sexual activity was "a monogamous relationship within the context of marriage;" and extramarital or premarital sex was likely to be "psychologically and physically harmful."[10] Since passage of the law, abstinence-only programs grew in the nation's schools. A 2002 survey found that 23 percent of high school sex-education programs were teaching abstinence-only curricula, compared to only 2 percent in 1988.[11]

But by the 1990s, neither schools nor parents held the exclusive franchise on sex education. The mass media already were eclipsing these traditional institutions. A national survey in 1997 found that more than half of high school boys and girls were learning about birth control, contraception, or pregnancy prevention from television, while two thirds of the girls and 40 percent of the boys had learned about these topics from magazines.[12] Scholars studying the impact of television on teens found that messages in entertainment programming could affect attitudes, expectations, and behavior.[13] Since the 1970s, a handful of nonprofit organizations had been collaborating with the creators of prime-time TV and soap operas to insert dialogue and storylines into programming, in order to educate youth about birth control, drunk driving, and other social issues.

Over the years these efforts grew into an entire infrastructure of "entertainment-education" initiatives, part of the landscape of the television industry.[14]

Advocates for Youth was one of the pioneers in the entertainment-education movement, establishing a Los Angeles office in the 1970s to encourage producers and writers to incorporate messages about birth control, abortion, and sexually transmitted diseases into their TV programs.[15] By the mid-1990s, the group's Media Project was an established presence in the entertainment community, conducting informational workshops, consulting on scripts, and handing out awards for responsible and "balanced" depictions of sex.[16] After passage of the abstinence-only legislation, Advocates for Youth's work in Hollywood became a key component in its political fight against the law. The group argued that young people had a right to a full spectrum of sexual information, including lifestyle options, birth control, abortion, and sexually transmitted diseases.[17] By collaborating with creators of popular TV programs, the Media Project sought to communicate directly to teens, circumventing the restrictive curricula that many schools adopted in the wake of the new law.[18]

With the growing number of entertainment programs created exclusively for teenagers, the project found a stable of willing and eager producers with whom to forge partnerships, developing ongoing storylines and characters that could help educate the loyal teen following. Project staff worked closely with producers of popular teen shows such as *Dawson's Creek* and *Felicity* to develop episodes dealing with teen sexual health and date rape.[19] The project's Web site featured detailed lists of specific program episodes that the group felt had done a good job of dealing with sexual issues. "Whether dealing with sexual abuse, contraception, or unplanned pregnancy or portraying strong parent-child communication or peer pressure resistance, the producers and writers of these programs have a right to be proud." The list included dozens of news and entertainment shows, with synopses of the episodes, as well as the ratings figures. For example, in a 2003 episode of *The Gilmore Girls* on the WB network, "Paris confides to Rory that she had sex for the first time, leading to a conversation about how the right time is different for everyone." The project commended a 2003 episode of *The Simpsons* in which Homer and Marge separate and Homer moves in with two gay guys, learning "to be more accepting of gays and lesbians." On ABC's *All My Children,* "JR gives Jamie a condom in preparation for him 'getting lucky' at a party."[20] Among the shows to receive the Media Project's SHINE Awards (for Sexual Health IN Entertainment) were: *Any Day Now, Sunset Beach, ER, The West*

Wing, Popular, Dawson's Creek, Will & Grace, Dateline NBC, That 70s Show, and *Moesha.*[21]

But addressing matters of sex in the context of prime-time television can be tricky business. Conservative media watchdog groups keep a close watch on television programs, taking their complaints to the government. In 2002, a storyline about oral sex (on which Advocates for Youth had consulted) was featured in the Fox Television series *Boston Public.*[22] Aired at 8:00 p.m., the program sparked an outcry from fifteen conservative groups, including Focus on the Family, Christian Coalition, and the American Family Association. The organizations petitioned the FCC to enforce its rules governing indecent content on television and radio. Though the commission levied no fines against the network, the steady drumbeat for government intervention into sexual content on television would grow.

While continuing to work with TV producers, Advocates for Youth also turned to the Internet, using its Web site not just as an educational vehicle, but also as an organizing tool in the fight against the abstinence-only policy. "End Censorship in America's Schools," students were urged. "Join other youth activists in the My Voice Counts! Campaign, as they raise their voices about the need for honest sex education in communities in the U.S. and abroad."[23] As more and more young people were going online to seek out resources and support for a range of health issues, those with limited knowledge of such matters at home or school now had unprecedented access to a wealth of information about sexual behavior, sexual lifestyles, and sexual health. Dozens of sites were set up to provide discussion forums on sexual issues, access to experts and community resources, as well as opportunities to be involved in the policy debate about sex education.[24] At the Live Teen Forum page of www.iwannaknow.org, created by the American Social Health Association, teenagers could communicate directly with a health specialist about their own sexual-health concerns, through e-mail or a toll-free phone number.[25] Rutgers University's Network for Family Life Education created www.sxetc.org, which quickly became a popular source of lively discussion and information among teens. The site included polls, surveys, and other interactive components to enable teenagers to voice their opinions on sex, sexually transmitted disease, and sex education policies. Sex, etc. also offered a downloadable "Teen Guide to Changing Your School's Sex Ed," with detailed instructions about building a local coalition, staging a community forum, drafting resolutions, and working with the press.[26]

Youth seeking information and support about sexual-identity issues could find a new world online. "For homosexual teenagers with computer access," wrote Jennifer Egan in the *New York Times* magazine, "the Internet has, quite simply, revolutionized the experience of growing up gay." For those "with the inhospitable families and towns in which many find themselves marooned," she explained, "there exists a parallel online community—real people like them in cyberspace with whom they can chat, exchange messages, and even engage in (online) sex."[27] Outproud.org, a Web site created by the National Coalition for Gay, Lesbian, Bisexual & Transgender Youth was one of hundreds of resources in this "parallel online community". It was set up "to help queer youth become happy, successful, confident and vital gay, lesbian and bisexual adults." The site offered "outreach and support to queer teens just coming to terms with their sexual orientation and to those contemplating coming out." Youth could find fact sheets and statistical information "to help make your case why supporting gay, lesbian, bisexual and transgender youth is important to your school." A Coming Out Archives provided a fully searchable database of hundreds of personal narratives, designed to "provide you with the benefit of the experiences of the millions of others who have found the right words on their own journeys. Sometimes things go well, sometimes they don't—whatever the results, they're here for you to see." Youth were also invited to share their own coming out stories on the Web site.[28]

The explosion of new media technologies has created both opportunities and challenges for health educators. The Internet and cable television can be used to circumvent the mainstream media, providing a relatively unfettered arena for addressing otherwise taboo topics. However, with the growing number of media available to teenagers, there is no guarantee that certain health messages will reach most or all teens. Media researchers have found that individual teenagers are customizing their own "sexual media diets," selecting from a growing menu of available TV programs, Web sites, music, and movies to suit their own needs, tastes, and desires for sexual information.[29]

"I Want My MTV!"

Executives at the Henry J. Kaiser Family Foundation took these trends into account when designing their public-education campaign on sexual health in the late 1990s. As Vicky Rideout, vice president and director of Kaiser's Program for the Study of Entertainment Media and Health, explained, the foundation's strategy was to "surround youth with a variety of messages in many different forms and styles."[30] With assets of more than a half billion dollars, the Kaiser Foundation was in a particularly unique and powerful position to carry out its goal.[31] The California-based philanthropic organization functions mainly as an "operating foundation," which means that, rather than just giving out grants to nonprofits, it can design its own large-scale public-education and research initiatives, often in partnership with other influential organizations. The foundation has been particularly effective at commanding the attention of the media, promoting widespread coverage of its research on such topics as health-care policy, women's health, HIV/AIDS, and minority health.[32]

Because television still plays a central role in the media diets of teens, part of Kaiser's effort has been aimed at the Hollywood creative community. The foundation followed the lead of other entertainment-education projects to encourage the television industry to use its programs as a way to educate youth about sexual-health issues. But with more resources than many nonprofits, the foundation has been able to develop a more comprehensive initiative for influencing entertainment television, combining its work to influence the producers of popular prime-time series with formal research that assessed the impact of its efforts on both the viewers and the programming. Foundation staff worked regularly with the producers of the NBC prime-time series *ER,* helping them insert storylines on a variety of health-related issues, including episodes that dealt with emergency contraception and sexually transmitted diseases. Follow-up surveys of viewers of these programs showed that the depictions helped increase awareness of the issues, prompted discussions among friends and family members, and in some cases helped people make decisions about their own health care.[33] For several years, Kaiser underwrote Advocates for Youth's Media Project in Los Angeles.[34] The foundation also has tracked the sexual content of entertainment television programming, conducting biennial studies to measure levels and kinds of sexual activity, in addition to depictions of "safe-sex" practices.[35]

To reach its target audience more directly, Kaiser partnered with popular teen magazines such as *Teen People, TM,* and *Seventeen,* working with editors to develop special features on sexual-health issues and reader surveys about sexual behavior.[36] The foundation's most ambitious effort for reaching young people is through its partnership with MTV. Launched in 1981, MTV is the number-one cable network among 12–24-year-olds, and the network has become "nearly synonymous with youth."[37] Its global reach (in 2006) includes more than 400 million subscribers in more than 164 countries.[38]

Sexual issues are front and center in the lives of the MTV Generation. The proportion of sexually active girls age 15–19 rose from 47 percent in 1982 to 55 percent by 1990.[39] Although rates of teen pregnancy declined somewhat since the high point in 1990, more than one million pregnancies still occurred in teenagers between ages 15 and 19, with nearly 30,000 in girls under 15.[40] And a quarter of sexually active teenagers contracted a sexually transmitted disease every year.[41] As far as teens are concerned, explained MTV's Jaime Uzeta, director of strategic partnerships and public affairs, sexual health is "public health #1."[42] For many social critics, however, MTV was part of the problem. Since the cable channel began, its 24/7 stream of graphic sexual images had sparked protests from parents groups and conservative Christians.[43] But for Kaiser, the pervasive sex on the cable network created an opportunity. Since MTV already was promoting sexual activity, it could be persuaded, with some financial incentives, to add responsible messages to the mix.

For years, nonprofit organizations seeking to use the mass media for their social-marketing campaigns often relied on the goodwill of the television networks and local stations (and encouragement by federal regulators) to provide free airtime. But many public-service announcements (PSAs) have been buried in the wee hours before dawn. With deregulation of the broadcasting industry in the 1980s, the number of PSAs that networks and stations were willing to run at all declined sharply.[44] Many nonprofits turned to paid advertising to get their messages on at a desirable time.[45] But the TV industry still was reluctant to air controversial PSAs. This was especially true with sexual issues.[46]

The Kaiser Foundation's strategy with MTV was a deliberate departure from traditional public-service campaigns. The foundation entered into a business relationship with the cable channel, offering financial and organizational support to sweeten the deal. This arrangement became the model for Kaiser's other media efforts over the years. "Kaiser has approached its entertainment media partnerships as business propositions with a philanthropic purpose," foundation executives explained. Kaiser crafts its agreements in "formal memoranda of understanding," offering an appealing package to its media partners that includes support for: "issues research; briefings for writers and producers, and other media staff; substantive guidance on message development; and funds to support program production and the creation of information resources for consumers." In return for this financial commitment from the foundation, media companies contribute "creative and communications expertise; on-air programming on the issues addressed by the campaign; and guaranteed placement of the PSAs and other content to reach target audiences." Kaiser's campaigns are "undertaken in much the same way that any commercial product would be marketed—by using the best creative teams to help develop compelling messages for the target audience and securing commitments that ensure that they are seen on the right television shows in the right time slots." Kaiser's "product" is not "sneakers or beer" however, but "awareness and prevention."[47]

The Be Safe campaign was launched in 1997 with a series of hip PSAs, a toll-free hotline, and the booklet "It's Your (Sex) Life," offering detailed advice and information about how to avoid getting sexually transmitted diseases.[48] In the first six months, 150,000 viewers called the hotline, and more than 100,000 of them requested a copy of the Kaiser booklet.[49] By 2002, the effort had been "rebranded" under MTV's existing Fight for Your Rights pro-social initiative with a new tag line added: Protect Yourself.[50] Fight for Your Rights: Protect Yourself soon became a recognizable brand on its own, woven throughout the TV schedule and Web site on the popular youth network. The campaign themes and messages became a pervasive presence on MTV, appearing in PSAs, interactive Web features, and print materials, as well as on-the-ground, grassroots activities. The Kaiser Foundation spent $440,000 on its partnership with MTV in its initial year.[51]

Though carefully framed as a public-health initiative, the campaign was not without controversy. In 1998, Christian right-to-life group Rock for Life, a division of the American Life League, attacked the Kaiser/MTV campaign for promoting sexual behavior among teens. "MTV has no business teaching kids about sexual relationships and promoting abortion," the group charged. "MTV is trying to take over the role of the parents and families and teach deadly values to kids under the disguise of looking out for their welfare." The group charged that by sending out informational materials directly to teens on abortion and contraception, it was attempting to undermine parents.[52]

Despite these initial complaints, the campaign not only continued, but also expanded. During the first five and a half years, more than sixty public-service ads were produced, airing more than 4,600 times.[53] The videos were much edgier, realistic, and graphic than anything that could run on broadcast television. Through research, the campaign also was able to identify specific subsegments of the MTV audience, tailoring PSAs to gays and lesbians, Latinos, females, and African Americans. One of the ads, for example, featured an inner-city Latino rocker speaking directly to the camera, trying to convince himself that he would know if he had a sexually transmitted disease. "If I was feeling weak, or had a rash downstairs—I'd definitely know something," he assures himself. But the voice-over announcer warns: "Most people with STDs show no symptoms at all. Get perspective. Get tested. For more information go to Fight for Your Rights @ MTV.com."[54]

Programming on MTV is an integrated multimedia effort, employing cross-platform promotion strategies that have become standard operating procedure for entertainment-media companies in the digital era. The network takes full advantage of the multitasking media habits of youth to extend the reach of its brand and ensure maximum exposure and ongoing relationships with its viewers. These same strategies were central to the Fight for Your Rights: Protect Yourself campaign. Branded sexual-health messages were woven throughout the MTV franchise. Through interactive media, teens could discuss sex and sexual-health issues candidly with experts as well as peers, without fear of parental interference. Public-service advertising and TV programs featured links to the special Web site created for the campaign, itsyoursexlife.com, where viewers could access a wealth of online resources, including: "e-PSAs"; a searchable nationwide database of HIV- and STD-testing facilities; an interactive guide to sexual health; a sexual-health news site; monthly features that provide the latest information about HIV/AIDS and other topics; and 24-hour message boards where viewers could meet and discuss sexual-health-related issues in an ongoing dialogue. Through a toll-free hotline run by the Kaiser Foundation, teens could receive the free "Sex Life" booklet, and be "connected immediately to a live operator at the CDC's National HIV/AIDS or STD hot lines or to their local Planned Parenthood clinic."[55]

In keeping with MTV's edgy, rebellious-youth image, the campaign encouraged its viewers to become engaged in sexual politics. The project's Web site featured a Take Action section, where MTV's other partners, including Advocates for Youth, offered opportunities to become involved in grassroots efforts. "Find out where you can be trained as a peer educator, learn how to take political action both locally and nationally, and get inspired by checking out what other young activists are doing," the Web site urged.[56] But Kaiser was careful to distance itself from these overtly political aspects of the initiative, frequently pointing out that it neither engaged in nor funded advocacy. In its report "Reaching the MTV Generation," for example, among the specials listed as part of the campaign were several MTV broadcasts that directly took on the controversial policies around abstinence-only education. One was the 2002 MTV/*Time* magazine special "Sex in the Classroom." The program dealt with abstinence-only and comprehensive sex-ed curricula in the nation's schools, and it included sections where "students, educators, and experts speak out." Though the foundation listed this broadcast among the campaign's programmatic accomplishments, it noted that the special was "produced independently of the Foundation, but aired as part of the Fight for Your Rights campaign."[57]

Dozens of TV programs were produced dealing with sexuality and sexual-health issues. During the 2002–2003 season, the network's documentary series *True Life* featured several episodes with sexual themes. In the episode "I Need Sex RX," camera crews followed several young people as they visited doctors and health-care centers, providing an up-close-and-personal view of one woman having her first gynecological exam and another getting tested for HIV. "It Could Be You" featured two young HIV-positive women on a road-trip around the country, meeting up with others who had tested positive for the disease. The special *9 Things You Need to Know before You're Good to Go* featured R&B/hip-hop star Tweet offering practical advice on how to have sex without getting an STD. The documentary *Dangerous Liaisons* examined the consequences of mixing drugs and alcohol with sex. And on *Live Loveline,* comedian Adam Carolla and MTV's sex doctor, Drew Pinsky, answered studio-audience and call-in questions about STDs. *MTV News* ran stories about teen pregnancy, STDs, HIV/AIDS, and other sexual-health issues. During the 2000 presidential election, the MTV special *Choose or Lose: Sex Laws* addressed such hot political issues as abortion, sex-ed policy, gay rights, and age of consent. The network's global reach enabled it to stage live, large-scale, international media events. The 2001 broadcast of *Staying Alive* 3, a special about young people and HIV/AIDS, aired in 150 countries, including South Africa, Kazakhstan, Russia, and China.[58]

On-air specials frequently were coordinated with "offline" community events, in partnership with local chapters of Planned Parenthood and other sexual-health groups. Youth also could take part in live, online "e-discussions" about the TV programs on the MTV Web site.[59] To kick off the 2002 campaign, MTV broadcast a live one-hour special called the "National Sex Quiz," promoting it in advance both on the air and online: "When it comes to sex, everybody thinks they know the score. You know everything you need to know . . . right? You sure? Prove it: Take MTV's 21-question pop quiz on sex and health, and then tune in to our live special on April 20 to find out if you know as much as you think (and get schooled if you don't!)."[60] More than 700,000 viewers participated in the event, according to MTV executives.[61] Inserts on "safer sex" were distributed in MTV's "Party to Go" CDs, as well as the *Real World* video, reaching 117,000 individuals.[62] As part of MTV's 1998 Campus Invasion concert tour, "sexual information" tents were set up at thirty different colleges and universities around the country, staffed by college health counselors who distributed free literature and condoms.[63]

But while TV specials and Web-site materials can offer fairly straightforward information on sexual health, incorporating the campaign's themes into unscripted reality shows sometimes can result in a muddled and confused message. With its rotating stable of eager young participants, *The Real World* has been a hotbed of topical issues confronting teens and young adults, making it a natural venue for pro-social messages. Over the years, the show has featured young people dealing with a variety of problems, including AIDS, sexual identity, and alcohol abuse.[64] According to Kaiser executives, sexual-health themes have been "placed" in the popular series.[65] Since it is a reality show, however, the issue cannot be written into scripts. Rather, when characters confront sexual issues on the show, the incidents can be linked to a more deliberate pro-social message.

So, when Trishelle went through her pregnancy scare, MTV put the young woman on an "online talk show," where she could debrief with her *Real World* viewers, accompanied by an expert to provide additional health information. But as Trishelle answered questions, it became clear that the fishbowl nature of the show itself may have contributed to her problem. When asked by the moderator why they did not use a condom when they had sex, Trishelle replied: "I think the first time we had sex it wasn't planned. And . . . the fact we had cameras on us, no one wanted to get up and get a condom. Then the camera crews would know that we were definitely having sex." The Planned Parenthood representative quickly intervened with her own advice: "The good news is most people don't have cameras on them when they're having sex. It's still a good idea to have condoms nearby if there's any chance you'll get involved in sexual behavior." After hearing the Planned Parenthood expert reel off a long list of birth control methods, the moderator asked Trishelle: "Did you consider going on the birth control pill?" The young woman replied that she had planned to go on the pill, but because she had to make a hasty move to Las Vegas in order to be on the show, she was having trouble getting her prescription transferred.[66]

The *Real World* incident underscored what some media critics see as a contradiction in the Kaiser/MTV initiative. The highly popular youth network is a perfect venue for reaching the audience that Kaiser seeks to influence, but embedding socially responsible messages into a channel known for its titillating and graphic sexual content may send a double message to young people. "It's tremendously ironic," Bob Thompson, director of Syracuse University's Center for the Study of Popular Television, commented to the press. "The campaign is working with a network whose programming features exactly the opposite messages." But Kaiser officials counter that teens and young adults already are heavy viewers of MTV, "so the bottom line is, are they better off with this information included or without it?" Kaiser's Vicky Rideout told the press. "And the answer is that those who get our materials are more likely to see a doctor, talk to their parents and use birth control as a result."[67]

Whether one agrees with the approach or not, the campaign's message appears to have gotten through. According to research released by the Kaiser Foundation, "a majority (52%) of all 16–24-year-olds in the country say they have seen sexual health ads on MTV, and a third (32%) say they have seen full-length shows." The research also found that "nearly two-thirds (63%)" of those who saw the campaign personally learned from it, and many told researchers they had become more cautious and careful in their own sexual behaviors, as a result of paying attention to the campaign.[68] The campaign has won numerous awards, including an Emmy and a Peabody award in 2004.[69]

"Knowing Is Beautiful"

Kaiser's campaign with MTV laid the groundwork for a much larger initiative, this time in partnership with the music network's parent company, Viacom. Launched in 2003, the KNOW HIV/AIDS campaign took full advantage of the massive holdings in the corporation's global media empire. Public-service advertising ran on all of Viacom's TV networks, radio stations, and billboards, and AIDS-related themes could be woven into Viacom-produced entertainment series.[70] Kaiser already had been working with Black Entertainment Television (BET) for several years on the public-service campaign Rap It Up.[71] The new partnership with Viacom (which purchased BET in 2000) enabled the foundation to extend its reach to the media conglomerate's other outlets, creating highly targeted messages aimed at African-American youth.[72] Launched in 1995, the United Paramount Network (UPN) had become one of the most popular TV channels among African-Americans.[73] KNOW HIV/AIDS PSAs were run during the network's Monday-night lineup of popular sitcoms, aimed at teen girls; a separate set of spots for boys aired on Thursdays during *World Wrestling Entertainment SmackDown*. Each series of PSAs was given its own gender-specific Web site.[74]

Like the MTV campaign, KNOW HIV/AIDS tapped into youth multitasking habits with messages that crossed platforms. In the case of entertainment programming, the Web site was used to promote the pro-social

TV episodes. The plot summary of UPN's sitcom *Girlfriends* was e-mailed to viewers to encourage them to tune in. "Maya, Lynn, Joan, and Toni have gotten real about HIV/AIDS," the message read.

The girlfriends learn that an old college friend (Kimberly Elise from *John Q*) has contracted HIV from her husband, who was also Joan's boyfriend in college and is now living with AIDS; Lynn, after attending a poetry slam, decides to refocus her documentary on sexuality to address AIDS; and Toni and her fiancé decide to take an HIV test. Be sure to watch the May 12 episode, when Lynn wraps up her documentary, which ends on a surprising and unsettling note.

Viewers were urged to spread the word among their peers. "Encourage your friends to watch," the message instructed, "send this email to your friends!" The missive included a link to an online preview of the show, as well as "three things you can do in your own community to get the word out. . . . 1. Call and/or forward this email to 10 friends to let them know about the show. 2. Make a night of it . . . host a 'house party' and watch the show with friends. 3. Get your neighborhood together . . . hold a potluck at your local church or community center and watch the show together."[75]

The foundation's focus-group research with African-American youth revealed that many of them did not want to be tested for HIV. Not only were they fearful of the results, but they also worried about being stigmatized just for going in for the test. To "normalize" testing, Kaiser and Viacom launched the series of public service advertisements, "Knowing Is Beautiful."[76] Produced by the Crispin Porter + Bogusky ad agency, the glossy, slick ads used hip-hop artists and themes to promote the positive aspects of getting tested for HIV. In the TV spots, narrated with poetry by hip-hop artist Common, four different characters were shown in various "moments of intimacy or contemplation," all sporting a small adhesive bandage that signifies they had gotten the AIDS test. Common's poetry, according to the campaign's description, "underscores the personal empowerment of having been tested for the disease." The spots were tagged with the KNOW HIV/AIDS Web site, where visitors could find local testing centers as well as additional information.[77]

Some media critics argued that the campaign's approach glamorized the disease. As columnist Ric Kahn commented in the *Boston Globe:* "By eschewing the scary and statistic-driven messages of typically dull public service announcements in favor of a style more reminiscent of a Gap ad's sensual sepia tones, hip-hop iconography, and an adhesive bandage from a blood test primped up as a beautiful flower-shaped HIV-test trademark the media blitz's participants are hoping to turn the clinical into the cool." The campaign might be "disguising the fact that, despite great medical strides, people are still dying from AIDS," he suggested, "dismissing the many who are still living but shredded by the side effects of their medications, from nausea to nightmares; even loosening safe-sex strictures with its elegant touch."[78]

Such isolated criticisms, however, did little to detract from the popularity of the initiative, which, like the Fight for Your Rights campaign, was highly acclaimed, winning Emmy and Peabody awards.[79] A Kaiser survey released in October 2004 showed that not only were a large percentage of African-Americans aware of the KNOW HIV/AIDS campaign, but a majority of viewers said they had learned important information from it, and it had influenced their own personal plans. Many reported taking steps to protect their health as a result of viewing the campaign materials.[80]

In 2003, the foundation began working with the United Nations to develop a multinational initiative on AIDS.[81] The following year, U.N. Secretary General Kofi Annan hosted a summit with CEOs from twenty-two media companies from around the world, including Kaiser's U.S. partners, as well as the heads of the British Broadcasting Corporation (BBC), the South African Broadcasting Corporation, Russia's Gazprom-Media, and Rupert Murdoch's Star Group Ltd. in India.[82] The Global AIDS Initiative spawned several national public efforts in other countries modeled on the Kaiser work with MTV and with Viacom in the United States. For example, Transatlantic Partners Against AIDS, a new nongovernmental organization (NGO) with funding from the Bill and Melinda Gates Foundation, partnered with Kaiser and more than thirty media companies in Russia to launch the Russian Media Partnership to Combat HIV/AIDS. In India the Sony Company agreed to integrate HIV messages into three episodes of its top-rated *Indian Idol*, a reality TV series modeled on the highly popular U.S. *American Idol*.[83]

The Kaiser Foundation's campaigns illustrate how large, well-funded organizations partnering with huge media conglomerates can take advantage of the economies of scale in an era of concentrated media, while at the same time targeting in very precise ways the demographic groups they need to reach. By bringing substantial financial resources and expertise to the table, the foundation has been able to gain exposure for its messages across dozens of multimedia platforms and outlets. Commenting on the KNOW/HIV Campaign, *Advertising Age* noted that "Kaiser secured a deal where it would invest $600,000 a year and in return, receive about $200 million in media time and space."[84] Kaiser's work also is emblematic of the blurring of content and promotion that has come to characterize so much of contemporary media. Like product placement, which integrates brands seamlessly into the narratives of television programs and movies, social-marketing campaigns increasingly are weaving their messages into the fabric of popular culture.

Telling "The Truth" to Youth

In 1994, two state attorneys general—Michael Moore of Mississippi, and Hubert "Skip" Humphrey III of Minnesota—filed suit against the major U.S. tobacco companies to recover the costs to their state Medicaid programs for treating tobacco-related diseases. Within months, all other states followed their lead. The negotiations became highly contentious, triggering a lobbying frenzy by the tobacco industry on Capitol Hill, dragging out for years, and pitting various sectors of the public-health community against one another.[85] Four years later, forty-six state attorneys general signed a historic $206 billion settlement with the four major tobacco companies.[86] Under the terms of the Master Settlement Agreement (MSA), the industry was forced to pull ads for its tobacco products from buses, subways, and other public-transit systems, as well as sports arenas, shopping malls, stadiums, and video arcades. Cartoon figures, including the notorious Joe Camel, were put into retirement. And ubiquitous brand-bearing merchandise—from hats to shirts to backpacks—were banned forever.[87]

The settlement also created a new nonprofit called the American Legacy Foundation. The DC-based organization was to be funded through annual payments of $300 million per year from the tobacco companies. The yearly support would remain at that level (adjusted up or down for inflation and declines in sales) until 2003, then reduced to $32 million annually through 2008.[88] With this money, the foundation would oversee a sustained nationwide public education campaign to reduce smoking among young people.[89] Throughout the 1990s, despite numerous public-education efforts targeted at teens, smoking rates remained alarmingly high. Cigarette smoking among high school students had increased during the 1990s, with more than a third using some type of tobacco, according to the CDC. Rates among younger children also were disturbing, with about 13 percent of middle schoolers using tobacco. Eighty percent of adult smokers start before they are 18, making teenagers the most vulnerable group at risk to begin smoking.[90] The unprecedented influx of funds made possible by the settlement enabled the new American Legacy Foundation to launch a bold national "counteradvertising" campaign, modeled on recent successes of public-health groups at the state level, with the goal of reversing these trends.

Controlling Tobacco

The Master Settlement Agreement was the culmination of more than thirty-five years of escalating battles between the tobacco industry and a growing army of public-health advocates, consumer groups, and government agencies. Both sides had learned important lessons. Beginning in the late 1960s and early 70s, in the midst of a burgeoning consumer and public-interest movement, new antismoking activist groups had formed, taking aggressive stances against tobacco.[91] These early efforts grew into a larger "smoking control" movement. Rather than focusing only on encouraging individuals to quit smoking, the goal was to change the "environment" that promoted and supported smoking. Advocates engaged in strategic, confrontational political actions that took direct aim at the industry producing and selling cigarettes. They also pushed a broad agenda of federal, state, and local laws, passing ordinances around the country to reduce smoking in the workplace, restaurants, and public buildings. Laws that increased taxes on cigarettes were able to discourage youth from purchasing the products, and in turn reduce smoking.[92]

Marketing also was a key target. In the 1960s, antismoking advocates successfully petitioned the FCC to require counteradvertising in the form of free PSAs, to rebut the millions of dollars spent by the tobacco industry to advertise its products on television. The resulting messages helped trigger a significant reduction in smoking rates. They also alarmed the tobacco industry, convincing its leaders to support a congressional ban that removed all cigarette commercials from television in 1971, rather than trying to withstand such an effective campaign to undermine them.[93]

In 1988, tobacco-control groups in California succeeded in a ballot initiative to increase the state surtax on cigarettes, earmarking a percentage of the revenue for a hard-hitting counteradvertising initiative.[94] "The campaign was like nothing the world had ever seen," recalled Stanton Glantz and Edith Balbach, in their firsthand case history, *Tobacco War*. "Rather than talking about the dangers of smoking, or even secondhand smoke, the campaign directly and explicitly attacked the tobacco industry." The first TV commercial was a parody of corrupt tobacco executives, huddled in a smoke-filled boardroom, cynically strategizing how to replace the thousands of customers they were losing either because they had quit smoking or had died from its effects. Faced with a multibillion dollar "problem," one of them explains, "We need more cigarette smokers. Pure and simple." In cold-hearted business language, he charges his cronies with recruiting "3,000 fresh volunteers every day." "Forget all that cancer, heart disease, stroke stuff," he orders. "Gentlemen, we're not in this for our health." The scene ends in a chorus of cynical laughter.[95]

These kinds of messages became part of a broad-based media-advocacy strategy to reframe the public debate over smoking through public shaming of tobacco companies for promoting disease and death.[96] Predictably, the California campaign prompted sharp criticism from cigarette makers. A spokesman for the Tobacco Institute labeled the ads "an unsavory assault on tobacco companies," and the industry began a long-term political effort to undercut the media campaign. Though the countercommercials continued for several years, relentless pressure by tobacco interests on the media and on state policymakers ultimately succeeded in disrupting and weakening the California antismoking initiative and forcing a "softening" of tone in the ads.[97]

Despite continuing opposition from the tobacco industry, the new style of counteradvertising became a linchpin in the growing number of media campaigns against smoking. Both Massachusetts and Florida launched efforts similar to those in California. The Campaign for Tobacco-Free Kids, a nonprofit policy-advocacy organization founded in 1994, incorporated the hard-hitting, public-shaming messages into its own policy advocacy campaigns, buying print ads that exposed unscrupulous "stealth marketing" by cigarette companies, and depicted innocent children as victims of purposeful tobacco-industry strategies that encouraged underage smoking.[98]

In 1997, the Florida Department of Health, which already had settled its lawsuit with the tobacco companies, used the funds to launch the truth® counteradvertising campaign. One ad featured an Academy Awards–type broadcast, where an honor was about to be bestowed for "most deaths in a single year." The nominees were suicide, illicit drugs, tobacco, and murder, each of which was represented by a man wearing a tuxedo. When tobacco was mentioned, the industry nominee enthusiastically applauded himself.[99] The ad campaign worked, contributing to a dramatic decline in smoking among middle school and high school students.[100] Florida's successful truth® model laid the groundwork for a much larger national effort.

But leaders in the industry also had learned much from their decades of combat with tobacco-control advocates. When it came time for the dealmaking over the Master Settlement Agreement, they insisted on language designed to reign in some of the vitriolic attacks that had come to characterize many of the antismoking media campaigns:

> The National Public Education Fund shall be used only for public education and advertising regarding the addictiveness, health effects and social costs related to the use of tobacco products and *shall not be used for any personal attack on, or vilification of, any person (whether by name or business affiliation), company, or governmental agency, whether individually or collectively.* (emphasis added).[101]

Executives at the American Legacy Foundation found themselves in a contradictory position. They were mandated to create a public-education campaign that would engage a target audience of teenagers already cynical about the flood of commercial messages and PSAs coming at them. Research showed that one of the most effective ways to "denormalize" smoking was to expose the manipulative strategies used by the tobacco industry to snare youth.[102] According to a study in the *Journal of the American Medical Association,* "Anti-tobacco advertisements need to be ambitious, hard-hitting, explicit, and in-your-face."[103] If the money were spent on "nice" ads that had little effect, the agency would be criticized strongly by the public-health community. But the type of advertising that had been shown to work with teens was precisely what the tobacco industry sought to suppress. For Cheryl Healton, a professor of public health at Columbia University, who was hired as the foundation's president and chief executive, this prohibition created a daunting challenge. As she wrote in a 2001 article in the *American Journal of Public Health,* "no existing countermarketing effort other than that of the American Legacy Foundation is required to operate under a clause proscribing vilification." Nonetheless, Healton vowed not to let the clause "impede the foundation's efforts to run an effective, hard-hitting countermarketing campaign that speaks the truth to teens and exposes the deceptive marketing strategies of the tobacco industry."[104] But the antivilification clause would remain a powerful legal weapon that the tobacco industry continued to wield as it closely monitored how the money extracted from its coffers was spent.

Unselling Smoking

Several years before the final MSA was signed, the CDC contracted with Columbia University to begin planning for the anticipated national campaign. The university began working with top advertising and youth-marketing experts, including executives from Motown, Procter & Gamble, and Teenage Research Unlimited.[105] Market research, using the latest psychographic methods, identified a precise target audience of teens between the ages of 15 and 16, who were the most vulnerable group needing immediate intervention. They were rebellious, "sensation-seeking" risk-takers who also sought ways to fit in. As the campaign rolled out, it would be designed not only to appeal to 12–17-year-olds, but also to tap into the deepest needs and desires of this specific "open-to-smoking" group.

With the money from the tobacco settlement in hand, the American Legacy Foundation partnered with top ad agencies, including Arnold Worldwide and Miami-based Crispin Porter + Bogusky, the agency that had spearheaded the successful truth® campaign in Florida. As Arnold Worldwide's creative director, Pete Favat, told the press, "the idea was to create a cool brand that means not smoking—that means a brand of rebellion against the tobacco industry."[106] The truth thus became its own brand, whose primary goal was to "unsell" smoking to teens.

From the outset, the campaign was designed to mirror the multiple strategies the tobacco companies themselves had used to target young people. Over the years, industry executives had denied repeatedly that they tried to recruit youth. But their efforts were well documented in the reams of internal memoranda and market-research reports uncovered during the discovery process in the lawsuits against the tobacco industry, as well as through widely publicized leaks from corporate whistleblowers.[107] In the thirty years since cigarette commercials had been taken off the air, the industry had compiled a new arsenal of alternative marketing weapons, many of them aimed at reaching young people. Companies shifted their ad dollars into magazines and billboards and devised a variety of innovative stealth strategies such as peer-to-peer street marketing, concerts, and branded publications. Tobacco brands also had become a ubiquitous presence in the lives of youth, appearing on everything from T-shirts to coffee mugs to mouse pads.[108] While the MSA forced the companies to give up some of these successful marketing tools, the antismoking opposition forces eagerly seized them and adapted them for their own campaigns.

The American Legacy Foundation also drew heavily from popular teen Web sites, such as Bolt, that were generating attention in the dot-com marketplace. The truth® was designed as a "multi-faceted and multi-platform" campaign, with young people themselves involved in every step of its design and implementation.[109] "Keeping it real, in the case of truth® is an actual mandate," observed *Clientology,* an ad-industry trade publication, "speaking to the trendsetting and opinion-leading teens and the teens who are most at risk, and keeping the stink of well-fed middle-aged white guys that hangs about many cool-seeking marketers away from the brand." In addition to a series of paid TV and print ads, the campaign developed a full complement of Digital Age youth-marketing tools, including a Web site (truth.com), branded merchandise, and peer-to-peer buzz marketing.

The Truth® Tour took the brand on the road, enlisting youth to "infect truth" among their peers across the country, making appearances at popular teen music and sports events.[110] A team of eighty young people volunteered for the summer 2001 Outbreak Tour, riding in large orange trucks to make over 400 stops during a four-month period from May to September in New York, Los Angeles, Chicago, and Atlanta. The state-of-the-art "truth® trucks" were fully equipped with teen pop-culture paraphernalia, including DJ decks, video monitors, Sega Dreamcast games and Internet access stations. "Look for truth® at concerts, festivals, parades, action sports events, skateparks and beaches," campaign promotional materials announced, promising to make appearances that summer at such hot teen venues as MTV's Beach House in Florida, Lil Bow Wow's Scream Tour, the Warped Tour and the Panasonic Shockwave Tour.[111]

The message of the campaign was a far cry from the scare tactics of the infamous fried-egg "this is your brain on drugs" PSA.

> In a savvy approach to our demographic segment's psychology, truth® ads do not "preach" at young people or tell them what—or what not—to do. Rather, the campaign conveys verifiable information in an edgy and sometimes humorous way that empowers teens to figure out the world for themselves and take control over their own lives. In a fast-paced, no-nonsense style, the advertisements feature young people who present stark facts about the death and disease attributed to tobacco use, the content of cigarettes, and the social costs of smoking.[112]

Promo magazine's Betsy Spethmann observed: "They don't tell kids not to smoke, exactly, they just make it really hip to not be sucked into that whole marketing scam."[113]

But this hip, no-holds-barred approach was controversial immediately. "The first ad campaign put us through a real test of our mettle," recalled Healton.[114] The foundation planned a high-profile launch on network television, with a series of innovative spots in prime time. But the project ran into trouble even before the first footage was shot. Broadcast TV networks were no strangers to controversy. Throughout their half-century history, their programming had triggered pressure campaigns from interest groups, advertisers, and the government.[115] Over the years, the networks had instituted a set of careful screening mechanisms to insulate themselves from pressure and to preempt outside problems. They were particularly averse to "advocacy advertising." When the first set of concepts for the antismoking commercials initially was submitted to CBS, NBC, and ABC, all three networks rejected the ideas, raising a host of "content" concerns. An unidentified official at CBS told the press that some of the spots might be "too grim and offending." In one ad, a teenager was to be shown actually smoking, prompting network executives to warn that such a message could "glamorize" the behavior that the campaign was "seeking to disparage."[116] In another spot, a teenager was asked to imagine applying an acne cream that ended up causing his entire face to fall off. The message was that cigarettes were the only product that, when used as intended, caused illness and death. Network officials worried the spot might unfairly implicate these other types of non-tobacco products, even if featured only for dramatic effect.[117]

But underlying these official explanations for rejecting the planned ads were other reasons that the networks were less inclined to talk about publicly. The tobacco industry was hovering over the campaign, inserting itself wherever it could to raise objections. Healton recalled that "a lot of pressure was put on the networks and on the board" of the American Legacy Foundation.[118] Some in the press speculated that the networks were balking because the same companies whose marketing strategies were being attacked in the antismoking campaign were big buyers of TV ad time for their other products. Philip Morris, for example, also owned Kraft Foods and Miller Beer, each of which had lucrative commercial deals with the networks.[119]

By early 2000, when revised commercials were in the can, it still was unclear whether the major networks would agree to run any of them. ABC and CBS had turned them down, but NBC was still considering. But several cable networks and teen-oriented channels (such as the WB) had agreed to air at least two of the spots, resulting in a "soft launch" of the $185 million campaign.[120] So by February, teens watching *Dawson's Creek* or *Buffy the Vampire Slayer* were subjected to what at first looked like yet another "extreme sports" soft-drink commercial. But the surprise ending packed a wallop. Three youth are on a bridge, about to jump into a ravine. Right before they take off, they each grab a can of "Splode" soda. After successfully making the plunge, the first two jumpers pull the pop-tops from their cans and allow the soda to spray into their mouths. But when the third jumper opens his can, it explodes, obliterating him in a ball of flames. The screen fills with the stark message: "Only one product actually kills a third of the people who use it. Tobacco."[121]

While this spot was edgy enough, there were others in the initial series that drew fire from tobacco companies. One of them was shot at the headquarters of the Philip Morris Company. Two teenagers enter the lobby of the corporation carrying an oversized briefcase marked "lie detector," and announce that they plan to deliver it to the marketing department in order to clear up any confusion about whether tobacco is addictive. Company security guards, obviously flustered by the unwelcome intrusion into business-as-usual, usher the teens out the door. In the other spot, a group of teens pulls up in front of the company's headquarters in a large truck and begins unloading hundreds of large sacks marked "body bag," throwing them into a huge pile next to the building. One of the teens shouts through a megaphone: "Do you know how many people tobacco kills every day?" Though neither of the ads identified the company as Philip Morris, the company responded angrily, charging the foundation with violating the vilification clause.[122] Several attorneys general who had negotiated the final settlement raised objections as well. To quell the controversy, the foundation temporarily pulled the ads. "This is a distraction from the goal: to stop 3,000 kids a day from becoming addicted to tobacco," explained Washington State's Attorney General Christine Gregoire, who led the negotiations and also chaired the American Legacy Foundation board. "It is not worth being distracted by one or two ads with others in the arsenal."[123]

The move prompted charges from some tobacco-control advocates that the foundation had caved in to pressure from the tobacco industry. "How can you run an anti-smoking campaign and not vilify the industry?" former U.S. Food and Drug Administration commissioner David Kessler told the press. "It would be better to not take the money if the industry is able to pull the strings and take control." Attorney General Michael Moore, who had spearheaded the legal fight against the tobacco industry, added: "I'm very uncomfortable with the tobacco companies being the censors of what kind of anti-tobacco message can be used to reduce teenage smoking." University of California at San Francisco's Stanton Glantz, a longtime antitobacco activist, had even stronger words. "I think the Legacy Foundation has basically destroyed itself," he pronounced. "The tobacco companies always threaten aggressive ads, and you have to stand up to them. You just cannot turn over control of your ads to Philip Morris."[124]

After substantial internal deliberations at the foundation and some retooling, the controversial spots eventually returned to the airwaves. Reversing its initial decision on the campaign, NBC agreed to run the "body bag" ad in its broadcast of the 2000 Olympics. A network spokesperson cited several reasons for the change, including the fact that some "elements" within the ad had been changed, the overall campaign appeared to be effective, and the ad in question had been running on cable channels without negative feedback. The American Legacy Foundation committed $15 million for airtime on NBC, MSNBC, and CNBC during the highly rated sports event. A Philip Morris spokesman declined to comment on the ad buy, but did say that the company did not "believe some of what they [ALF] have done has been consistent with the spirit" of the settlement agreement.[125]

Two new commercials were launched during the 2001 Super Bowl. They were as hard-hitting as the first group of spots, though there was no specific connection, directly or indirectly, to any particular company. One ad questioned the industry's claims that it was changing its practices. A former smoker, speaking through an "electrolarnyx," reminds viewers that the industry's deadly and addictive products had not changed, nor had the consequences of using them. The other ad showed an average American speaking with great pain about the death of his 46-year-old wife from lung cancer, as photos of her family flash intermittently throughout.[126] The spots aired without any major protest from tobacco companies.

Having successfully weathered its first year, truth® began to garner awards.[127] Initial research also indicated that the initiative had succeeded in attracting the attention of its teen target audience.[128] No doubt the controversies themselves had helped raised its profile. In the meantime, the campaign harnessed the Web as a tool not only for promoting the truth® brand online, but also for mobilizing young people against the tobacco industry. The TV and print ads carried the Web site address, where teens could find additional information about tobacco-marketing strategies, including internal documents released through the court cases. The site also encouraged teens to engage in their own activism against cigarette-marketing campaigns. For example, teens were challenged to monitor whether tobacco companies were keeping their promise not to put their ads into magazines read by significant numbers of underage youth. To document abuses, the campaign sent its volunteers out into the field, instructing them to tear out any cigarette ads they found in youth-oriented magazines and turn them in at events that were part of the truth® tour, where they could be redeemed for truth®-branded merchandise.[129] The paid truth® ads, including some of the most controversial, also could be accessed on the Web site, along with an array of other features that had become common on teen Web sites, including games, newsletters, chat rooms, and e-mail. The site also served as a recruiting tool for volunteers to join the yearly Truth® Tour.

The truth® campaign set up a special Web site for grassroots activists, many of whom had received funding from the tobacco settlement through their respective states.[130] With an edgy, urban design, streettheory.org featured teens in the back seat of a car with their hands and feet sticking out, the caption reading: "Fueling a tobacco-free generation." Billed as an online community and resource hub for smoking-control activists, the Web site offered links to activist groups around the country "to share ideas about strategies, resources, local interventions, and activism ideas that are utilized in their own communities." In addition to campaign tactics and facts and figures about the tobacco industry, activists could download posters, bumper stickers, and other tools for their own local campaigns.[131]

But while the Internet could be a valuable tool for engaging directly with youth, the foundation faced special challenges there as well, with the vilification clause a constant concern. "In one of the first board meetings," recalled Cheryl Healton, "there was a discussion—not heated, but intense—about whether we should have a mechanism on our Web site where young people could write 'Web letters' to the tobacco industry." Such features were commonplace on youth-activism Web sites, taking advantage of the interactive properties of the Internet to encourage and facilitate citizen participation.[132] However, in this case, explained Healton, "there was a concern that under the MSA, if we facilitated their communication with the tobacco companies, we would be held accountable for what these young people said." She argued strongly that youth had a First Amendment right to express themselves to these companies and that the foundation should facilitate and encourage them. This was also consistent with one of the primary goals of the campaign, which was designed not only to educate youth, but also to engage and involve them, encouraging them to take ownership of the cause. The board ultimately decided to set up the mechanism, designing it in such a way that youth could send their Web letters directly to the tobacco companies without the foundation ever seeing them.[133] But because of concerns over the vilification clause, the foundation began careful monitoring of the message boards on the site.[134] These cautious steps to constrain youth communication on the Web site stood in sharp contrast to the free-spirited and unfettered ways in which young people were using the Internet for self expression. But the foundation was in a tough position. On the one hand, it wanted to tap into teenage rebelliousness and harness the energy against the aggressive and manipulative marketing strategies of the tobacco industry. On the other hand, it had to be careful about perceptions that its campaign was inciting uncivil behavior by young people. And while other social marketers might have been able to use the new online medium to engage with youth outside of public scrutiny, every move made by the truth® campaign was watched closely by tobacco-industry executives and their lawyers.

In the summer of 2001, the campaign released a new radio ad, featuring a recording of a phone call made by a teenager to the Lorillard Tobacco Company. Identifying himself as a dog walker, the youth explains that he has some "quality dog urine" for sale and hopes the company can use it, since it is an ingredient used in cigarettes. The spot was aimed at highlighting chemicals found in cigarettes and cigarette smoke, such as ammonia and cyanide, based on documents that had been uncovered during the litigation. "There are nearly 600 chemicals tobacco companies report they add to cigarettes. This is something we think teens should know," Cheryl Healton explained later to the press.[135] The tobacco company was outraged. Charging that the commercial was "false and misleading," and asserting that dog urine was not an ingredient in cigarettes, Lorillard filed a complaint with the Federal Communications Commission.[136] Lawyers urged the commission to bar radio stations from airing the spot, arguing that it violated an FCC ban on broadcasts of recordings that were made without the knowledge or permission of one of the parties. The foundation countered that the FCC ban on secret taping did not apply to organizations producing public-service spots, but only to broadcasters themselves, in order to keep shock jocks from embarrassing unsuspecting listeners.[137] Nonetheless, the controversial ad soon disappeared from the airwaves.[138]

But a new series of truth® commercials promoting the same theme debuted a few months later during the annual MTV Video Music Awards. In "Ratman," a young man wearing a giant rat costume crawls out of a subway in New York's Times Square, where he greets passers-by with a hand-scrawled sign bearing the warning: "There's cyanide in cigarette smoke. Same as in rat poison." Another ad, "Bio-Suits," featured a group of teenagers attired in bright orange biohazard-style suits roaming through a major tobacco conference in search of tobacco-industry executives. Their goal is to offer protective suits to participants at the meeting so that people can protect themselves from the harmful chemicals found in cigarettes. A third spot—"Ammoniaide"—shows two teenagers who have decided to "take the idea of the lemonade stand to a new level at a major tobacco conference." What they offer is not lemonade but a new drink, "Ammoniaide," which they have flavored with ammonia "to enhance the drink's taste—borrowing the idea from tobacco industry documents that say that ammonia is added to cigarettes for flavor." All three spots ended with the tagline from the previous summer's Outbreak Tour: "Infect truth®." "These new ads will excite teens to spread the truth® about Big Tobacco and the toxic substances Big Tobacco adds to an already deadly substance, tobacco," Healton announced.[139]

Meanwhile, University of Michigan's annual Monitoring the Future survey showed that smoking among American teens had declined sharply, a dramatic reversal of the disturbing increases of the 1990s. "These important declines in teen smoking did not just happen by chance," the study's director told the press. "A lot of individuals and organizations have been making concerted efforts to bring down the unacceptably high rates of smoking among our youth," he explained, giving specific credit to the American Legacy Foundation's campaign.[140]

But the tobacco industry was not willing to allow the successful antismoking campaign to proceed unchallenged. In January 2002, Lorillard, labeling the entire truth® campaign "unscrupulous" and "deceitful," gave notice to the press and public that it believed the ads violated the vilification clause of the settlement agreement, indicating its intent to take punitive legal action.[141] Faced with an impending lawsuit, the foundation made its own preemptive move. Filing suit against the Lorillard Tobacco Company in a Delaware state court, the foundation called for a declaration exempting it from enforcement under the tobacco settlement. Its argument was that while the nonprofit organization had been established by the settlement, it had not been a party to the original agreement and its constraining language.[142] Within days, Lorillard sued the foundation in the company's own state of North Carolina, charging that the dog urine ad was false and that it violated the settlement agreement. "There are many ways to educate young people," a Lorillard spokesman told the press. "You can be edgy. You can be different. But you can't personally attack or vilify a company or its employees." Healton responded that the ads had "not vilified or personally attacked any person or any tobacco company. We have educated young people about the addictive and deadly consequences of tobacco use. And we will continue to do so."[143]

The Lorillard lawsuit not only took issue with an individual ad in which it had unwittingly played a part, but also aimed to put a stop to the entire truth® campaign. During the next four years of litigation, tobacco lawyers continued to gather ammunition to support their case, monitoring each new ad for evidence of "vilifying" content. They also made it clear to the press that if the judge ruled in their favor, they would ask that the

American Legacy Foundation be dissolved, the funds returned to an escrow account, and a new organization set up in its place.[144] The lawsuit made the campaign's Web site a particular target, charging that the foundation was encouraging "personal attacks" on the tobacco company's employees by helping the public send "harassing and vulgar" e-mail messages.[145] Since the foundation had not kept the letters that young people had sent to tobacco executives through the Web site, its lawyers had to get copies of them through the trial's discovery process in order to counter the tobacco company's charge. According to Healton, they found that Lorillard had been "very selective" in choosing which e-mail messages to use in its case. Ninety percent of the messages were about teens' personal experiences with the harmful effects of cigarettes, with accounts of parents and other family members who had lost their lives to tobacco—"very sad" to read, she recalled.[146]

In their arguments before the court, lawyers on both sides haggled over the precise definition of vilification, wielding dueling dictionaries to support their respective positions. Industry attorneys argued that "to vilify" was "to debase, degrade, or lower the target's standing." The foundation countered that the word meant "to make false or abusive statements," maintaining that because the content of its ads was factual and the tone humorous or satirical, it had not violated the agreement.[147] Ultimately, a Delaware judge ruled against the tobacco company, deciding in August 2005 that the truth® campaign did not violate the terms of the settlement agreement. "There are no scurrilous and vitriolic attacks" in the ads, the judge told the press. "There is no cruel slander. There is no public ridicule or calumny." Foundation executives were jubilant. "This decision will save hundreds of thousands of lives," Healton commented to the press, and "will allow our foundation to continue to tell American youths the facts about tobacco and the industry that markets its products to them."[148]

Buoyed by the successful resolution of its court case, the foundation also could take official credit for the positive impact of its public education efforts. An independent study published in the March 2005 issue of the *American Journal of Public Health* documented that the truth® campaign had led to significant declines in youth smoking rates. It also found that the more youth were exposed to truth® messages, the faster the decline in smoking.[149] "The truth® campaign works because it uses the exact same techniques the tobacco industry uses to attract kids," Matthew Myers, president of the Campaign for Tobacco-Free Kids, told the press, noting that the campaign's independence from the tobacco industry had been a key to its success.[150]

But just as the campaign emerged victorious from its long and draining court case, it faced serious financial challenges. Under the terms of the agreement, funding had been guaranteed only for five years. Any continued support from the tobacco industry would be forthcoming only if the companies that were party to the original agreement continued to maintain more than 99.05 percent of the U.S. market share.[151] As *Advertising Age* had noted as early as 2001, "It now appears that the payments won't continue after March 2003 because of share growth by independent tobacco companies that were not part of the original pact."[152] Anticipating its loss of income, the foundation had developed a number of strategies for diversifying its base of support, seeking corporate sponsors and strategic partnerships, and holding awards dinners and other fundraising events.[153] The foundation met with officials at the Kaiser Family Foundation in early 2005 to explore an effort modeled on its sexual-health campaigns with Viacom. Some industry insiders remained skeptical, however. "Legacy's efforts could prove Sisyphean in nature because media companies may find it impossible to ignore Big Tobacco's purchasing power," *Ad Week's* Wendy Melillo speculated. "Any partnership with Legacy is likely to be frowned upon by the tobacco industry."[154]

For the American Legacy Foundation to continue its edgy, hard-hitting, highly successful efforts to reduce teen smoking, it would need to secure sufficient funding to pay for them. To help with these efforts, former surgeons general, U.S. secretaries of health, and CDC directors formed a blue-ribbon Citizen's Commission to Protect the Truth. One of its tactics was to file friend-of-the-court briefs in lawsuits against the tobacco industry, urging judges to include the foundation as a recipient of any monies awarded by the courts.[155] There was no guarantee, however, that the tobacco industry would back away from its own aggressive efforts to reign in the truth®.

Calling All "NOISEmakers"

If new marketing strategies could be effective at fostering healthy behaviors among youth, perhaps they also could address the vexing problem of civic disengagement. As Robert Putnam documented in his 2000 book *Bowling Alone,* declines in democratic participation by youth were part of a much deeper pattern of "social malaise" affecting broad segments of the American public. But among young people, the indicators were particularly troubling.[156] While the national rate of voter participation had dropped by 9 percent among all age groups over a twenty-five-year period, the decline among youth age 18–24 was double that figure.[157] Youth voting rates had remained in a steady decline, beginning with a promising 55 percent voter turnout in 1972, and plunging to a dismal 36.1 percent by 2000, with a slight spike in 1992.[158] Not only were youth less likely to vote, they also were less interested in political discussion and public issues than either their older counterparts or young people of previous decades.[159] While many young people felt strongly about issues that affected them directly—drunk driving, depression, teen suicide, guns at school, and drugs—most were not inclined to take political action to remedy these problems.[160] And though youth were volunteering in record numbers, they had not made the connection between community service and political participation.[161] "Put simply," wrote Michael Delli Carpini of the Pew Charitable Trusts, "America's youth appear to be disconnecting from public life, and doing so at a rate that is greater than for any other age group."[162]

By the end of the century, foundations were pouring large sums of money into think tanks and university research institutes to try to find out why youth were not engaged.[163] Numerous public and private initiatives were launched to rally America's youth to action.[164] In 1997, General Colin Powell convened a national Presidents' Summit for America's Future, featuring former presidents Gerald Ford, Jimmy Carter, and George H. W. Bush, as well as First Lady Nancy Reagan. The high-profile Philadelphia event brought together youth organizations from around the country, joined by thirty governors, one hundred mayors, and dozens of business leaders. In addition to launching a new nonprofit, America's Promise, the summit called upon all participating organizations and institutions to make "specific, measurable commitments . . . to turn the tide for America's Youth."[165]

In response to the summit's challenge, nonprofit group Save the Children decided to launch a new organization to engage youth.[166] Diane Ty, a marketing executive at American Express, came in to develop and implement the project. With the help of such experts as celebrity pediatrician T. Berry Brazelton, Save the Children initially explored the creation of a large, grassroots membership organization that could do for children what AARP was doing for seniors by galvanizing a broad-based constituency of young people as advocates on behalf of children and youth policies. Ty drew up a business plan for a new organization, with elements comparable to those of AARP, including membership cards and a magazine.[167]

But in the age of the Internet, such plans seemed a little anachronistic. With increasing numbers of young people online, it made more sense to launch a Web site. Hundreds of nonprofits around the country already were using the Web to reach out to youth. Though fragmented and not always in the foreground of the glitzy new dot-com culture, these online civic ventures were becoming a significant presence. Most arose from the existing nonprofit associations, institutions, and organizations that make up civil society, though some grew out of private-sector initiatives, government programs, or cross-sectoral collaborations. They ranged from small, locally based efforts such as YouthLink of Hampton, Virginia, (yl-va.org) to large-scale, national programs like Youth Service America (ysa.org). Some were entirely virtual, made possible by the technology that enabled an individual or group to establish a Web site even in the absence of a "bricks-and-mortar" base of operations. Others reflected the efforts of long-standing, real-world fixtures that only recently had added online components to their programs for youth (e.g., the Girl Scouts at gsusa.org, or the YMCA at ymca.net). Still other sites spanned international borders, overcoming language and cultural barriers, and drawing youth from around the world into online discussions and projects (e.g., Taking ITGlobal.org and the World Bank Institute's World Links for Development at worldbank.org/worldlinks/english/index.html).[168] It was far too early to assess whether these initiatives were having any impact. But they did suggest something about the Internet's potential as a tool for civic and political involvement.

In the spring of 1999, Ty had a "seminal meeting" with Dan Pelson, cofounder of Bolt.com. "This was the first time I was exposed to the power of the Internet," she recalled. She and her partner, Liz Erickson, began devouring every book and article they could about the new online marketplace, soon reaching the conclusion that an exclusively Web-based initiative would be the best way to go. Through one of Save the Children's new board members, they were introduced to Tom McMurray, a Silicon Valley venture capitalist who had been a pioneer in the dot-com market, and an original partner in Yahoo! "He gave us the confidence that our business plan would work," Ty said, urging the pair to begin building a prototype for the Web site.[169] The plan called for initial start-up funding to come from foundations and major individual donors, and partnerships with many of the Silicon Valley companies that were expecting large profits in the rapidly growing dot-com marketplace.

In developing a Web site to promote civic involvement among teenagers, the team faced a number of challenges. While voluntarism among teens was on the rise—spurred in part by the growth of service-learning programs in schools—80 percent of them were not consistently volunteering.[170] In

1999, the David and Lucille Packard Foundation gave the new group a $150,000 planning grant, which included funds for focus-group research with teens to learn more about what would motivate them to become involved in voluntarism and activism. The research identified a "huge opportunity to change the image of service," Ty remembered. Involvement in service, considered by many teens to be too "goody two-shoes and nerdy," suffered from an obvious image problem. But once youth were engaged in the issues, they wanted to help.[171] To counter the "nerdy" image of activism, the design of the site had to be vibrant and "hip," capitalizing on the fact that fun and accessible features were driving young people to the Web. Featuring "interactivity, instant access to information, unique insights about other kids, and easy involvement," the Web site, YouthNOISE (YN), strove to create the right online dynamic for this particular audience. As the focus group report explained, "The opportunity is to create an entirely new identity for activism by 'shocking' kids with a cool layout and graphics that aggressively counter the 'boring' stigma and blows the dust off of old notions of activism and uses every opportunity to *flex its 'progressive' muscle.*"[172]

Ty and her associates relied on state-of-the-art tools of market research to determine what would appeal to teens and how best to approach them. The project's focus-group research confirmed what experts on adolescent development already knew; most teens lived in their own isolated bubbles. Teens' self-obsession was itself a "cause," one that teens already were pursuing with "consuming interest, deep conviction, and tenacious energy."[173] So one challenge was to find ways to hook teenagers. The Web site would have to offer them something that meshed with their prevailing mind-set, tapped into their needs and desires, and fit within their predominately narrow, self-focused perspective.[174]

Another challenge was finding ongoing financial support for the new venture. The business plan called for corporate sponsorships for the Web site's contests, surveys, and other features. For example, the Just 1 Click feature gave teens an easy way to authorize an online donation to a selected cause. When a teen clicked on the icon, the sponsoring company would donate a few cents to the featured issue. To encourage participation, companies would supply their products as prizes. Companies would donate the money and help underwrite the site, providing needed income to YN in exchange for the opportunity to reach a very valuable audience demographic. Teens would be urged to e-mail their friends with the link to the Just 1 Click Web page, using the Web's unique capacity for viral marketing to maximize not only teen participation in online philanthropy, but also the reach of the corporate logo within the teen market.[175]

In forming these alliances with corporations, YouthNOISE was tapping into the growing interest in "cause marketing," a practice employed by an increasing number of companies to link their products to causes and issues in order to build customer appreciation and loyalty. Considered a "win-win" strategy, cause marketing benefits both the commercial and the non-profit partners. Corporations sought association with well-known causes, and nonprofits stood to gain higher visibility from their connection to a corporate trademark. Companies were willing to pay nonprofits considerable sums of money for the association, with many creating specific lines in their marketing budgets for cause marketing.[176] "Linking a brand or product to a charity or a philanthropic organization has become one of the hottest ways to build customer appreciation and loyalty," explained a December 2000 *Wall Street Journal* article. "The donor companies believe that even though they share part of their margin with a nonprofit, they still make more money by increasing the volume of sales."[177] The strategy became particularly valuable to corporations seeking to market their products and services to teenagers. As a 2000 Cone/Roper study reported, "when price and quality are equal, 89 percent of teens report they would be likely to switch brands to one associated with a good cause." Cause marketing was considered a reliable way to influence girls' purchases, the report explained, "and girls are more likely than boys to tell their friends about companies that support causes."[178]

YouthNOISE garnered support from a variety of sources to secure the funds needed for its Web site. Save the Children's board gave the new project $250,000, which provided funding for additional staff and enabled Ty to hire twelve teenagers to help refine and rename the original concepts and site features in order to "bring the Web site to life."[179] The Packard Foundation, the Surdna Foundation, and the AOL Time Warner Foundation provided grants to the new enterprise, which also secured gifts from such high-profile show-business celebrities as musicians Bonnie Raitt and Graham Nash. In addition to some corporate cash contributions, YN developed partnerships with a handful of prominent new-media and technology companies that promised in-kind support or deeply discounted products and services. These included Vignette, InfoPop, WebTrends, IBM, and AOL Time Warner.[180] U.S. Interactive, a Web-services company, agreed to provide $1.4 million in pro bono services to build the YN site.[181]

Plans were developed for further intense fundraising from Silicon Valley companies.

The work of conceptualizing and building the new nonprofit online venture was proceeding smoothly and the new Web site was being readied for a high-profile launch in the fall of 2000. But its intended debut fell victim to a larger set of events, as the highly inflated dot-com market went into its dramatic slide.[182] This turn of events hit YouthNOISE hard. U.S. Interactive suddenly went out of business, with only 60 percent of its work on the youth Web site finished.[183] As a consequence, the launch was canceled, fundraising strategies had to be revisited, and the project was forced to regroup.

It took nearly a year to finish the construction of the site. In early 2001, YN staff began working with one of its media partners, *Seventeen* magazine, on a cross-promotion strategy timed to coincide with the rescheduled launch of the YN Web site, which was targeted at girls age 13–18. With the help of Save the Children, the magazine commissioned a feature article profiling the lives of four Afghan refugee girls exiled in Pakistan and fighting for their right to be educated. The article offered a rare glimpse for its female teenage readership into the treatment of women in Afghanistan, a place most U.S. teens knew little about. The magazine referred readers to the YouthNOISE Web site. "There's a simple but productive way to show your support of girls' education in Afghanistan," explained the magazine: "(1) Visit Youthnoise.com and click on Just-1-Click. (2) Register. (3) Click once a day. If only 3,000 people click daily for 60 days, YouthNOISE will meet its goal of 200,000 clicks and will then contribute $10,000 to Save the Children's Afghan education programs, through which 3,000 refugee girls will get much-needed Peace Packs full of books and supplies." The Just 1 Click program also offered prizes to three participants picked at random, with the grand-prize winner receiving a *Seventeen* gift basket and two runners up receiving jewelry.[184] The YouthNOISE site finally was scheduled to launch as the magazine's September issue hit the stands.

As it turned out, YouthNOISE's online premiere coincided with one of the most historic and tragic events in recent American history, a defining moment for the Digital Generation. As millions of Americans began their day the morning of September 11, their world was irrevocably transformed by the actions of nineteen terrorists. First word of the attacks sent people scurrying to their television sets, where many remained transfixed for hours. In the aftermath of the plane crashes into the World Trade Center and the Pentagon, more and more people turned to the Web to participate in local, national, and global discussions of the events, to donate money, and to volunteer for a myriad of collective efforts. Phil Noble of Politics Online tagged the Internet the "People's Channel," providing many services not available through other media, and doing "what it does best—communicating and connecting."[185]

For those personally caught up in the attacks, it was a means of communications when others failed. People stranded in the World Trade Center towers sent e-mails and instant messages to their loved ones; Blackberrys and pagers came through when mobile phones and land lines failed. Hundreds of online groups formed to do all the things people wanted to do—reach out to each other, share their grief, search for friends and loved ones.[186]

For teens and young adults, the Internet played a particularly central role during the crisis. Many turned to their own online communities, communicating with their peers across the country to make sense of the suddenness and severity of this national tragedy. "It started almost immediately," reported the staff of WireTap, an online youth magazine run by AlterNet (alternet.org/wiretapmag). "Youth went online to join message boards and email their friends about the events as they happened. Some, like those in New York, were writing in states of fear and shock and posting their words where others could hear about what they were going through."[187]

On the day of the attacks, with Washington, DC, under siege, YouthNOISE offices were evacuated, and all the staff and interns were sent home. But work was able to continue as a "virtual staff" was reassembled quickly, communicating through e-mail and instant messaging from home. One of the summer interns, who already had returned to college, became the ad hoc coordinator of online communication, keeping track of where each IM "buddy" was and making sure everyone remained in touch with one another. Working all day and into the night, they were able to put up content by the next day. They had not planned to launch their e-mail newsletter, NOISEnews, so soon but decided to go with it anyway, hastily pulling together the first issue.[188] With increased traffic on the site, other new features were added to address the crisis, including an educational piece on tolerance ("Islam, Muslims, Arabs, and Intolerance"), stories of youth bravery and commitment ("Teen Survivors of '93 Attack Lend a Hand") and opportunities for YouthNOISE teens to make their own contributions. For example, through the Memory Chain Memorial, they could write to other teens who had lost parents or family members. Site visitors also were encouraged to write to policymakers. The Web site later reported that NOISEmakers used the Change the Rules section to write more than 2,500 letters to President Bush.[189]

Over the next several weeks, YouthNOISE, along with a growing network of online youth projects, provided an unprecedented forum for political debate, voluntarism, and artistic expression. Young people around the country and the world were able to engage in intense discussion about a variety of deeply felt issues, including nationalism, patriotism, capitalism, the global economy, race, ethnicity, and religion. In the uncensored forum of the Web, youth were outspoken in expressing their anger, their fears, and their opinions.[190] On the Youth Radio site (youthradio.org), one audio clip began with youth reaction to the attacks, but the discussion quickly turned to other issues, such as the United States's role and behavior as an influential actor in a complex global economy.[191] The Web also served as a digital billboard where youth could post their artwork, cartoons, photos, video, and poetry to express their reactions to the events.[192] On HarlemLive.org, launched in 1996 for the youth of Harlem, hundreds of photos and quotes gathered from on-the-street interviews documented the shock of people in the Greenwich Village area of New York. DoSomething (dosomething.org) offered ways that youth could take action—in their schools and communities—to help prevent post-9/11 prejudice and discrimination against Arabs and Muslims.[193] MTV.com provided continuing news coverage with information on how viewers could become involved in the relief effort, and a link to the American Psychological Association to help teens cope with loss. JustResponse (justresponse.org), an online information clearinghouse dedicated to finding a "just and effective" response to 9/11, offered pro-peace information and other online resources. Based around articles written by political and religious peace activists as well as students, the site addressed such topics as U.S. policy options and their implications, possible roles for the United Nations, questions about the "just" nature of a war on terrorism, and implications of a war for constitutional rights.[194]

The 9/11 experience offered a glimpse of the role that the Internet could play in the civic and political lives of young people. But it was by no means a magic bullet. As experts pointed out, this new medium—with its unparalleled ability to deliver content, foster discussion, and circumvent traditional gatekeepers—could become an invaluable resource for those youth who already were politically engaged. However, for those uninterested in public life, it would take a great deal more to encourage their involvement.[195]

In January 2002, the Center for Information & Research on Civic Learning and Engagement commissioned a survey of American youth age 15–25 to assess how the events of September 11 had affected them. There was little doubt that the terrorist attacks had been a defining moment for the generation coming of age at the turn of the century, as significant for these youth as the assassination of President John F. Kennedy had been for the Baby Boomers. But it remained unclear whether the experience of this tragedy would translate into increased involvement in public life. As the survey found, the terrorist attacks and the war on Afghanistan "appear to have influenced the way young adults *feel*—about the government, their communities, and—in theory—about their own civic and political involvement." Youth were "more trusting of government and institutions" since 9/11. And there had been significant upsurges in "reported interest in community and issue organization involvement." But there was little evidence that the experience had "*yet* impacted young adults' community or political *behavior*." Even with heightened interest in government, youth were not showing greater interest in voting.[196]

But as political scientist Anna Greenberg wrote in a 2003 issue of *American Prospect*, this was a generation with great potential to make a contribution to American politics. "Political professionals usually dismiss Generation Y because it votes at a much lower rate than older Americans," she pointed out. "Yet, even at this depressed rate," Greenberg reminded her readers, "voters under 25 years old will constitute between 7 percent and 8 percent of the electorate in 2004," rivaling other "coveted swing groups" in size. "Today's youngest voters," Greenberg noted, "were raised during the heady 1990s, a time of seemingly endless dot-com possibilities, as well as social projects such as AmeriCorps that were championed by the nation's political leadership. Volunteer programs blossomed and flourished on college and high-school campuses. . . . To write them off politically is to risk someone else mobilizing a sleeping giant."[197] With another presidential election on the horizon, however, considerable effort would be required to awaken this "sleeping giant."

7 Peer-to-Peer Politics

In the months leading up to election day in 2004, dozens of new Web sites began popping up on the Internet, all with the same goal. The veteran Rock the Vote was joined by a new army of online activists with strange and provocative names—PunkVoter.com, League of Pissed Off Voters, and Vote or Die!—in the biggest battle for the youth vote in U.S. history. Voter-mobilization campaigns linked up with pop culture, enlisting familiar icons from the media world in the all-out effort. Borrowing from cutting-edge digital-marketing strategies, they refashioned social-networking software and other personalized peer-to-peer communications tools into political weapons. Some took to the streets, enlisting mobile technologies to coordinate and orchestrate on-the-ground "smart mob" battles.[1]

While the Internet had begun to play an increasingly prominent role in campaign politics, 2004 marked the first truly high-tech election. An explosion of political Web sites and a burst of innovation took center stage, capturing the public imagination. A new generation of bloggers did an end run around the news media, scooping stories, whipping up controversies, and forcing issues onto the political agenda. Even the voting process itself became high-tech. The debacle of hanging chads and missing votes four years earlier had spawned a new generation of digital voting devices to replace outdated mechanical machines.[2] The election became a crucible of heated, frenetic activity, as strategists experimented with every possible means to further their goals, inventing new software, revamping existing digital tools for new uses, and hurriedly putting them into use, sometimes before the kinks had been worked out.[3] As the race tightened and election day neared, a frenzy of new efforts mobilized, combining traditional grassroots get-out-the-vote tactics with hastily devised online innovations.

The voter-mobilization strategies were part of a broader cyberactivism movement, as young people seized the Internet and other digital technologies to promote a variety of political causes.[4] Youth joined the growing ranks of "citizen journalists" launching their own online media to use the Internet as a powerful megaphone. The easy availability of software produced an explosion of blogs, online publications, and streaming-video outlets, enabling young activists to add their voices to the cacophony of conversations taking place in cyberspace. Some cyberactivists mobilized to fight over the very future of the Internet itself, engaging in "online civil disobedience" and taking on powerful corporations in public-policy battles over the control of digital media.

Rocking the Vote

Founded in 1990, Rock the Vote was a pioneer in the youth-vote movement, serving as a prototype for a series of similar efforts throughout the decade and well into the twenty-first century. Through its association with MTV, Rock the Vote was one of the first organizations to tap directly into the new youth media, creating a powerful blend of pop culture, activism, and marketing that was tailor-made for the Digital Generation. The movement's highly visible campaigns featured a succession of popular musicians, from Iggy Pop and Madonna in the early 1990s to the Dixie Chicks in 2004. Rock the Vote enlisted thousands of volunteers across the country to reach out to youth in popular venues—at rock concerts and media events, and aboard the Rock the Vote and MTV Choose or Lose buses. Rock the Vote's Web site pulsed with vibrant interactivity, beckoning visitors to e-mail their congressional representatives, join the Street Team, and show up at the nearest "meet-up." In the months leading up to the 2004 election, almost half of the country's 18–24-year-olds visited the Rock the Vote Web site.[5]

While the organization has become best known for its mission to increase democratic participation among young people, Rock the Vote was created as a self-defense strategy by the music industry, in response to a pressure campaign over the content of record albums during the 1980s. Foreshadowing the V-chip wars of a decade later, the battle over music labeling involved some of the same political players, and followed a similar pattern of advocacy-group campaigns, government pressure, and industry compliance. The Parents Music Resource Coalition (PMRC) was no ordinary parents group; its founders were wives of powerful Washington officials. Tipper Gore was married to Al Gore, then a senator, and Susan Baker to James Baker, treasury secretary. Allied with well-known membership organizations, including the National PTA and the American Medical Association, PMRC launched a campaign against "suggestive" and "offensive" song lyrics in the mid-1980s. Through its high-level connections, the groups wielded the power of the press and the strong arm of Congress to cast the public spotlight onto the shrouded world of youth music culture, reading raunchy lyrics into the congressional record and releasing lists of "offensive" recordings to the media.[6] Just as congressional actions in the 1990s were to create a climate of pressure for the television industry, hearings on Capitol Hill during the 1980s ultimately forced the music industry to agree to self-regulatory labels.[7] In 1985, the Recording Industry Association of America (RIAA) entered into an agreement with PMRC to create warning stickers for albums that contained references to suicide, violence, drugs, sex, and alcohol.[8]

But some members of the music industry strongly opposed the agreement, arguing that RIAA's deals threatened free expression. Nor did the labeling pact prevent further criticism of music lyrics. By 1988—an election year when Al Gore ran in the Democratic presidential primary—PMRC was again on the march, this time over what it saw as inadequate industry compliance with the existing labeling agreement. With this rising tide of criticism continuing to plague the music business, industry insiders told the trade press they were feeling "disenfranchised" and "powerless." Several record executives decided to take more strategic action in "self-defense." As Jeff Ayeroff and Jordan Harris of Virgin Records saw it, the push to regulate the music industry was coming mainly from a "dated constituency" of parents, lawmakers, and other older Americans, while the consumers of record albums—primarily young people—were not involved. What the music industry needed, they thought, was some way to mobilize its customer base to fight back. Since the battles were played out in the halls of Congress, one of the most important elements of this strategy would be to create a campaign to encourage more young people to vote.[9] With the youth vote in steady decline, assaults on youth music culture could be a perfect cause around which to reengage political participation.

In 1990, Ayeroff and Harris convened an industry meeting in L.A. to "organize a voter registration campaign aimed at young record-buyers and concertgoers," marshalling the marketing expertise and financial coffers of the music business for the effort.[10] The plan was to employ a variety of popular music venues to reach youth, including music retail stores, MTV shows, and concerts, "to use people who we know have an influence with our audience—namely all the musicians and performers on MTV."[11] "If we can make stars of Madonna and Paula Abdul and M.C. Hammer," organizers told the press, "we can get people voting."[12] Rock the Vote was launched at the September 1990 MTV Music Video Awards show, with the support of a broad array of music artists, celebrities, and companies.[13] One of its first campaigns was to push for passage of the National Voter Registration Act.[14]

In mid-1992, Senator Al Gore (D-TN), chosen as Bill Clinton's vice-presidential running mate, was in a rather awkward position when he went to Hollywood to solicit entertainment-industry support for the ticket. His record on Capitol Hill was scrutinized closely, and some executives asked for assurances that no further actions against the music business would come from Washington if Clinton and Gore were elected. By this time, however, the threats had subsided. PMRC had pulled back from its confrontational stance, repositioning itself as a clearinghouse for information about music labeling.[15] While conservative critics continued to rail at the music industry for corrupting the values of youth, Democratic candidates backed off of that segment of the entertainment industry, and Tipper Gore receded into the background on culture issues.[16]

Ironically, the very group that had been created to counter Tipper Gore's campaign against music lyrics may well have helped elect her husband. The 1992 election was Rock the Vote's first all-out, star-studded voter-registration effort. In partnership with MTV, the group enlisted an army of celebrities in a series of media events, concert tours, and public-service messages to promote its message. "We got two options," rapper Ice-T chanted in a series of unconventional PSAs on MTV. "Either vote or hostile takeover. I'm down with either one. We're youth; we have to change things." As *Time* magazine observed, pop star Madonna "literally wraps her otherwise scantily clad body in the American flag and cries out 'Vote!' to the staccato rhythms of her hit song, Vogue, ending with the admonition, 'if you don't vote, you're going to get a spankie.'" These compelling appeals were complemented by more conventional grassroots, get-out-the-vote tactics. MTV collaborated with the League of Women Voters to produce and distribute a user-friendly guide to voter registration in all fifty states. The music network also assigned 24-year-old reporter Tabitha Soren to cover the election from a youth perspective, airing regular segments as part of a new Choose or Lose campaign.[17]

Presidential candidates in the 1992 election took their campaigns to MTV, late-night talk shows, and other unconventional venues as a strategy for circumventing traditional news and reaching youth voters directly.[18] Bill Clinton played the saxophone on the *Arsenio Hall Show* and appeared on an MTV youth forum, where he fielded questions on a range of contemporary topics.[19] In contrast, President George W. Bush avoided the youth cable network until the end of the campaign. When he finally consented to be interviewed by Soren, "he did so while drinking a cup of coffee, exuding body language of discomfort and even contempt. By comparison, Clinton looked like the hippest of hepcats," recalled music industry executive Danny Goldberg in his book *Dispatches from the Culture Wars*.[20]

By election day, Rock the Vote and its partners in the youth-vote effort claimed to have registered 350,000 new voters, taking credit for helping to reverse a twenty-year decline in youth voter turnout.[21] The Clinton-Gore ticket garnered the majority of voters age 18–24.[22] In January, Rock the Vote hosted a glitzy inaugural ball for the new president and vice-president. A few months later, the group celebrated another victory at the signing of the new National Voter Registration Act on the White House lawn. Cofounder Jeff Ayeroff was among the celebrants, referring to his new political group as a "rock & roll rifle association."[23]

Rock the Vote's privileged position within the music industry gave the nonprofit unique access to a stable of popular performers and celebrities, who were eager to link their names with the cause. Its close partnership with MTV placed the group at the forefront of youth media culture and enhanced the music network's legitimacy as a social force. Its yearly awards events earned it a prominent place among the other show-biz spectacles of glitz and glamour. Corporations seeking to reach the youth market enthusiastically jumped on the Rock the Vote cause-marketing band-wagon, inserting their brands into its high-profile campaigns. Pepsi under-wrote $1.2 million of a Rock the Vote television program that aired on both Fox and MTV. Reebok supported a campus tour, selling T-shirts and cups with the Rock the Vote logo at its stores, and funneling a portion of the proceeds back to the nonprofit.[24]

Throughout the next decade, the nonprofit continued its campaign to mobilize young voters, while other groups—such as World Wrestling Enter-tainment's Smackdown Your Vote!—modeled their efforts on Rock the Vote's successful blend of pop culture and politics.[25] Employing state-of-the-art techniques, the nonprofit enlisted the help of M80—the same youth-marketing company that ran guerilla street-marketing campaigns for the Backstreet Boys, 'N Sync, and other popular musical groups.[26] Rock the Vote also began expanding its agenda beyond voting to encompass a variety of hot-button liberal political issues particularly relevant to youth.[27] In addition to advocating free expression, a core goal from the beginning, the group sought to engage young people in issues such as education, violence, health care, the environment, discrimination, and money.[28] The group also expanded its brand into additional media outlets beyond MTV, seeking partnerships with ESPN, Telemundo, BET, WB, Fox, and other networks with large youth audiences.[29]

Rock the Vote quickly achieved a level of credibility and influence that trumped that of many other nonprofits attempting to engage youth. Foundations viewed the nonprofit as a worthwhile investment for their charitable giving, and a direct connection to the populations they sought to help.[30] With the high-stakes 2000 presidential election on the horizon, the Pew Charitable Trusts awarded the group more than $3 million over a two-year period beginning in 1998, to support an intensive campaign "to promote civic engagement among young adults."[31] This influx of funds enabled the group to launch a massive campaign, harnessing the power of both traditional media and digital technologies in a renewed get-out-the-vote effort. But while the group boasted gains in youth-voter registration, the overall voter turnout by young people was disappointing.[32] The controversy surrounding the long, contentious vote-counting process in 2000 was hardly inspiring for youth, so the group used the experience to call for further activism. "Young people have long suspected that something is wrong with the political system," Rock the Vote's Web site told its visitors. "The 2000 presidential election proved that the electoral process is flawed. The time has come to defend the most fundamental American right—the right to vote."[33]

Youth as e-Citizens

The outcome of the 2000 election, and the disappointing youth turnout, helped spawn numerous new initiatives aimed at increasing the youth vote. Some projects were housed in the ivy halls of the nation's universities, as foundations invested large sums of money to find out why youth were not voting and to develop innovative ways to reengage them. In 2001, the Pew Charitable Trusts funded a new Center for Information & Research on Civic Learning & Engagement at the University of Maryland.[34] Two years later, with another national election on the horizon, Harvard University's Institute of Politics, working with other colleges and universities around the country, established a National Campaign for Political and Civic Engagement.[35] Polls and focus groups were conducted, producing a spate of new primers, guides, and fact sheets, all available for instant downloading from the Internet, about how to reach and engage young voters.

For example, CIRCLE released the fact sheet "Young People and Political Campaigning on the Internet."[36] And George Washington University's Graduate School of Political Management issued the special report "Campaigning to the Internet Generation."[37]

A growing number of advocacy groups, political parties, and youth organizations began going online to spread the word about youth voting and to engage young people directly, many offering online voter-registration links. The nonprofit Youth Vote Coalition served as a portal to numerous local and national youth-vote initiatives, providing statistics on youth voting; links to voting, academic, state, and federal Web sites; a state-by-state voters guide; legislative updates; a digest of news on civic participation; and listings of offline events.[38] Both major political parties had their own youth-vote initiatives. The GOP launched the Young Republicans' Online Community Network as well as a Web site for the College Republican National Committee. The conservative America's Future Foundation, "a network of America's next generation of classical liberal leaders," offered a political commentary "Webzine" called Brainwash. Democratic Party Web sites included the College Democrats of America and the Young Democrats of America. The unconventional Republican Youth described itself as "a nationwide network of Republican students and young professionals who believe in developing a generation of Republican leaders who are pro-choice, pro-environment and pro-fiscal responsibility."[39]

By the next presidential election in 2004, these online efforts were joined by dozens more, creating what political scientists W. Lance Bennett and Michael Xenos called a "youth engagement Web sphere" on the Internet that was far larger and more sophisticated than any before.[40]

Moving On and Meeting Up

The digital tactics that would come to define much of the 2004 election had their roots in the burst of innovation at the heart of e-commerce. Moveon.org was one of several "political-technical hybrid organizations" that would play a significant role. Created by Wes Boyd and Joan Blades, two Silicon Valley software developers who used the profits from their company to fund the venture, MoveOn drew heavily from the experimental business models of the dot-com era. As Garance Franke-Ruta wrote in *The American Prospect*, "the Internet boom created a new base of wealth free from long-standing allegiances or deep involvement in traditional political circles and a new generation of individuals steeped in the boom years' free-agent, entrepreneurial, startup mentality."[41] MoveOn.com burst into the public arena during the 1998 scandal over President Clinton and Monica Lewinsky and the ensuing impeachment proceedings, launching a Web-based "flash campaign" that flooded Capitol Hill with 2 million e-mail messages and more than 250,000 phone calls, urging lawmakers to "censure and move on."[42]

In many ways, MoveOn's strategy was emblematic of a new kind of "armchair activism," requiring little more for democratic participation than the simple mouse clicks and minimal data entry involved in a routine e-commerce transaction. As Boyd explained, MoveOn took advantage of the essential "stickiness" of the Internet to develop a following of loyal members who, through the instantaneous interactivity of the new medium, could be directed to take concerted, collective action aimed toward a specific goal. "We don't look at our work as persuasion or education," he explained, but people can be motivated to act if you provide services to them. "If they hear from you, they will sign up and stick with you." Creating these dynamic, ongoing, and loyal relationships required continual "servicing."[43] The liberal nonprofit chose its campaigns carefully, focusing on "populist issues, ones that have real, broad resonance, and are easily understood," where there is a "disconnect" between public opinion and government action.[44] It also used the Internet not only to direct mass actions, but also to solicit input on policy positions and ad campaigns, and to organize and orchestrate "offline" political action by its members—ranging from candlelight vigils against the war in Iraq, to meetings with congressional members in local districts, to grassroots screenings of filmmaker Michael Moore's movie *Fahrenheit 9/11*.[45] During its first two years, political scientist Michael Cornfield observed, "MoveOn matured from a record-setting publicity magnet into a unique breed of pressure group: a citizen portal that blends the community spirit of grassroots movements with the sophisticated tactics of a PAC."[46]

As the Bush Administration made its public case for an invasion of Iraq in late 2002, MoveOn made plans to run a protest ad in the *New York Times*. Within twenty-four hours the group raised the $70,000 it needed, and over a three-day period was able to garner more than $400,000 for the cause.[47] In February 2003, the group orchestrated a "virtual march on Washington," mobilizing hundreds of thousands of people to send simultaneous messages—by e-mail, fax, and telephone—to the Senate and the White House, opposing the impending war on Iraq. Supporters registered online to join the protest, which was billed as a "way to influence policy without leaving your living room." The protesters jammed the switchboard on Capitol Hill and forced Senate offices to hire additional staff for the day to handle the volume of phone calls.[48] Later that year, MoveOn sponsored an online contest for an anti-Bush ad to run on TV during the Super Bowl. The Internet solicitation produced 1,500 entries, all of which could be viewed on the group's Web site.[49] Two particularly virulent ads, which compared Bush to Adolph Hitler, prompted outcries from critics that MoveOn had crossed a line into hate mongering. The group countered that it had not endorsed the spots, while acknowledging that they had been "in poor taste."[50] After MoveOn raised the $2 million to buy airtime for the winning commercial, CBS refused to air the ad, citing its policy against issue-advocacy ads. The group responded by launching a national campaign to buy time on individual TV stations around the country. For the first time, observed *Wired* magazine, "the most advanced campaign weaponry, the 30-second attack, was put directly into the hands of the activist base." By this time, the magazine noted, "opposition to the war was merging with the interest in the presidential campaign, and the MoveOn list ballooned to some 2.3 million names," adding that the group had raised more than $29 million, most of it from small individual donors. Convinced of the powerful potential for MoveOn's special blend of online organizing and fundraising, philanthropist George Soros pledged $2.5 million in matching money to the nonprofit.[51]

Meetup.com was another online innovation that would prove valuable to political-campaign efforts. Its original purpose was not for political organizing, but rather for encouraging the kinds of communities of interest that were essential for building "social capital."[52] Meetup founder Scott Heifferman was a young online-marketing entrepreneur who had made a fortune creating flashing banner ads. After reading Robert Putnam's *Bowling Alone*, he developed a new Web site that would help provide the social glue for reconnecting an apathetic public.[53] Like other social-networking software, Meetup enabled individuals with an interest in any number of arcane activities—from quilting to trading cards to cockapoos—to find like-minded friends in their own local communities. Unlike an online community however, where relationships are almost entirely virtual, Meetup was designed to facilitate real-world connections by providing listings of meetings in cities and towns across the country.[54]

Online Political Citizens

For years, young people had been turning away from traditional news sources, to the great angst of newspapers, network news divisions, and academic experts. By 2000, more than a third of Americans under thirty relied on late-night comedians for their news, and nearly 80 percent of youth were learning about politics from comedy programs such as *Saturday Night Live* or nontraditional outlets such as MTV.[55] While many people were alarmed at these trends, some viewed them more optimistically. As Bruce Williams and Michael Delli Carpini wrote in the *Chronicle of Philanthropy*:

> A Jay Leno monologue satirically pointing out the political ignorance of the general public, a scene from *Law & Order* exploring racial injustice in our legal system, an episode of *The Simpsons* lampooning modern campaign tactics, or an Internet joke about Bill Clinton that generates discussion about the line between public and private behavior can be as politically relevant as the nightly news, maybe more so.[56]

However, a study four years later by the Pew Research Center for the People and the Press found that individuals who learned about politics from entertainment TV programs—whether young or not—were "poorly informed about campaign developments." In contrast, the study found, "those who learn about the campaign on the Internet are considerably more knowledgeable than the average, even when their higher level of education is taken into account." And while young people continued to abandon traditional news sources, the study found, they increasingly turned to the Internet for their political information, with approximately one fifth of 18–29-year-olds getting their campaign news from online sources.[57]

The 2004 election also saw the emergence of a new category of Internet users, which researchers at George Washington University labeled "online political citizens." Though not exclusively youth, this cohort of Internet-savvy political participants included a significant number of young people, with 36 percent of them age 18–34, compared to 24 percent of the general public. A large majority of them (44 percent) had not been politically involved before and had never "worked for a campaign, made a campaign donation or attended a campaign event." These highly charged, politically engaged individuals eagerly embraced the full array of new online political tools available to them. "They visit campaign Web sites, donate money online, join Internet discussion groups, and read and post comments on Web logs," the study noted. They also "organize local events through Web sites such as Meetup.com or donate money to their causes on sites such as Moveon.org or grassfire.org." They "use campaign Web sites as hubs" and "depend heavily on e-mail to stay in touch with the campaigns, receive news stories and muster support."[58] On the cutting edge of technological innovation, the new breed of Digital Age citizens, suggested the researchers, "may be harbingers of permanent change in American politics."[59]

Democratic primary candidate and Vermont Governor Howard Dean became the poster child of Internet politics, as organizers and supporters alike seized the power of the Web to forge new strategies and tactics. Joe Trippi, Dean's campaign manager, became a guru of this new style of youth-oriented online politics. In his book chronicling the heady days of the campaign, Trippi wrote enthusiastically of the technological weapons that were assembled into a political arsenal, branded with the candidate's name, and proudly bandied about to the press. After supporters began using Meetup to find local "Deaniacs" in their own towns, campaign organizers quickly developed their own software to augment the online resource, developing Get Local tools to enable people to enter a zip code and find the closest Dean meeting.[60] "And in the open-source tradition," Trippi wrote, "we put the software out there for people not only to use, but also to improve, which they invariably did." The campaign launched its own DeanLink software, modeled on the popular Friendster social-networking Web site, to give "Dean supporters the chance to meet others like themselves." DeanLink made it possible to keep track of the people with the largest social networks, encouraging them to enlist their friends in the political effort. The campaign also made full use of blogs, which already were playing a prominent role in campaign politics. "The blogosphere was where we got ideas, feedback, support, money—everything a campaign needs to live," Trippi recalled. "And the first stop for people who wanted to get involved was often the official Web log, 'Blog for America,' . . . where the online campaign began its translation to the real world." Through the viral marketing power of the Internet, "other bloggers would write about the campaign everyday, quickly spreading the word online, offering commentary, and sometimes second guessing campaign strategy . . . the bulk of the daily blogging about the Dean campaign, like the campaign itself, came from grassroots organizers."[61] At the outset of the campaign, blogs publicized the first Meetup gatherings, which drove curious Internet users to the campaign Web site, where they were encouraged to sign up and donate.[62] Though Howard Dean failed to win his party's nomination, his highly visible campaign created a buzz in the media and served as a model for other Internet-based efforts.

Vote or Die

In the period between the 2000 and 2004 elections, 14 million young people became eligible to vote—3.5 million new eligible voters per year. The population of young citizens had grown to levels not seen since the early 1980s, according to CIRCLE, and was expected to continue growing.[63] This expanding youth block was "politically up for grabs," according to a bipartisan survey conducted in January 2004. "American youth are nearly evenly split when it comes to relating to political parties, and perhaps of greatest importance, most currently say they are independents." Because members of this new generation were still in the process of developing their own political identities, they were "susceptible to appeals from both political parties."[64]

Election year 2004 saw the rise of what the *Boston Herald* called a "dizzying array of voter-mobilization efforts."[65] A key factor driving this unprecedented level of activity was an important lesson from the most recent federal election—that a very small number of votes could be the deciding factor in determining the winner.[66] The youth-vote campaigns were particularly intense in the battleground states of Ohio, Wisconsin, Florida, and Oregon. "This year," observed *Billboard* magazine a few days before election day,

> the youth vote is not only being racked, it's being rapped, punked, mobbed and even smacked down. Generation Y voters have been the target of an unprecedented campaign by rockers (Bruce Springsteen with Vote for Change and Christina Aguilera with Declare Yourself), hip-hoppers (Russell Simmons and the Hip-Hop Summit Action Network) and wrestlers (Hurricane of WWE's Smack Down Your Vote).[67]

Foundations, music-industry celebrities, corporations, and wealthy donors poured enormous sums of money to infuse new energy into existing youth-vote projects and to spawn new ones. Dozens of efforts were launched, each with its own brand—from Redeem the Vote to Punk Voter to the League of Pissed Off Voters. As their coffers filled with this new influx of money, the campaigns commandeered every possible grassroots and mass-media weapon at their disposal to get the word out. Their tactics included the conventional and the New Age. They set up tents on college campuses across the country, hailing passers-by with an opportunity to register to vote between classes. They launched massive advertising campaigns, filling television screens and billboards with their appeals. And they drafted into their battles a new set of digital tools that already were a daily part of young people's lives. As Nat Ives observed in the *New York Times*:

All the vote marketers are searching out their targets with a sprawling set of marketing strategies, like sending interactive text messages to cellphones, selling tie-in merchandise like $20 designer t-shirts, creating Web logs and producing performances by everyone from the Rza of Wu-Tang Clan fame to the Rock n' Roll Worship Circus.[68]

The New Voters Project, an initiative launched by state Public Interest Research Groups (PIRGs) in late 2003, and sponsored by the Pew Charitable Trusts and the George Washington University Graduate School of Political Management, had an election-cycle budget of $10 million. Working with more than a half-dozen prominent partners, including Rock the Vote and MTV, the project was focused on massive get-out-the-vote tactics in six grassroots states.[69] Cast the Vote, which had garnered $173,000 in foundation funds, and $150,000 in in-kind donations—including billboards in Yankee Stadium and Times Square—ran PSAs on movie screens and staffed voter-registration tables in the lobbies of movie theaters.[70] The Youth Vote Coalition was operating with a $660,000 budget and had amassed a coalition of 106 national groups.[71]

Following in the tradition of Rock the Vote, many of the efforts had show-biz, pop-music connections, inserting the cause into the foreground of contemporary youth culture. Founded in 2004 by liberal TV producer Norman Lear, Declare Yourself aimed "to energize and empower a new movement of young voters to participate in the 2004 presidential election," with a budget of over $9 million.[72] The group boasted an impressive group of corporate sponsors, including AXA Financial, Yahoo!, Clear Channel, Friendster, and Tower Records. It also enlisted a who's who of pop-culture celebrities—including Leonardo DiCaprio, Kirsten Dunst, Reese Witherspoon, and Peter Sarsgaard—for a series of college-campus tours, concerts, and TV shows, as well as a blitz of advertising.[73] "We're approaching a cause as a brand," explained an ad-agency executive involved in the campaign, "it's not any different than any corporate American company . . . it's all about creating a brand of passion for consumers."[74] To promote its Declare Yourself/Yahoo! Online Voter Registration Drive, pop-singer icon Christina Aguilera was displayed on a massive billboard on Hollywood's Sunset Boulevard with her mouth sewn shut, along with the slogan "Only You Can Silence Yourself."[75]

Music was the common link among many of the youth-vote initiatives, as a record number of musicians threw themselves into the cause and new groups were formed to encourage more participation by artists. "The idea was pretty simple," explained the Web site for Music for America (MFA), an organization launched in 2003. "There are bands out there who would like to bring a positive political message to their fans. There are fans who want to help spread that message. All we need to do is hook them up and provide some good materials."[76] Billing itself as a "peer-to-peer, decentralized, youth mobilization movement," MFA sponsored a series of live concerts around the country, amassing local grassroots volunteers in every city, using blogs and other Internet tools to promote the events and link individuals together.[77] As one observer noted, "many of the musicians involved with these organizations claim that they understand better than anyone how young people feel about politics, mostly because they too are voting for the first time in 2004."[78] "I've been thinking about and talking about voting for a long time," said Ani DiFranco in a June 2004 article in *Billboard* magazine, explaining why she had decided to launch her Vote Dammit Tour to target young people in the swing states. "Unlike my anarchist friends, I think it's a pretty good idea. I think we've tried not voting, and that doesn't work."[79]

The Hip Hop Summit Action Network (HSAN) had been around for three years before the 2004 election, launched by rap-music mogul Russell Simmons with a series of events headlined by hip-hop musicians and aimed at urban and Hispanic youth. Like Rock the Vote, its original goal had been a response to growing onslaughts against the popular music form.[80] Partnering with Rock the Vote, Smackdown Your Vote, Choose or Lose, and others, HSAN launched a renewed get-out-the-youth-vote campaign for the upcoming election, vowing to register "two million more in 2004."[81] In July 2004, rap star, Sean "P. Diddy" Combs announced his new voter mobilization group, Citizen Change, with a compelling slogan that reflected the high-stakes nature of the upcoming election: Vote or Die! The new organization featured its own A-list of musical celebrities, including Snoop Dogg, Jay-Z, and 50 Cent. Offering a raft of Vote or Die! T-shirts to his fans, Combs was making plans to take his cause directly to both upcoming nominating conventions, and to "serve as a wake up call to young and minority voters as it turns up the heat on the 2004 election." Promising a "get out the vote campaign on a scale and style that has never been seen before in America," the musician vowed to "blanket every space that young people travel in with images that they relate to with its powerful 'Vote or Die' message."[82]

Like the counterculture movement decades before, the blending of music and politics was a powerful concoction that resonated with young people. It was also an effective strategy for reaching out to segments of the youth population who felt disconnected from public life. In the case of African-American youth, for example, "Hip Hop is an avenue that validates and credentials politics and civic engagement," explained a report by pollsters Lake, Snell, and Perry. "There are few things this cohort holds in higher esteem than the culture," the report said, noting that 88 percent of African-American voters found hip-hop credible on politics, with 48 percent regarding it as very credible. "Hip hop not only brings excitement and attention to the cause," the report explained, "it also brings validation and may be a requisite for messaging among young African American voters."[83]

For the music industry, fans helped infuse traditional marketing campaigns with a loftier mission. Warner Music Group slapped voter-awareness stickers on its CDs, with links to its Web site, which in turn linked to other youth-vote Web sites. Warner also added "vote" message tags to its TV and radio advertising as well as the promotional and marketing material used by its grassroots and street-marketing teams.[84]

Some campaigns were aimed at a broad audience, others at more narrow demographic segments of the youth population. The Advertising Council's PSA campaign to "Fight Mannequinism" encouraged 18–24-year-olds to "stay involved with their communities by doing what they can, when they can. Whether it is by voting in local elections, volunteering in their spare time, or just reading the newspaper and discussing current events with their friends." Modeled on the truth® antismoking campaign, the project used the Web and TV spots to reach youth by humorously showing "what happens when people become inactive and aren't involved—they turn into mannequins."[85] L.A.-based Voces del Pueblo (voices of the people) targeted Latino youth "who are most likely to opt out of participating in the electoral process."[86] The Black Youth Vote project partnered with BET for a black college tour, with a budget of $5 million.[87] Redeem the Vote aimed to register "people of faith regardless of party affiliation, or personal political beliefs, but as a matter of Christian principle."[88] Gay and lesbian youth activists from both political parties organized their peers to vote. The National Stonewall Democrats launched the Stonewall Student Network; the Log Cabin Republicans organized a leadership forum and a series of campus outreach efforts.[89]

While the majority of youth-vote efforts claimed to be bipartisan, the rhetoric that many youth used reflected an anger targeted specifically at the policies of the Bush Administration. Kristin Jones, writing in the *Nation,* profiled a new generation of get-out-the-vote youth groups fueled by strong anti-Bush sentiment.[90] "This spring, with an eye on mobilizing angry punks," she wrote, Punk Voter was using "hard-edged, partisan tactics," noting that the group had enlisted musicians Jello Biafra, NOFX,

Alkaline Trio and Authority Zero in its Rock Against Bush tour, drawing sold-out crowds in California, Oregon, Washington, Nevada, and Arizona. The League of Pissed Off Voters (part of Indyvoter.org) was "one of the first groups to try to establish a voting bloc specifically on the basis of being young and angry," Jones observed. The group's book *How to Get Stupid White Men out of Office: The Anti-politics, Un-boring Guide to Power*, coedited by Adrienne Brown and William Wimsatt, was a grassroots primer for translating anger into action. The introduction offered a contradictory message that was emblematic of the conflicting attitudes of many young progressives, urging followers to vote Democrat while acknowledging that "Democrats are not our friends."[91]

The Internet was a central part of all these campaigns, not only providing each effort with a direct means for reaching its target audience, but also fostering collaboration among the groups, and forging virtual coalitions through links and cross-promotion strategies. This online fluidity enabled visitors to travel across Web sites quickly and effortlessly, gathering information, communicating with others, and joining whatever effort matched their interests and passions. Most of the Web sites offered a link to online voter-registration, providing a form of instant gratification unparalleled by any other means. On the Rock the Vote site, for example, a Register to Vote tab linked to a pop-up window with a voter-registration form that visitors could print and mail to their state elections office, under the slogan "Fill it and print it, lick it and mail it."[92] Many campaigns promoted peer-to-peer viral-marketing efforts, mimicking the strategies of commercial marketers. Through social-networking software, the online world also served as a powerful enabler for thousands of "offline," real-world efforts, from concerts to rallies to protests.

Moveon.org, already a trailblazer in Internet organizing, launched Click Back America, a "college click drive," in March 2004 to raise $1 million from students around the country for an advertising campaign attacking the Bush presidency.[93] By August the effort had morphed into its own branded campaign, MoveonStudentAction.org, organized by two students from Brandeis University, Ario Rabin-Havt and Ben Brandzel. Working with the liberal public-relations firm Fenton Communications, the group launched a series of high-profile efforts to focus attention on the dangers to young people posed by the Bush Administration. Raising money online, MoveonStudentAction bought space in the *New York Times,* where the group published an open letter to President Bush. Signed by 65,000 young people, the letter warned that the Bush policy in Iraq was leading to a post-election military draft and demanded an exit strategy to end the war. The ad also ran in 155 college newspapers in the battleground states.[94]

Though MoveonStudentAction did not have a particularly compelling, graphically rich Web site, it developed innovative tools to mobilize peers. "What is distinct about our effort," Brandzel later explained, "was that it was 'grassroots driven,' using the online media to generate action, but with 'no personality at the center'." Rather, the Web site itself was the hub. A "Voter Multiplier" page on the MoveonStudentAction site invited members to upload their friends' names and e-mail addresses—from their PalmPilot, Outlook, or Facebook programs—in order to create their own "personal precincts." With a few strokes of a key, each individual could contact hundreds of friends instantly, e-mailing them personalized messages—from "virtual doorhangers" to online voter-registration links to election-day reminders to cast their ballot, along with directions to the right polling place.[95]

Other social-networking sites spawned their own political counterparts. James Hong and Jim Young had created the successful online dating site, HotOrNot, attracting a large following of 18–24-year-olds. For the election, the duo launched a new site called VoteOrNot. The venture was based on the same principle as "connector" marketing efforts such as Procter & Gamble's Tremor that friends could do a much better job of influencing each other than impersonal advertising messages could. To attract people to the site, the sponsors offered a $200,000 sweepstakes that would be split between the winner and the person who had referred him or her to the site. Members who joined VoteOrNot would be linked to another Web site where they could register to vote. Launched over Labor Day weekend 2004, VoteOrNot signed up more than 100,000 people before the end of October.[96]

The success of Friendster also inspired a political clone, though not an official offshoot of the original social-networking site. Political Friendster was created by Stanford University student Doug McCune, who wanted to "do something that involved the election." So he came up with the idea of using a Web site to illustrate the connections among politicians. "I just had the idea that since it was such a familiar concept for kids my age that using that concept to apply to politics would strike a chord," McCune told the *New York Times*. The site worked like a twenty-first-century version of C. Wright Mills's book *The Power Elite,* identifying who a politician's "friends" were and exposing political connections. Instead of posting their own profiles to the site, visitors would post information about politicians and then link them to the other people they knew about. For example, clicking on Hillary Clinton revealed that one of her "friends" was Wal-Mart founder Sam Walton and, in turn, that one of Walton's "friends" was President George H. W. Bush, who had awarded the discount-store magnate the Medal of Freedom in 1992. The site, which mimicked the style of Friendster, billed itself as a "parody" of the original. To preempt any copyright problems, McCune posted a disclaimer: "If you're from Friendster and want to sue me, then take a deep breath, calm down, laugh a little bit and chill out."[97]

Branded Activism

Rock the Vote was the most recognizable organization in the youth-vote movement, taking its mobilizing model to a new level during the 2004 election. Through its trademark fusion of consumerism and citizenship, the group created a highly charged campaign designed to penetrate every sector of youth culture. Four corporate sponsors—Dr. Pepper/Seven Up, Unilever's Ben & Jerry's Homemade Ice Cream, Motorola, and Cingular Wireless—paid $1 million each to support the Voter Registration Bus and Concert Tour. The money from these companies made up 35–40 percent of the nonprofit's $7–8 million budget for the year, with the group's awards event bringing in another $1 million and foundation and individual donors making up the rest. Rock the Vote's position in the music industry enabled it to draw from headline bands—including the Dixie Chicks, Alanis Morissette, Snoop Dogg, and the Dave Matthews Band—who provided their services free, performing at fifty-six tour stops between June and November.[98] The message also was spread through a variety of television channels, magazines, and radio outlets, including MTV, Comedy Central, the WB, and the E cable network.[99]

Equipped with the latest state-of-the-art features, the Web site, rockthevote.com, served as the hub of this maelstrom of preelection activity, linking with the growing number of youth-vote initiatives in a synergistic network of online relationships.[100] The site offered numerous ways for individuals to get involved, tailored to a variety of interests, including Chick Vote and Rap the Vote. By registering online, members could join the Rock the Vote Street Team, linking up with others in their communities to become part of the army of volunteers who were registering new voters at concerts, clubs, and campuses across the country.[101] Donations could be made easily with a click of the mouse. Just another click was necessary to purchase Rock the Vote gear, including "Give a Shit" T-shirts and branded thongs.[102] Yet another click would jump to Amazon.com, where the latest CDs by Rock the Vote music-award winners were for sale.[103] Youth also could participate in the Rock the Vote blog, to learn "what Capitol Hill is saying and find young people's response."[104] The nonprofit went to elaborate means to spread the Rock the Vote brand throughout the Web, including free downloads of banners and radio ads, as well as links to its voter-registration page, available to "anyone and we mean EVERYONE." Groups and individuals even could import the online voter-registration tool and rebrand it for their own Web site. "This unprecedented network of thousands of tools," the Web site predicted, "will make the NEW vote, the SWING vote, and bring 20 million GenNext voters out to the polls this year."[105]

The Rock the Vote brand was also propagated through software applications, wireless technologies, and commercial Web sites that married activism and advertising. In partnership with a company called Meca, the nonprofit created Rock the Vote Communicator, a branded version of instant messaging, offering "six available Rock the Vote–themed skins" that were "designed to appeal to the elusive 18–24 voter demographic."[106] The joint venture enabled teens and young adults to chat with their friends and exchange political opinions. It also served as an organizing tool for Street Teams. "As the election draws closer and voter-registration closes," explained one trade publication, "street team volunteers will guide the discussion towards moving to the polls and making sure that their newly minted political activists follow through by pulling the lever of their choice on November 2."[107] The nonprofit linked up with the popular MySpace.com, tapping into its technology and youth user base. "MySpace.com's social-networking platform," a Rock the Vote spokesperson explained to the press, "will exponentially open up communication among young people to Rock the Vote's political tools and street teams." As part of the agreement, MySpace.com agreed to "actively promote Rock the Vote throughout the network," creating a profile of the group, promoting its affiliated musicians, and incorporating a link to its online voter-registration page into the MySpace Home page. The joint effort developed a variety of tools "to inspire, organize and mobilize young people to vote," including MP3s, photos, and buddy icons.[108]

Rock the Vote's mobile project was modeled on several successful "smart mob" political efforts in other countries, including campaigns by activists in Spain the night before the March 2004 elections, where "the spread of text messaging mobilized some thousands of people who congregated in front of the political party running the country, Partido Popular, in just a couple of hours."[109] Rock the Vote's version of these campaigns, however, was an integrated cause-marketing venture with Motorola. To launch Rock the Vote Mobile, Motorola sent e-mail messages to its thousands of cellphone users, attaching a video that featured Rachel Bilson, star of the popular Fox TV show *The O.C,* inviting young people to sign up online for the campaign. As an added incentive, the company offered sweepstakes with prizes that included Ben and Jerry's ice cream and Motorola handsets. Through this opt-in process, youth could be plugged into a constant stream of interactive content and activities through their cell phones.[110] Biweekly polls were able to "take the pulse of 18–30-year-olds on top-of-mind topics from education and economics to job creation and the war on terrorism," campaign materials explained, and a regular feature asked voters which candidate was "likely to get their vote on election day."[111] Users could also receive "wake-up calls" and ring tones from Rock the Vote musicians, enter election-related contests, and participate in a variety of text-messaging surveys. Undecided voters could take the "candidate match" survey. After answering ten questions on issues such as the war, the environment, and the economy, they would receive a text message with the name of the candidate who best fit their own values and interests. When asked by one skeptical reporter about the neutrality of such a quiz, especially when administered by a liberal group, a Rock the Vote spokesman responded with assurances that it was "being extraordinarily careful about how the questions are drafted . . . we have a team of attorneys review them to be sure the questions are unbiased, and we link with outside sources to give more information."[112]

Through its ongoing partnership with MTV's Choose or Lose campaign, the nonprofit sponsored a "PRElection." The unique effort blurred the lines between music fandom and citizen participation by combining a mock online election with real-world voter-registration. Using special forms approved by the Federal Election Commission, young people were able to cast their votes for president in an MTV.com poll, while at the same time registering to vote in the upcoming real election. Once they had registered for the PRElection, the fans could enter weekly and monthly sweeps and gain access to "exclusive music and videos at MTV.com." Prizes included: "a trip to the MTV Beach House; a July date with an MTV VJ; tickets to the August Video Music Awards; and appearances on *Total Request Live.*" By June, MTV had 15,000 registrants and its two spring shows had garnered the highest ratings ever for the network's *Choose or Lose* programming."[113]

These combined efforts enabled Rock the Vote to attract an unprecedented number of people to its Web site. A post-election memo tallied the results:

> In January 2004, our site saw 3.4 million hits; by July, we had reached 8 million hits per month. In October, the Rock the Vote Web site had 27.4 million hits from people registering to vote, learning about the issues, and finding ways to get involved. In all, we received an incredible 190 million hits for the 2004 election cycle . . . more than 45% of 18–24-year-olds visited our Web site in the months leading up to the election.[114]

More than 120,000 people joined the Rock the Vote Mobile campaign.[115] "We made over 200,000 contacts to this list in the final days of the campaign," the group explained, "including celebrity voice mails that explained how to find a polling place through the Web or through an automatic patch-through to 1800MYVOTE1." Online voter registrations totaled 1.2 million.[116] The election-year initiative also generated a sizable database for the nonprofit.[117]

With the 2004 election, the music industry's original plan to mobilize its consumer base, hatched nearly two decades earlier, had come to full fruition. Rock the Vote had forged a new model for democratic participation, one that merged the roles of fan, consumer, and citizen in the youth media culture of the Digital Age. As other groups followed in Rock the Vote's path, music was fully integrated into their mobilization efforts, with major labels providing funds, lending their artists to the movement, and incorporating get-out-the-vote slogans into their own sales campaigns. In some cases, music celebrities themselves led the way, modern day troubadours who stirred their young fans into action. This powerful merger of pop culture and politics was also the perfect cause-marketing vehicle for corporations, who were able to link their brands to the hope of democratic renewal.

The Reengaged Generation

Two months before the election, CIRCLE teamed up with MTV to survey youth voters between the ages of 18 and 29. The research found that, compared with survey results during the 2000 election, when youth-vote turnout remained low, more than twice as many young registered voters were paying "a lot" of attention to the campaign—as much, the researchers noted, as they were in 1992, when the youth turnout had spiked.[118] This optimistic assessment was shared by several other polls that were closely monitoring the youthvote in the final months and weeks leading up to election day.[119]

But when the polls closed, initial news accounts were disappointing. "This was not the breakout year for young voters that some had anticipated," reported the Associated Press. Despite the enormous outlays of money and time, it appeared that voter turnout among youth between the ages of 18 and 24 was about the same proportion of the electorate that it had been in 2000.[120] This first account of the returns was picked up by other news media in the early reporting on the election, playing into the conventional news frame of youth apathy and cynicism. But CIRCLE's Mark Lopez knew that this interpretation could not be right. He and his colleagues immediately began crunching numbers, working through the night, and calling reporters to correct the story.[121] Part of the problem was that the statistics were confusing. As the *San Francisco Chronicle* tried to explain to its readers a few days later, after speaking with CIRCLE staff: "participation among the nation's 40 million 18 to 29 year olds was up—to 20 million, compared with 16.2 million in 2000. But so was voting across the board. With a total voter turnout greater than 120 million, the much ballyhooed youth voters turned out to be 1 out of 10, which is just about exactly the percentage they were four years ago."[122] But even these complex explanations didn't tell the final story, which could not be determined until six months later, when the U.S. Census Bureau released its official results. When it did, the findings were dramatic. Voter turnout among youth had reached the highest level in more than a decade. "The increase in turnout by the youngest voters, age 18–24, was higher than any other age group," CIRCLE explained, "making it a significant and disproportionate factor in the overall jump in the number of Americans going to the polls last fall." The turnout rate among voters under age 25 had jumped 11 points, from 36 to 47 percent between 2000 and 2004, while the overall voter turnout rate increased by about 4 points, from 60 to 64 percent.[123] Commenting on these numbers, CIRCLE labeled the young voters "the Reengaged Generation."[124]

In a fact sheet released a few months later, CIRCLE suggested that "the confluence of extensive voter outreach efforts, a close election, and high levels of interest in the 2004 campaign all worked to drive voter turnout among people to levels not seen since 1992." But the researchers expressed some caution in reading too much into these results, adding that "it remains to be seen if this increase in voter turnout in 2004 is part of a new trend, or is instead a spike like that in the 1992 election."[125]

While it was difficult to predict whether the level of engagement among youth would remain this high, it was clear that other trends during the 2004 election were likely to continue. Calling it a "breakout year for the role of the Internet in politics," a report by the Pew Internet & American Life Project found that 75 million Americans "used the internet to get political news and information, discuss candidates and debate issues via e-mail, or participate directly in the political process by volunteering or giving contributions to candidates."[126]

Freeing the Culture

If music was a touchstone for the youth-vote movement, it played a far different, but no less important, role in the "free culture" movement. These activists were also avid fans, and some were musicians themselves. (Many were participants in the 2004 get-out-the-vote efforts.) But rather than joining hands with the industry, they organized against it, taking on the large corporations that controlled much of youth popular culture. At the heart of their battles was a passionate belief that young people should be creators and shapers of a new participatory free culture, rooted in the inherent capacities of the Internet. Like the rest of their generation, these activists had grown up with digital technology, internalizing the ease of use, accessibility, freedom, and constant connectivity that went along with them. And like their contemporaries in the youth-vote movement, they were able to seize the new digital tools as weapons for their political efforts, in this case to challenge the public policies and corporate interests that were influencing the future direction of the Internet itself.

Most of these young activists were still in high school in 1998, when the Digital Millennium Copyright Act (DMCA) passed Congress. The legislation had been hotly debated within the closed circle of K-Street Washington lobbyists and a handful of public interest groups. But with little mainstream press coverage for this arcane, inside-the-Beltway issue, the public was largely unaware of the law. The DCMA was a response to corporate fears that control of copyrighted material was being undermined by a new generation of digital technologies that made it easy to download, distribute, and change content. Increasingly, copyright protections were built directly into the software of many commercial applications, designed to thwart such activities.[127] The new law criminalized the production and dissemination of technologies that could circumvent these encoded copy-protection devices, imposing penalties as high as ten years in prison or $1 million in fines for willful violations of the provision.[128]

The full implication of the DMCA, which took effect in 2000, was not apparent immediately. As digital media became increasingly personalized, many young people maintained a strong sense of intimacy and ownership in their relationships to new technologies. What they did in their daily lives seemed far removed from distant policy matters.[129] But the widely publicized lawsuits by the recording industry placed youth in the middle of a hotbed of controversy and debate. Suddenly the long arm of the law was reaching not only into their online experiences, but also into their homes and schools.

The cofounders of Downhillbattle.org—Holmes Wilson, Tiffiniy Cheng, and Nicholas Reville—were college students when the lawsuits against Napster first made headlines. All of them shared a passion for both music and technology. As they watched the public debate over Internet file-sharing, they became increasingly frustrated and angry over what they saw as one-sided coverage by the press. They believed reporters were ignoring the role that the music industry itself was playing by failing to respond to the changing needs of its customers. They also thought that the negative coverage of file-sharing had ignored the public benefits of the popular practice. In their view, sharing music online could give people equal access to a "treasure trove of culture," instead of allowing radio disk jockeys and record companies to be gatekeepers.[130]

The activists took to the Web, where they could present their case "in a clear and simple and funny way" and tap into the viral nature of Internet communication to spread the word and reach a "ton of people." One of their key tactics was to create "stunt pages" in order to "throw a rock in the debate." Like other cyberactivists, the group figured out that with a minimum of technical skill, it was possible to put up a Web site overnight, mimicking the style and content of corporate sites so that search engines would take users directly to the spoof site.[131] Their first stunt page was iTunes Is Bogus. The parody was similar to many of the tactics in the anti-tobacco truth® campaign, following the design of the official iTunes Web page, but carrying a message that attacked the company. A headline at the top read: "iTunes Music Store. Facelift for a Corrupt Industry." The rest of the page carried a series of essays, framing much of the argument in consumer-oriented language. "Let's start simple," the first essay began, "the iTunes Music Store is not a good value for customers." While iTunes claimed that people could buy entire CD albums for $8–12, the site explained, this was much more than CDs would cost at Amazon or eBay, where they could be picked up used for $5. "If you don't care about liner notes," the Web site advised, "you can burn the CD from a friend for 25 cents and send the musician a buck, and you can always use iTunes to rip it onto your computer or mp3 player."[132] This kind of irreverent tone, advocating "digital civil disobedience," and providing explicit instructions for acting out against copyright law, was characteristic of Downhill's approach.[133] To augment their online strategy, the activists began pitching stories to reporters, directing them to the Web site. "Because there was a real void and the need for another point of view," the group recalled, "the press was often willing to print quotes from our three-day old organization." As a result, Downhill Battle was able to gain national press exposure.[134]

What a Crappy Present was another of the group's stunt pages. This "antiadvertisement for CDs," featured a photo of a little girl opening a Christmas present under the tree, her face a mass of disappointment at the CD inside. The Web page offered advice to children who found themselves in similar situations, along with information for their clueless parents. "Kids today are so good at downloading music from the Internet," the site explained, "that most of them already have all the music they like on their computer, or if they don't have it yet they can get it in 10 minutes." Launched during the highly publicized lawsuits against families whose children were accused of illegally downloading music, the spoof site offered instructions on how to avoid legal problems: "If your family turns off 'sharing,' downloading songs is 100% safe." It also advocated consumer boycotts, reminding parents, "when you buy major label CDs you're paying companies to sue families and marginalize independent music." The advice to kids was much more explicit and subversive. The Web site displayed step-by-step instructions for purposeful disobedience, illustrated with pictures of a child carrying them out. "Try to find the receipt," it suggested. "A parent's wallet or purse is a good place to start looking." After that, "Get yourself to the mall and return the CD," it advised. "Even if you don't have the receipt, some places will give you store credit (especially if you act real sad)." Finally: "Find the biggest pack of CD-Rs you can get for the price of the CD (usually 25 or 50). Now you're back in charge of your music. Rock on!"[135] According to Downhill Battle, one million people visited the Crappy Present site.[136]

The activists also used the Web to orchestrate collective actions against the music industry. Its Grey Tuesday Internet campaign earned the small upstart group widespread recognition within the online activist community, as well as mainstream press coverage. It began when a disk jockey took music from the Beatles's *White Album* and remixed it with tracks from hip-hop artist Jay-Z's *Black Album*, producing a new hybrid *Grey Album*. As scholar Sam Howard-Spink explained in his case study of the Grey Tuesday campaign, these "mash-up" or "bootleg" albums were created by cultural artists who remixed elements from existing musical pieces together into a new genre of hybrid works. With origins dating back to the early days of hip-hop, the creation of "sample-based" or "remixed" recordings had accelerated with the advent of digital technologies, online networks, and file-sharing software.[137] But while increasingly popular among a growing number of music aficionados, remixing ran up against the interests of powerful music corporations. In the case of the *Grey Album,* EMI Records and Capitol Records, the companies that owned the copyright on the sound recordings of the *White Album,* threatened legal action against the DJ and "anyone who sold or distributed the *Grey Album.*" Downhill Battle swiftly moved into action, staging an online protest, and offering free downloads of the album on its Web site. As Internet activist groups had done a decade before to protest the Communications Decency Act, more than 400 Web sites participated in Grey Tuesday, turning their sites "grey" for a day, with nearly 200 of them hosting their own downloads of the controversial album. Activists at Downhill Battle used the event as a way to publicize their concerns over copyright law, generating attention from major news outlets, including the *New York Times,* MTV, and the BBC.[138]

Downhill Battle soon extended its efforts beyond these attention-getting stunts. Like many of their generation, the activists had a passion and facility for using new digital technologies to create and distribute their own work.[139] Sharing the zeal of other open-source advocates, the group began developing and promoting technologies for creating a do-it-yourself "free culture."[140] It established a nonprofit Participatory Culture Foundation and launched new software applications, such as the Broadcast Machine, available for free to anyone wanting to develop online "peer-to-peer television."[141] Advertising the software on its Web site, the group promised a new vision for future media: "You (and any other individual or organization) will be able to publish full-screen, high-quality video to thousands and potentially millions of people at zero cost. . . . We are offering free support to organizations that are interested in starting channels. . . . it will take you less than one hour to set up your own channel."[142] Through these peer-to-peer channels, "kids can get TV from one another rather than from Viacom."[143]

A truly participatory culture, however, required more than creating innovative software. It also depended on a legal and technological infrastructure that would support peer-to-peer communication, production, and distribution. Policies such as the DMCA threatened this kind of "open architecture" upon which the Internet had been founded. The law's impact on music distribution was a concrete illustration of the way in which corporate practices and government regulations could affect the fluidity and openness of digital media. With its aim to "create a decentralized music business and a level playing field for independent musicians and labels," Downhill Battle joined a growing number of music activist groups that were attempting to change the way music was created and distributed.[144]

Organizations such as the Future of Music Coalition (FMC) saw digital technologies as a way to loosen "the stranglehold of major labels, major media, and chain-store monopolies." Founded in 2000 by a group of independent musicians, FMC's mission was to "address pressing music-technology issues and to serve as a voice for musicians in Washington, DC, where critical decisions are being made regarding musicians' intellectual property rights without a word from the artists themselves."[145] These groups were allied with other advocates engaged in policy debates over the future of digital media. Public Knowledge was established to fight for a "vibrant electronic commons" in the digital landscape, participating in congressional debates and regulatory proceedings over highly technical policies such as "broadcast flag" and "open access" to cable broadband platforms.[146] The Electronic Frontier Foundation, by this time one of the oldest Internet-policy organizations, also was active in the intellectual-property debate, describing itself as "a modern group of freedom fighters" engaged in defending "the vast wealth of digital information, innovation, and technology that resides online."[147]

Many of the activists were deeply influenced by the writings and teachings of Lawrence Lessig, a law professor whose widely popular books, *Code, The Future of Ideas,* and *Free Culture,* provided the intellectual underpinnings for a growing free-culture movement.[148] His books earned him a wide following within the general public and among intellectuals. A popular speaker on the college lecture circuit, Lessig was able to translate legalistic jargon into compelling and vivid prose, providing illustrations that resonated particularly well with the experiences and values of the Digital Generation. While a strong supporter of legal protections for copyright, he provided a well-documented set of arguments that the current direction of intellectual-property regulation in the United States was threatening not only the Internet, but also the larger culture. "Capturing and sharing content," he explained, "is what humans have done since the dawn of man. It is how we learn and communicate." But capturing and sharing through digital technology is different, explained Lessig:

> You could send an e-mail telling someone about a joke you saw on Comedy Central, or you could send the clip. You could write an essay about the inconsistencies in the arguments of the politician you most love to hate, or you could make a short film that puts statement against statement. You could write a poem to express your love, or you could weave together a string—a mash-up—of songs from your favorite artists in a collage and make it available on the Net.[149]

The problem with the new DMCA, Lessig pointed out, was that it went too far, essentially undermining the very structure and operation of the Internet, which had been built on openness, sharing, and the notion of individuals building on each other's work. The DMCA, as well as other digital copyright-protection schemes, prevented many legal uses of content, shutting down the opportunity for the kind of sharing and building on other people's work that had been essential to the growth and enrichment of cultural experience. "We come from a tradition of "free culture," Lessig explained. "A free culture is not a culture without property, just as a free market is not a market in which everything is free. The opposite of a free culture is a 'permission culture'—a culture in which creators get to create only with the permission of the powerful, or of creators from the past."[150]

Lessig also put some of these ideas into practice by setting up the Creative Commons, a nonprofit that enabled copyright holders to create "flexible licenses" that would set the terms under which others could use their work, and thus offer an alternative to the rigidity of current copyright law.[151] He called for a more balanced approach to copyright that would foster competition and innovation in the distribution of music and other content, without harming copyright holders.[152] Lessig was joined by intellectuals and policy advocates calling for open access to broadband technologies, open-source software, and other proposals to promote full participatory culture in the Internet Age.[153]

Lessig's teachings also inspired the creation of a new youth organization dedicated to spreading the free-culture message among college students. The group Freeculture.org began in 2003 as a small club of Internet enthusiasts at Swarthmore College. Like Downhill Battle, the group's first online stunt vaulted it into the national spotlight. A hacker got into the computer system at the Diebold Corporation, maker of high-tech paperless voting machines. Among the thousands of e-mail messages retrieved were some embarrassing internal communications suggesting that there were serious problems with the new system. Immediately, the messages spread throughout the Internet, prompting the company to issue cease-and-desist letters to a number of Internet service providers, claiming that the information was protected under the Digital Millennium Copyright Act. When Nelson

Pavlovsky and Luke Smith learned of the e-mail messages, they posted them on their Web site, and encouraged other student activists around the country to do the same.[154] They also contacted the Electronic Frontier Foundation, working with the organization's lawyers to sue Diebold for abusing copyright law to suppress freedom of speech.[155] The suit attracted national press and congressional attention. Congressman Dennis Kucinich (D-OH) called on the House Judiciary Committee to conduct an investigation. Within three days, Diebold announced it would no longer try to stop the distribution of its memos on the Web.[156]

Buoyed by their victory, the student activists decided to launch a more ambitious effort. They bought the domain name freeculture.org, invited Lessig to Swarthmore to give a talk, and started promoting chapters at other colleges around the country.[157] Lessig placed a link to the new organization on his own Web site, where his book *Free Culture* could be downloaded for free.[158] Freeculture.org began working with other groups, including Downhill Battle, on a series of Internet campaigns to promote free speech, open-source software, and less restrictive copyright laws.[159]

By 2005, freeculture.org organized its first summit of like-minded activists. Billing itself as an "international student movement," the group's Web site listed a growing network of nearly two dozen local campus chapters.[160] "Through the democratizing power of digital technology and the Internet," reads the group's mission statement, "we can place the tools of creation and distribution, communication and collaboration, teaching and learning into the hands of the common person—and with a truly active, connected, informed citizenry, injustice and oppression will slowly but surely vanish from the earth."[161]

Whether such lofty and ambitious goals ever could be fully realized was uncertain. But these youth Internet activists—in both the 2004 election and the free-culture movement—were demonstrating an investment in digital technology that went beyond their role as consumers. They were taking ownership of the new media as tools for the practice of citizenship.

8 The Legacy of the Digital Generation

In its 2006 New Year's issue, the trade publication *Kidscreen* offered predictions about the future of the media culture for children born that year. Drawing on the expertise of youth marketers, the magazine's crystal ball revealed a scenario of ubiquitous digital media that liberate and empower. "2016's ten-year-olds will be smack in the middle of Generation Next," predicted Harris Interactive's John Geraci. "Using their bedrooms as command central," kids will move freely about in cyberspace, interacting with playmates across the street and around the world. Technology "will permeate just about every aspect of a child's world," as today's portable gaming systems, cell phones, PDAs, and laptops "merge into one mondo device." Asked how they saw the future, children in the focus groups offered similar visions, with one boy exclaiming, "toys will be floating around!!!!" These digital devices of the future, say the experts, will accompany young people wherever they go in their daily sojourns, assuring the next generation of young people the constant connectivity to friends and family that will have been part of their lives since birth.[1]

None of this should seem surprising; most of these trends are already underway. In the little more than a decade since the World Wide Web was launched, the media system has been altered in profound ways, and likely will continue on its technological course into the future. Children are growing up in an immersive media culture that has become a constant and pervasive presence in their lives. While it is still too early to understand its full impact, the record so far suggests it will remain a significant personal, social, and political force. The events of the past ten years offer a glimpse of the multiple and varied ways that technology has been woven into the day-to-day experiences of the Digital Generation. With the Internet's unprecedented instantaneous access to vast global resources, youth have been able to seek out information on any number of topics, find like-minded peers, and connect with online communities. Many young people have taken ownership of the myriad new tools available to them, harnessing them for communication, personal expression, and exploration. Digital media have become indispensable allies in the quest for identity, training wheels for social interaction. Online journals, social-networking software, and Webcams enable new levels of self-reflection and documentation of young peoples' inner lives. Many youth have eagerly embraced the Web as an electronic canvas to showcase their writing, music, artwork, and other creations to the infinite audiences of cyberspace. The tech-savvy among them have found the Internet to be a laboratory for experimentation and innovation. For young activists, digital media are powerful outlets for the practice of "personalized democracy."

Not all youth are engaging with new digital media in the same ways, and some may not even see themselves as part of the Digital Generation.[2] Despite repeated government promises during the 1990s to "bridge the digital divide," a significant number of children and teens remain disconnected from the Internet.[3] While public policies have helped ensure greater access through schools and libraries, they have not yet erased the troubling gap between children with access at home and those without it.[4] As media scholar Ellen Seiter notes, "The growing importance of the Internet has created a new disparity across class lines in children's access to skills, social networks, and intellectual resources. While children with high household incomes enjoy speedy, pervasive access to technology at home and at school (either private or affluent public school districts), others struggle to compete for intermittent access to slow machines that are outdated and erratic."[5] These persistent and troubling inequities only serve to underscore the central position that technology holds in our society. Those young people without access to the Internet—not to mention broadband service and mobile devices—are considered to be seriously disadvantaged, cut-off from opportunities, unskilled for future work, and disconnected from peers.

And yet the same medium that is increasingly considered an essential utility continues to stir anxiety and confusion. The Internet has become a flashpoint for public debate as we struggle to sort out the conflicting and contradictory roles it is playing in children's lives. It challenges assumptions, confounds categories, erases boundaries, and, in the end, defies easy classification. It can serve as a powerful research tool, enhancing knowledge and teaching critical skills; but it can also be a seductive and addicting online playground, mesmerizing young people in all-encompassing interactive environments. It is an immense public sphere, a forum for robust youthful debate on critical issues of the day; but it is also a sophisticated marketing machine, whose ubiquitous surveillance devices can track one's every move. It can be an engine of collective action, sparking civic participation and mobilizing political movements; but it also can be a digital cocoon that promotes and encourages self-obsession and isolation. For youth in trouble, the Internet can serve as a vital safety net, where professional help is available just one click away; but it also can be a dark and dangerous underworld, where predators and pornographers lie in waiting.

These complex and contradictory images of the Internet mirror the multiple and conflicting roles played by young people themselves—both directly and indirectly—in the public-policy debates and the marketplace during the formative period of the digital era. The early commercialization of the Net as the driving force in the hot high-tech stock market boom placed the Digital Generation in a critical R&D role, testing new products and forging new practices in the early fluid period of online experimentation (a role they continue to play today). The online market's focus on "digital kids" merged with the official government policy to connect children to the Information Superhighway.[6] Congressional efforts in the early 1990s to shield children from pornography, indecency, and predators cast a far different light on the Digital Generation, whose vulnerability was repeatedly highlighted.[7]

In many ways, the transition to the Digital Age has called into question our contemporary conceptions of childhood.[8] The public discourse has produced conflicting notions of the child—from high-tech guru, to innocent victim, to avid consumer—based on competing institutions and interests. Policymakers have relied on child-development research to designate age categories in need of special protective interventions. Market researchers also have drawn from the developmental literature to legitimate their own constructions of new childhood demographic categories. Marketers have been so successful in this area that the concept of "tweens" is now commonly accepted in the public mind, both in the United States and abroad.[9] The press—with its steady stream of labels—"My Space Generation," "Multitasking Generation," "Wired Generation"—often has characterized all young people in monolithic and simplistic terms, defining them almost exclusively on the basis of technology. So much media and market attention has been placed on middle-class children and teens that the behaviors, needs, and concerns of poor children, rural children, and children of color—the communities with the least access to sophisticated new technologies—have been largely overlooked.

Some social critics, including the late Neil Postman in his book *The Disappearance of Childhood*, have lamented the media culture's role in dissolving the boundaries between children and adults.[10] To some extent, such concerns have been heightened in the digital era. For example, the Internet makes it possible for children and teens to interact on an equal footing with adults in a virtual environment where one's identity and age can be kept secret. But at the same time, contemporary media culture, through the proliferation of specialized children's TV channels, Web sites, and youth brands, has purposefully cultivated a separateness between the realm of childhood and that of the adult world, as part of the further segmentation of the marketplace.

The concept of empowerment has been a recurring theme, taking on different meanings depending on who uses it. Marketers frequently speak of children and teens as powerful and autonomous agents in the new digital economy, and the relentless engine of market research issues a continual flow of information to the press promoting this image. In the face of looming concerns over online safety, the Clinton Administration, in concert with the industry, promised parents that they, too, could be empowered by new filtering technologies that would help them regain their authority against the onslaught of a rapidly exploding media culture that threatened to upset the balance of control in their own families.

The persistent public concerns about dangers in cyberspace deflected attention away from the rapidly developing online marketplace and from a series of legislative actions (including the Telecommunications Act of 1996 and the Digital Millennium Copyright Act) that profoundly influenced the power structure of the electronic media. Even the campaign to establish online market safeguards for children, which ultimately resulted in passage of the 1998 Children's Online Privacy Protection Act (COPPA), was shaped by the prevailing public obsession with Internet safety.

Continuing Culture Wars

The public-policy debates during the 1990s left a legacy of "parental empowerment tools," including an astonishing array of rating systems, filtering-software products, and labeling schemes. There are also dozens of private initiatives aimed at cordoning off sections of the expanding digital media landscape, either as areas forbidden to children, or as gated communities set aside to ensure their safety. For example, AOL and other Internet service providers and search engines have created mechanisms that allow parents to set the levels of access and activity for their children, according to age.[11] Commercial filters, such as NetNanny, CyberPatrol, and CyberSitter, have been integrated fully into the online landscape, armed with a variety of tools to allow parents to block children's access to chat, instant messaging, and other worrisome online pastimes. Some also include software that blocks pop-up ads and cookies (while allowing advertising messages aimed at children, as well as links to commercial product sites).[12] A Pew Internet & American Life survey found that more than half of Internet-connected families with teens are using some kind of filter, though their overall effectiveness remains unclear.[13]

But while these strategies functioned quite well, during the first decade of the digital era, at preempting or undermining government legislation to control online content, they have not been able to thwart continuing efforts by Congress and the executive branch to push for child-protective measures on the Internet. The U.S. government passed a law in 2003 creating a new "dot-kids" domain, an official subsection of the .us domain, described as a "child-friendly" safe zone for children under 13, the online equivalent of the children's section of the library. The ACLU objected to this law because of the undue influence of the government in determining what kinds of content would be acceptable, although the group has not tried to challenge the legislation in the courts.[14]

Civil-liberty groups were successful at blocking implementation of the 1998 Child Online Protection Act (COPA), which would have made it a crime for commercial organizations to allow access by minors under 17 to sexual material deemed "harmful to minors."[15] The U.S. Justice Department launched a legal defense of COPA, as parties on both sides fought it out for years in the courts. In 2004, the Supreme Court upheld the injunction but sent the case back down to a lower court, instructing it to evaluate whether technological advances would provide adequate protections and eliminate the need for the law.[16] As the case was readied for new hearings, the Justice Department staged a series of high-profile sting operations, arresting dozens of online child predators in the United States and abroad (including one embarrassing arrest of an official at the U.S. Department of Homeland Security).[17] The news media suddenly were filled with a spate of frightening stories about the dangers to teens of social-networking sites such as MySpace.com, sparking renewed parental concerns.[18] In a highly controversial move, the Justice Department issued subpoenas to search engines and Internet service providers, including Google, Yahoo!, and America Online, demanding they turn over data about online searches, hoping to use the information as evidence that it was still easy for children to access pornographic and indecent content online.[19] The incident shed light on the vast stores of personal information collected on the Internet, raising more worries about online privacy. "This issue is going to come up over and over again," warned an official at the nonprofit Electronic Frontier Foundation. "I don't think this should make anybody very comfortable about the future. Google still has this stuff and people will still try to seek it."[20]

After its highly contentious and well-publicized launch in the late 1990s, the V-chip remained a well-kept secret for years, a predictable outcome given the confusing hybrid age-and-content labels (e.g., TV-14 DSL), and the industry's failure to promote the new system. Nearly seven years after the TV ratings were implemented, a survey by the Kaiser Family Foundation found that while parents had some familiarity with the ratings icons—which about half of them used to guide their families' viewing—only 15 percent of parents were using the V-chip itself to block out programs, and only four out of ten consumers who bought new TVs were even aware that the sets came equipped with the device.[21] "Right now, the V-chip is virtually unused," one industry executive told the press.[22]

But when the TV industry found itself under attack over alleged indecent programming, the obscure V-chip was dusted off and reintroduced to the American public through a series of highly publicized campaigns. With Janet Jackson's infamous "wardrobe malfunction" during CBS's airing of the 2004 Super Bowl, broadcasters once again were under siege. In the ensuing weeks and months, the incident mushroomed into a much wider public controversy. Washington political pundits debated the issue on the Internet, in the print press, and on the dozens of television talk shows and 24-hour news channels that peppered the media landscape in the multichannel era of cable and satellite TV. The FCC began slapping fines on dozens of other broadcasters and jawboning industry leaders to clean up the airwaves. The U.S. Congress held a series of public hearings on broadcast indecency. The National Association of Broadcasters hastily organized a summit, calling together network executives, producers, advocacy groups, and advertisers to address the problem. The industry created a new print ad and PSA campaign to promote the V-chip. Television representatives proudly invoked it in their congressional testimony. And corporate lawyers referenced it in their formal comments at the Federal Communications Commission.[23]

Even the recalcitrant NBC Network, which had refused to sign the 1997 agreement adding the content descriptors, changed its mind in the face of renewed government and public pressure. In a somewhat disingenuous explanation, a network spokesman told the press, "When content descriptors were first implemented, we questioned if they would cause more confusion than they would help. We're in an environment where the number of channels has grown exponentially. It's increasingly hard for viewers to know what's on all those channels. We want to give them more information, and our research shows that they are interested in choice and control, not in being passive."[24] In fact, the television industry had decided to enlist the much-maligned chip as a device for deflecting public criticism and preempting government regulation. As *Broadcasting & Cable* observed, the dire warnings about the V-chip a decade earlier did not materialize, and it had been easily institutionalized into TV programming practices without any serious disruptions. Noting that "the sky didn't fall when the chip became a reality," the trade publication explained that the controversial requirement "ultimately relieved networks of pressure to alter content." As one network executive told the press, "When it comes to a choice between censorship and the v-chip, we'll take the V-chip."[25]

However, even these highly publicized defensive measures did not prevent further escalation of this new round of culture wars. The indecency controversy played right into Republican political strategies during the 2004 presidential election, keeping the issue in the headlines for many months. Two years after the Super Bowl incident, the issue was still hotly debated on the Hill and in the press. More congressional hearings were held, and dozens of legislative proposals were introduced, aimed at increasing the fines against broadcasters and holding individual artists responsible for uttering indecent words. The uproar over indecency also triggered renewed political attacks over television violence, with Congress ordering the FCC to consider additional content controls in that area as well.[26] Right-wing groups, allied with mainstream consumer organizations, pushed for new policies over cable television, requiring operators to offer individual channels on an "a la carte" basis, so that consumers could pick and choose which ones were acceptable to their families.[27]

While the Clinton Administration had worked with liberal child-advocacy, health, and education groups (including the Center for Media Education) on its children-and-media public-policy initiatives, much of the contemporary debate is being spearheaded by conservative organizations that have gained ascendancy and support during the Republican Bush Administration. The most prominent is the Parents Television Council, a nonprofit founded in 1995 "to ensure that children are not constantly assaulted by sex, violence and profanity on television and in other media."

PTC uses a multipronged strategy to promote its goals. Like Advocates for Youth and the Kaiser Family Foundation, the Parents Television Council has established relationships with Hollywood writers and producers in its efforts to encourage conservative family values in television programming.[28] But PTC is mainly a media watchdog group that pressures government agencies, networks, and sponsors in order to "stem the flow of harmful and negative messages targeted to children." Boasting a grassroots membership of more than one million people, PTC has been the primary force behind much of the campaign over television indecency, using its Web site and e-mail newsletters to generate sponsor boycotts and letter-writing campaigns at the FCC. Dismissing the V-chip ratings, PTC offers its members a more elaborate online alternative "Family Guide to Prime Time Television."[29] The group also launched an ongoing campaign against MTV, releasing an analysis of the network's programming in 2005, and charging the channel with "blatantly selling raunchy sex to kids."[30]

The Digital-Marketing Paradigm

With the culture wars raging on, children and youth are again at the center of a booming digital marketplace. "As marketers continue to shift spending from traditional to digital media," noted *Media Week*, "experts expect spending to surge between 20% and 30%" above the estimated $12 billion in advertising revenues during 2005, making 2006 "the year of behavioral targeting."[31] An article in the *New York Times* explained that "the major media companies would like, above all, to find ways of reaching the younger audience," quoting one industry executive who told the press, "They're open to trying new things, and they have more time on their hands."[32] Among the hottest properties are MySpace.com, Xanga.com, Facebook.com, and other social-networking sites, which have quickly amassed millions of eager young followers, while at the same time stirring alarm among parents. The News Corporation's purchase of MySpace for $580 million is a harbinger of further consolidation trends in the online industry, as advertisers continue to seek out the youth market.[33] With the explosion in podcasting and personalized videos, marketers are moving further onto these platforms as well, investing their ad dollars in youth new-media ventures such as MTV's Overdrive, which enables consumers to view live performances, music videos, and celebrity interviews on their iPods, and to create their own multimedia "playlists" that can be e-mailed to friends.[34]

Marketing continues to pervade young people's media experiences. While the roots of today's digital marketplace date back to the middle of the twentieth century, new interactive technologies have created capabilities that alter the media marketing paradigm in significant ways, extending some of the practices that have already been put in place in conventional media, but more important, defining a new set of relationships between children and corporations. The Children's Online Privacy Protection Act (COPPA) offers some safeguards for children under 13. A mandated review of the FTC rules in 2005 found that they had succeeded in limiting the amount of personal information collected from children online, while striking "an appropriate balance between protecting children's personal information online and preserving children's ability to access content."[35] As one industry insider told *Advertising Age*, "When some people heard about [the parental permission requirement], they basically jumped out of the lake, ran away and climbed up a tree. They simply shut out anybody under 13."[36] However, increased market pressures could encourage more creative strategies for getting around the law, which will require continued vigilance from consumer groups.[37]

In teen media, contemporary marketers have succeeded in completely integrating content, advertising, data collection, and sales. Digital technologies make it possible to track every move, online and off, compiling elaborate personal profiles that combine behavioral, psychological, and social information on individuals, and aggregating that data across platforms and over time. Marketing strategies are designed to work on a deep level, tapping adolescents' basic developmental needs, anxieties, fears, and sense of identity. Companies now can forge intimate ongoing relationships with individuals, recruiting teenagers into market-research enterprises and enlisting them to promote products to each other, while operating outside the purview of parents and family. Marketers also are making it easy and fun for young people to serve as their own personalized ad agencies, incorporating brands into their digital artwork and distributing their work virally on the Internet, cell phones, and iPods. These trends no doubt will continue and grow in the coming years, becoming even more sophisticated with further innovations.

But the expansion of commercial culture in children's lives also has triggered a backlash from a growing number of consumer groups. Commercial Alert, cofounded by longtime consumer advocate Ralph Nader, vows to "keep the commercial culture within its proper sphere, and to prevent it from exploiting children and subverting the higher values of family, community, environmental integrity and democracy."[38] The group has called on Congress to investigate some of the new media marketing practices targeted at children, including Disney's plans in 2005 to offer wireless cell-phone service to 8–12-year-old children. It also petitioned the Federal Trade Commission to investigate the buzz marketing strategies of Procter & Gamble's Tremor.[39] In 1999, Commercial Alert sent a formal complaint to the American Psychological Association (APA), urging the professional association to write to its members, "denouncing the use of psychological techniques to assist corporate marketing and advertising to children."[40] In response to the challenge, the APA created a Task Force on Advertising and Children, which released a report in 2004 cautioning psychologists against lending their help to advertisers, and calling for legislation that would restrict advertising targeted to children 8 years old and younger.[41] A coalition of child advocates, including Children Now and the APA, was successful in convincing the FCC to place some modest limits on interactive marketing to children in emerging digital TV broadcasting.[42]

The Boston-based Campaign for a Commercial-Free Childhood (CCFC) is leading another coalition of health-care professionals, educators, and advocacy groups in a national effort "to counter the harmful effects of marketing to children through action, advocacy, education, research, and collaboration."[43] In 2006, CCFC, along with the veteran health-and-nutrition organization Center for Science in the Public Interest, announced joint plans to file suit against Viacom's Nickelodeon and Kellogg, charging that the two corporations "are directly harming kids' health since the majority of food products they market to children are high in sugar, saturated and transfat, or salt, or almost devoid of nutrients," and urging the court to stop the companies from "marketing junk foods through Web sites, toy giveaways, contests, and other techniques" to children under the age of eight.[44]

Concerns over childhood obesity have triggered a number of government and private actions, focusing renewed attention on children's marketing.[45] If current trends continue, health experts warn, this generation of children may be the first in modern history that will not live as long as their parents.[46] In 2004, Congress directed the Centers for Disease Control and Prevention to study the role of marketing in food consumption among children. The CDC commissioned the Institute of Medicine (IOM) to conduct the research and make recommendations. IOM's 2005 report *Food Marketing to Children and Youth: Threat or Opportunity?* assessed the available research on the impact of food marketing on children. Its committee of experts included health professionals, media scholars, and marketing representatives. The report's carefully worded findings concluded that, "among many factors, food and beverage marketing influences the preferences and purchase requests of children, influences consumption at least in the short term, and is a likely contributor to less healthful diets, and may contribute to negative diet-related health outcomes and risks among children and youth."[47] Among the report's recommendations was a strong warning to the food industry to change its advertising practices. "If voluntary efforts related to advertising during children's television programming are not successful in shifting the emphasis away from high-calorie and low-nutrient foods and beverages to the advertising of healthful foods and beverages," the report said, "Congress should enact legislation mandating the shift on both broadcast and cable television."[48]

But the IOM report issued no such directives concerning food-marketing practices in new media. In fact, one of its most telling discoveries was that there was virtually no academic research on digital-marketing strategies, despite the fact that food companies have been shifting much of their marketing efforts to the Internet, cell phones, product placement in film and television, and other new venues.[49] A quick survey of recent promotional campaigns by food and beverage companies could serve as a compendium of contemporary digital-marketing strategies. Burger King's "Subservient Chicken," an interactive "viral video" featuring a giant chicken that responds to user commands to dance, jump, or play dead, has become such an online phenomenon that, in less than two years, it has attracted 17 million visitors, who spent an average of six to seven minutes interacting with the ad.[50] At Viacom's online teen channel the N, visitors can use the Web site's Video Mixer to create their own ads for Skittles and e-mail them to their friends, part of the trend toward "user-generated advertising."[51] At spicyparis.com, a Web site sponsored by fast-food chain CKE Restaurants, popular celebrity Paris Hilton can be seen washing a car as she downs a Carl's Jr. Spicy Burger. The "interactive playground" also offers gaming, sweepstakes, and giveaways, designed to keep visitors online as long as possible, and collect information for personalized database marketing.[52] Procter & Gamble launched a campaign with Nickelodeon's popular show *Fairly Odd Parents*, putting codes in Pringles Snack Stacks packages that enabled kids to go to Nick.com and insert the code numbers to watch the show, vote for their favorite flavor, and play games.[53] A study by the Yankee Group found that mobile phones are beginning to reach "critical mass as a marketing tool," with both McDonalds and Kellogg Co. conducting text-messaging trials, by placing access codes on 250 million take-out bags and 80 million cereal boxes.[54] And snack-food companies are taking advantage of the database capabilities of search-engine marketing. A full-page ad in *Advertising Age* displayed a happy Frito-lay executive standing in front of a huge display of Cheetos, with Chester Cheetah leaping into the air. "Frito-Lay has added a special ingredient to its snack line," the copy read. "It moves chips off shelves and into grocery bags. . . . It's the Yahoo Marketing Engine."[55]

Much of the formal studies of consumer socialization, which began during the contentious 1970s public-policy debates over children and advertising, have in recent years focused primarily on supporting marketers, rather than assessing the impact of advertising on children's behavior.[56] Research by Juliet Schor, Susan Linn, and other scholars has helped raise public awareness of contemporary practices in the children's marketplace, but a great deal more ongoing work will be needed to stay apace of the rapidly expanding market.[57]

Some advocates have called for a ban on all marketing to children, in both conventional and new media, citing similar laws in other countries.[58] Whether these proposals can pass constitutional muster remains to be seen, in light of First Amendment freedoms granted to advertisers in a series of recent court cases. In the meantime, there are interventions that can and should be made in the children's digital marketplace, ones which can be achieved through a combination of government regulation and industry self-regulatory guidelines. For example, practices such as buzz marketing, viral marketing, and immersive advergaming, to the extent that they operate without clear disclosures to consumers, may well qualify as "deceptive" under existing laws.[59] At the very least, marketing practices and research methods should be disclosed clearly to the public.[60] In the area of food marketing, where there is a critical public-health crisis, strong regulatory intervention is needed to protect children and teens from manipulative, intrusive personalized practices. For all age groups, government protections are needed to protect consumers from abusive data collection and profiling practices in the new media.[61] To help young people better navigate the new digital consumer culture, we need comprehensive media and consumer education in the schools, with a curriculum that is sufficiently flexible to stay current in this rapidly changing marketplace.[62] Ultimately, it will take a combination of government policy, responsible industry self-regulation, public education, and activism to ensure that marketers treat young people fairly in the Digital Age.

However, while it is still possible (and necessary) to establish safeguards against some of the most egregious marketing practices targeting children, the overall dimensions of this highly commercialized interactive-media culture will remain in place. As the pro-social campaigns of the Kaiser Foundation, the American Legacy Foundation, and YouthNOISE demonstrate, digital-marketing strategies have become increasingly essential tools for health professionals, civic organizations, and educational groups seeking to reach and influence young people. The Kaiser Foundation continues to be a leader in the field, staying abreast of cutting-edge practices in youth marketing, and encouraging other nonprofits to retool their strategies to fit the new imperatives of the Digital Age. In March 2006, the foundation hosted a special seminar on "New Media and the Future of Public Service Advertising," featuring marketing expert Joseph Jaffe, author of one of the latest how-to books, *Life after the 30-Second Spot: Energize Your Brand with a Bold Mix of Alternatives to Traditional Marketing.*[63] The foundation also released a report profiling several recent social-marketing campaigns that use new media techniques to reach youth audiences. For example, the VERB campaign, launched in 2002 by the U.S. Centers for Disease Control and Prevention, uses text messaging on cell phones to encourage physical exercise among tweens. The Above the Influence campaign, launched in 2005 by the White House Office of National Drug Control Policy and the Partnership for a Drug-Free America, employs a range of digital tactics to help teenagers "deal with the competing pressures in their lives that lead to using marijuana and other illicit drugs, drinking alcohol, and engaging in sexual activities." The project's Web site invites teens from around the county to contribute their own podcasts, videos, poetry, drawings, and graffiti art that express "their feelings about negative and positive influences on their lives."[64]

In the coming years, youth will continue to be the targets of a cacophony of messages from diverse and sometimes competing institutions, as for-profit companies and nonprofit organizations step up their efforts to "become part of the communication structure" of the Digital Generation.

Fulfilling the Promise

The transition to the Digital Age provides us with a unique opportunity to rethink the position of children in media culture, and in society as a whole. Even though many business and cultural practices are in place already, there is still enough fluidity in the emerging media system for actions to help guide its future. What we need now is a national and international conversation, informed by research, on how digital technologies can best serve the needs of children and youth. If done right, this process could lead to a series of public and private initiatives that would help ensure that this potential is fulfilled. Youth should be involved directly in that dialogue as key stakeholders, innovators, and leaders.

In the United States, much of the research on children and television has been driven by public-policy debates over the relationship between consumption of content (either prosocial or antisocial), and behavior. For example, thousands of scientific studies (and considerable federal money), as well as countless articles in the popular press, have been devoted to assessing the impact of televised violence on children's attitudes and actions.[65] Though these concerns remain important, the growth of digital media requires us to expand our thinking to encompass a view of children not only as consumers, but also as producers and contributors to the media culture and the larger society.

In the first decade of this new era, market and political forces shaped much of the public discussion, generating considerable heat but shedding little light on our understanding of the new media culture. Formal research on the relationship between children, youth, and digital technologies lagged behind market research. More recently, nonprofit organizations, foundations, and academic institutions have begun to fill in the gaps of our knowledge. The Pew Internet & American Life Project and the Kaiser Family Foundation, for example, have been tracking media-usage patterns by children and adolescents for the past several years, offering independent assessments of the role that digital media are playing, and providing valuable information to the press, the public, and the academy that can also serve as a baseline for long-term trends.[66]

To deepen our understanding of the new media culture, its institutions, and its varied and complex roles in the lives of children and youth, however, we need a more comprehensive multidisciplinary effort, combining the contributions of historians, sociologists, anthropologists, and economists. "Rather than see children as the object of media effects," explains Sonia Livingstone, professor of social psychology at the London School of Economics, in her book *Young People and New Media*, "they are instead seen as actors in the household and community, coconstructors of the meanings and practices of their everyday lives."[67] In recent years, a number of scholars and research institutions have undertaken studies from this broader, multifaceted perspective. For example, MIT Professor Henry Jenkins examines how children and teens are using the Internet to acquire skills, knowledge, and self-confidence to equip them with the attributes needed to participate in the world around them.[68] In a collection of essays titled *Girl Wide Web: Girls, the Internet, and the Negotiation of Identity*, researchers present in-depth case studies of how girls use the Web to explore sexual identities, carve out their own cultural spaces, and engage in artistic expression.[69] As some members of the Digital Generation become scholars themselves, they are contributing further to our understanding of this new cultural experience.

To ensure that scholarship stays abreast of rapid developments in the digital-media industries, we need adequate ongoing funding from both the private and public sectors. The National Science Foundation, along with several major private foundations, has been supporting a consortium of university-based Children's Digital Media Centers, where scholars conduct investigations of how children and teens use chat rooms, instant messaging, and online games.[70] The John D. and Catherine T. MacArthur Foundation's project on Digital Media and Learning is investing $50 million over five years to fund "research and innovative projects focused on understanding the impact of the widespread use of digital media on our youth and how they learn."[71] In addition to studies of the uses of new media by young people, there is an urgent need for serious ongoing examination of the institutions that are creating this digital culture, both its commercial and its noncommercial sectors. In an era when academic institutions increasingly are forced to seek outside support for their work, it is critical that research on new media and children be free of undue influences, either by industry or narrow political interests. Several prominent policymakers have championed the cause for additional funding to support research on children and new media. But with the continued focus on Internet safety and harmful media effects, in both the press and the policy arena, there is a real danger that federal funds may come with political strings attached, narrowing both the research agenda and the public debate.[72]

With so much policy focus on protecting children, not enough attention has been paid to nurturing the attributes that children need to grow through adolescence and into adulthood. As the University of London Professor David Buckingham has argued, instead of trying

> to reinforce the boundaries between childhood and adulthood, and to confine children more securely within them, we need to face the fact that those boundaries are increasingly being crossed and blurred in all sorts of ways. And rather than leaving children isolated in their encounters with the "adult" world of the contemporary media, we need to find ways of preparing them to cope with it, to participate in it, and if necessary to change it.[73]

The new media hold particular potential for fostering greater civic and political engagement by young people. The hundreds of youth civic Web sites created in the last decade offer abundant opportunities for community involvement and political expression.[74] The 2004 youth-vote initiatives suggest that the Internet can be an effective mobilizing tool in electoral politics. And the online efforts of Downhill Battle, Freeculture.org, and other cyberactivist groups demonstrate how some youth have seized these new digital tools to promote policy goals and to create and distribute their own cultural products. The real test, however, is whether such forms of participation can be extended to a broader segment of the youth population, and sustained beyond the occasional bursts of activity surrounding extraordinary events, such as the 9/11 tragedy or high-stakes national elections.

With the support of foundations and government agencies in the United States and abroad, some scholars have begun to explore the potential of the new media as a force for civic and political renewal among youth.[75] These efforts could grow into a broader movement.

Experts on civic engagement and youth development point to a crucial set of skills that youth need in order to become engaged as citizens. These include communications skills such as active listening, taking turns, and public speaking; intellectual skills such as critical and reflective strategies for processing information, formulation and expression of opinions, perspective taking, and principled reasoning; leadership and organizational skills such as organizing meetings, understanding and tolerance for diverse points of view, bargaining and compromise in group decisions, public problem solving, and coalition building; and civic-specific skills such as contacting public officials.[76] Digital media offer many tools for developing these skills. But there are also opposing tendencies in the new media that may serve to undermine its democratic potential. For example, the capacity for collective action, community building, and mobilization are unprecedented. But the move toward increasingly personalized media and one-to-one marketing may encourage self-obsession, instant gratification, and impulsive behaviors. And while digital media offer an incredible resource for mobilizing citizen action, the intertwining of political activism and brand marketing—so much in evidence in the youth-voter mobilization campaigns—ultimately may undermine these efforts.[77]

As we consider the policy issues for the Digital Age, the goal of fostering a healthy, democratic media culture for young people must be a top priority. This requires engaging in a variety of policy issues that traditionally have not been part of the children's-media policy agenda. For example, issues such as network neutrality, digital-rights management, and spectrum allocation are fundamental to the preservation and growth of an open and participatory digital media system for the twenty-first century.[78]

The true democratic potential of the new media can never be fulfilled unless everyone has the ability to participate.[79] While significant gains have been made in narrowing the digital divide, further efforts are needed, in both the private and the public sectors, to ensure universal access across existing lines of class, race, ethnicity, and gender.

Notes

Chapter 1

1. Andrea L. Foster, "Recording Industry Says It Will Sue Hundreds, Including Students, Starting This Month," *Chronicle of Higher Education*, September 12, 2003, A30.

2. Steve Lohr, "In the Age of the Internet, Whatever Will Be Will Be Free," *New York Times*, September 14, 2003, sec. 4, 1.

3. Frank Ahrens, "RIAA's Lawsuits Meet Surprised Targets," *Washington Post*, September 10, 2003, E1.

4. Timothy W. Maier, "Arresting Kids for Downloading Music," *Insight on the News* 20, no. 5 (2004): 29.

5. Pew Internet & American Life Project, "Music Downloading, File-sharing and Copyright: A Pew Internet Project Data Memo," July 2003, http://www.pewinternet.org/pdfs/PIP_Copyright_Memo.pdf.

6. Though Napster was forced into bankruptcy over copyright violations, the company subsequently reinvented itself as a for-pay music downloading service. Wikipedia, "Napster," http://en.wikipedia.org/wiki/Napster.

7. Pew Internet & American Life Project, "Music Downloading, File-sharing and Copyright: A Pew Internet Project Data Memo," July 2003, http://www.pewinternet.org/pdfs/PIP_Copyright_Memo.pdf.

8. Kevin Maney, "Music Industry Doesn't Know What Else to Do As It Lashes out at File-Sharing," *USA Today*, September 10, 2003, 3B.

9. Jessi Hempel, with Paula Lehman, "The MySpace Generation," *Business Week*, December 12, 2005, 86.

10. Don Tapscott, *Growing up Digital: The Rise of the Net Generation* (New York: McGraw-Hill, 1998), 1.

11. Tapscott, *Growing up Digital*, 33.

12. Austin Bunn, "The Rise of the Teen Guru," *Brill's Content*, July/August 2000, 64–69, 123–129. Remarkably, the majority of teens and their parents are in agreement on this point: 64 percent of online teens claim to know more about the Internet than their parents, and 66 percent of parents agree. Amanda Lenhart, Lee Rainie, and Oliver Lewis, *Teenage Life Online: The Rise of the Instant-Message Generation and the Internet's Impact on Friendship and Family Relationships* (Washington, D.C.: Pew Internet & American Life Project, 2001), 11, http://www.pewinternet.org/pdfs/PIP_Teens_Report.pdf.

13. Michael Lewis, *NEXT: The Future Just Happened* (New York: Norton & Co., 2001), 87–117.

14. Joseph Menn, *All the Rave: The Rise and Fall of Shawn Fanning's Napster* (New York: Crown Business, 2003); Trevor Merriden, *Irresistible Forces: The Business Legacy of Napster and the Growth of the Underground Internet* (Oxford, UK: Capstone Publishing, 2001).

15. Howard Rheingold, *Smart Mobs: The Next Social Revolution* (Cambridge, MA: Perseus), 160.

16. Clive Thompson, "The Virus Underground," *New York Times Magazine*, February 8, 2004, 28.

17. Henry Jenkins, "Introduction," *The Children's Culture Reader* (New York: New York University Press, 1998), 9.

18. David Thomas, "M for Violence, Mature Content: The Story Behind a Video Game Star," *Denver Post*, August 17, 2003, EE3; Kieran Nicholson, "Harris Wrote of Massacre Plan: Journal Entries on Columbine Predated Complaint to Sheriff," *Denver Post*, December 5, 2001, A1.

19. Kurt Eichenwald, "Through His Webcam, a Boy Joins a Sordid Online World," *New York Times*, December 19, 2005, A1.

20. While Internet service providers (ISPs) have vowed to shut these sites down, the article found that by searching "pro-anorexia" on Google, more than 30,000 links came up. January W. Payne, "No, That's Sick. Pro-Anorexia Web Site Authors Claim the Condition Is a 'Lifestyle Choice,'" *Washington Post*, Tuesday, September 14, 2004, F1.

21. Laura Sessions Stepp, "Offline and Still in Touch with Away Messaging," *Washington Post*, July 9, 2004, C1. In addition to such life-and-death encounters, the press is full other troubling stories about the negative impacts of the Internet on young people. Some college educators worry that students are using instant messaging to pass electronic notes to each other during class lectures. Lisa Guernsey, "In the Lecture Hall, a Geek Chorus," *New York Times*, July 24, 2003, G2. Others are concerned about the growing popularity of social networking Web sites like Theface-book.com. Calling it "the Swiss Army knife of procrastination," they fear that

obsession over the online pastime is creating new forms of "Internet addiction" among students. Peter Applebome, "On Campus, Hanging Out by Logging On," *New York Times*, December 1, 2004, B1. Some high schools in the Washington, DC, area have banned the use of school e-mail on such sites, fearing that teens are revealing too much about themselves on the Internet. Tara Bahrampour and Lori Aratani, "Teens' Bold Blogs Alarm Area Schools," *Washington Post*, January 17, 2006, A1. A *New York Times* article reported that the popularity of text messaging on cell phones was causing many young people to rack up hundreds of dollars in debt without a clue that their new addictive habit was costing so much. Lisa W. Foderaro, "Young Cell Users Rack Up Debt, One Dime Message at a Time," *New York Times*, January 9, 2005, A1.

22. "Although it is widely assumed to be true," media scholar Ellen Seiter points out, "no research proves that familiarity with computers and the Internet increases learning and competitiveness in the job market." Ellen Seiter, *The Internet Playground: Children's Access, Entertainment, and Mis-Education* (New York: Peter Lang, 2005), 6.

23. The survey found "a remarkable continuity in the belief that the Internet is a useful and even critical component of a child's education while at the same time it gives youngsters access to content with troublesome values." Joseph Turow and Lilach Nir, "The Internet and the Family: The Views of Parents and Youngsters," in Joseph Turow and Andrea L. Kavanaugh, *The Wired Homestead* (Cambridge, MA: MIT Press, 2003), 201.

24. Ellen Wartella and Nancy Jennings, "Children and Computers: New Technology—Old Concerns," *The Future of Children, Children and Computer Technology* 10, no. 2 (2000): 31.

25. "The Administration's Agenda for Action," http://www.ibiblio.org/nii/NII-Agenda-for-Action.html. As Marsha Kinder observed, the Clinton administration fostered what *Time Magazine* called the "cult of the child." As she pointed out, "An 'education president' who came into office only two years after passage of the 1990 CTA, Clinton's campaign rhetoric both in 1992 and 1996 was strategically structured around kids' culture." Marsha Kinder, ed., *Kids' Media Culture* (Durham, NC: Duke University Press, 1999), 16. J.F.O. McAllister, "The Children's Crusade," *Time*, August 25, 1997, 36.

26. National Telecommunications and Information Administration, "How Access Benefits Children: Connecting Our Kids to the 'World of Information': The Telecommunications and Information Infrastructure Assistance Program," http://www.ntia.doc.gov/top/publicationmedia/How_ABC/How_ABC.html.

27. Eliabeth Corcoran, "Mapping Out the Information Superhighway; New Report Details Ways to Get Nation Plugged In," *Washington Post*, December 18, 1995, F17; "Internet in a Box Joins Jones Intercable Project to Deliver Easy Internet Access

over Cable Television: Schools in Alexandria, Va. First Sites in Nation to Gain Broadband Cable Connectivity to Information Highway," PR Newswire, February 14, 1994.

28. Net Generator, "Welcome," http://www.netgenerator.com/html/home_facts.html.

29. "The Digital Generation; How Young People Have Embraced Computers and the Internet," in *A Nation Online: How Americans Are Expanding Their Use of the Internet,* National Telecommunications and Information Administration, http://www.ntia.doc.gov/ntiahome/dn/html/Chapter5.htm (accessed August 25, 2005).

30. Income level clearly plays a role in determining Internet access, according to the report. Only 73% of teenagers from homes earning under $30,000 a year reported using the Internet, while 90% of teens from families earning more than $30,000 per year were connected. Teens from families in the highest income levels (more than $75,000) were online at the highest rate (93%). Amanda Lenhart, Mary Madden, and Paul Hitlin, *Teens and Technology*, Pew Internet & American Life Project, July 27, 2005. P. 1. http://www.pewinternet.org/PPF/r/162/report_display.asp.

31. Mark Mather and Kerri L. Rivers, "City Profiles of Child Well-Being: Results from the American Community Survey," A KIDS COUNT Working Paper, October 2005, p. 3. http://www.aecf.org/kidscount/pubs/city_profiles.pdf.

32. "Today, half of all marriages end in divorce, and more than half of all children will spend some time in a single-parent family. About 69 percent of youth currently live in a family with two parents." From Joy G. Dryfoos, *Safe Passage: Making It through Adolescence in a Risky Society* (New York: Oxford University Press, 1998). See Judith S. Wallerstein and Sandra Blakeslee, *Second Chances: Men, Women, and Children a Decade after Divorce* (Boston: Houghton Mifflin, 2004).

33. "Over the last twenty-five years," Sylvia Ann Hewitt explains in her 1991 book, "the proportion of mothers in the paid labor force has tripled and the number of children growing up without a father has increased by a factor of two." These trends, she argues, have produced a "deficit that is increasingly threatening the well-being of children." From Sylvia Ann Hewlett, *When the Bough Breaks: The Cost of Neglecting our Children* (New York: Basic Books, 1991). See also David A. Hamburg, *Today's Children: Creating a Future for a Generation in Crisis* (New York: Times Books, 1992); and Arlie Russell Hochschild, with Anne Machung, *The Second Shift: Working Parents and the Revolution at Home* (New York: Avon Books, 1997).

34. Patricia Hersch, *A Tribe Apart: A Journey into the Heart of American Adolescence* (New York: Ballantine Books, 1998), 13.

35. Victoria Rideout, Donald F. Roberts, and Ulla G. Foehr, "Generation M: Media in the Lives of 8–18 Year Olds," Henry J. Kaiser Family Foundation. Executive Summary, 9–11, 24–25. http://www.kff.org/entmedia/entmedia030905pkg.cfm.

36. Mireya Navarro, "Parents Fret That Dialing Up Interferes with Growing Up," *New York Times*, October 23, 2005, Sec. 9, 1.

37. For example, see the National Institute on Media and the Family Web site at http://www.mediafamily.org/about/index.shtml.

38. The Kaiser study found that two-thirds of children have a TV in their bedrooms, half have a VCR/DVD player and a video game console and nearly a third have computers in their bedrooms. Victoria Rideout, Donald F. Roberts, and Ulla G. Foehr, "Generation M: Media in the Lives of 8–18 Year Olds," Kaiser Family Foundation. Executive Summary, 9–11.

39. Bob Thompson, "Guys and Digital Dolls," *Washington Post*, April 14, 2002, W08.

40. "Hedonomics," *Brandweek*, December 13, 1999.

41. "Forrester Technographics Finds That Young Consumers Are Internalizing Net Rules," August 10, 1999. http://www.forrester.com/ER/Press/Release/0,1769,158,FF.html (accessed August 17, 1999).

42. "A New Media Landscape Comes of Age," Executive Summary, *Born to Be Wired: The Role of New Media for a Digital Generation.* Commissioned by Yahoo! and Carat Interactive. Research conducted by HarrisInteractive and Teenage Research Unlimited. July 23, 2003. From the author's personal files.

Chapter 2

1. I was one of the participants at this event, invited to warn marketers about some of the ethical and political issues raised by their practices.

2. Institute for International Research, promotional pamphlet, "Youth Marketing Mega-Event 2003, June 22–25, 2003, Anaheim, California."

3. For a recent examination of contemporary children's market research trends, see chapter 6, "Dissecting the Child Consumer," in Juliet B. Schor, *Born to Buy: The Commercialized Child and the New Commercial Culture* (New York: Scribner, 2004).

4. See Susan Linn, *Consuming Kids: The Hostile Takeover of Childhood* (New York: The New Press, 2004). Juliet B. Schor, *Born to Buy: The Commercialized Child and the New Consumer Culture* (New York: Scribner, 2004).

5. Daniel Thomas Cook, "The Other Child Study: Figuring Children as Consumers in Market Research, 1910s–1990s," *The Sociological Quarterly* 41 (2000): 487–507. See also Daniel Thomas Cook, *The Commodification of Childhood: The Children's Clothing Industry and the Rise of the Child Consumer* (Durham, NC: Duke University Press, 2004).

6. Lizabeth A. Cohen, *Consumers' Republic: The Politics of Mass Consumption in Postwar America* (New York: Alfred A. Knopf, 2003), 318–319.

7. Thomas Hine, *Rise and Fall of the American Teenager* (New York: Bard, 1999), 232.

8. Ibid.

9. Eugene Gilbert, *Advertising and Marketing to Young People* (Pleasantville, NY: Printer's Ink Books, 1957).

10. See chapter 7, "Culture: Segmenting the Mass" in Cohen, 292–344. See also Joseph Turow, *Breaking up America: Advertisers and the New Media World* (Chicago: University of Chicago Press, 1997).

11. Norma Odom Pecora, *The Business of Children's Entertainment* (New York: The Guilford Press, 1998), 16–17.

12. Cook, "The Other Child Study," 500–501.

13. Ibid., 501. McNeal's books include: *Children as Consumers: Insights and Implications* (Lexington, MA: Lexington Books, 1987); *Kids as Customers: A Handbook of Marketing to Children* (Lexington, MA: Lexington Books, 1992); *The Kid's Market: Myths and Realities* (Ithaca, NY: Paramount Market, 1999).

14. As Cook writes, McNeal and other proponents of the marketplace "helped make this kind of research morally appropriate and ethically palatable by the conception of a little consumer nascent in the child." Cook, "The Other Child Study," 502.

15. As Norma Pecora explained: "Like early radio, children's television was first sponsored on a sustaining basis; programs were defined by the networks as promotional material. Of the programs offered during the first 4 years of television, over half received some support from the networks. The child audience was used to encourage the sales of television, selling parents on the educational and prosocial possibilities of the new technology. Once most households had a television set, there was no longer a need to attract a young viewing audience, sustaining program production was no longer practical, and stations went after the audience with income—adults." Pecora, *Children's Entertainment*, 31. See also Joseph Turow, *Entertainment, Education, and the Hard Sell: Three Decades of Network Children's Television* (New York: Praeger, 1981).

16. The emergence of research on children as consumers in the 1970s came about as a direct result of public policy debates over children's television advertising. But as Deborah Roedder John points out, the research generated out of these concerns also led to "a new generation of researchers and an emerging field of study pertaining to children as consumers." Deborah Roedder John, "Consumer Socialization of Children: A Retrospective Look at Twenty-Five Years of Research," *Journal of Consumer Research* 26 (1999): 183–213. See also Dale Kunkel, "The Role of Research in the Regulation of U.S. Children's Television Advertising," *Knowledge: Creation, Diffusion, Utilization* 12, no. 1 (1990): 101–119.

17. Cook, "The Other Child Study," 502. See also Stephen Kline, *Out of the Garden: Toys and Children's Culture in the Age of Marketing* (New York: Verso, 1993).

18. John McDonough, "Its 'Kids First' Philosophy and 'Let Kids Be Kids' Mantra Drive the Mighty Nickelodeon Engine," in special advertising section, "Celebrating 25 Years of Kids First!" *Advertising Age*, March 15, 2004, N1–N28.

19. This ad appeared in a number of trade publications during the early 1980s, at the height of intense bidding for local cable franchises. Author's personal files.

20. In 1980, advertisers spent $110 million to buy time on the three broadcast networks for children's commercials. By the early 90s, the figure had jumped to more than $170 million. Selina S. Guber and Jon Berry, *Marketing To and Through Kids* (New York: McGraw-Hill, 1993), 130. See also Juliet B. Schor, *Born to Buy: The Commercialized Child and the New Consumer Culture* (New York: Scribner: 2004) 39–68.

21. Lawrie A. Mifflin, "Growth Spurt Is Transforming TV for Children," *New York Times*, April 19, 1999, sec. A.

22. Kathryn C. Montgomery, "Digital Kids: The New Online Children's Consumer Culture," in Dorothy Singer and Jerome Singer, eds., *Handbook of Children and the Media* (Thousand Oaks, CA: Sage Publications, 2001), 635–650.

23. Mifflin, "Growth Spurt."

24. Karen Flischel, "Listening to Children in the U.S.: the Important Role of Research in the Development of Nickelodeon's Programme Strategy," *Admap*, April 1995. Accessed online at the World Advertising Research Foundation, http://www.warc.com.

25. Dale Russakoff, "Marketers Following Youth Trends to the Bank," *Washington Post*, April 19, 1999, A1.

26. Ibid.

27. Laura E. Wendt and Petra Sonderegger, "Bunking with Your Customers: A Weekend Retreat with Kids," *World Association of Research Professionals*, November, 2002. Accessed online at the World Advertising Research Foundation, http://www.warc.com.

28. Linda Simensky, "The Early Days of Nicktoons," in Heather Hendershot, ed., *Saturday Morning Censors: Television Regulation before the V-Chip* (Durham, NC: Duke University Press, 1998).

29. Russakoff, "Marketers." Lessley Anderson, "Mixing Teen Cool with E-commerce Savvy," *The Industry Standard*, June 17, 1999, http://www.cnn.com/TECH/computing/9906/22/teen.idg/index.html; Susan Stellin, "This Web Site Rocks! (It's Turning a Profit)," *New York Times*, December 3, 2000, sec. 3). Figures for children and teen spending vary, depending on the sources. But experts all agree on the rate of growth in the marketplace. For example, according to a 1998 article in *American Demographics*, "teenage spending has been a green machine for the past 44 years,

according to Lester Rand, president of the Rand Youth Poll. Spending by teenagers increased every year from 1953 to 1996 despite eight recessions and the drop in the size of the group when Generation X was in its teen years." Matthew Klein, "Teen Green—Teenage Spending," *American Demographics* 20, no. 2 (1998): 39.

30. Guber and Berry, *Marketing*, 9.

31. U.S. Department of Labor, "Work Experiences Differ between Girls, Boys," July 7, 2000.

32. Guber and Berry, *Marketing*, 8–9.

33. McNeal, *The Kids Market*, 14.

34. Ibid., 21.

35. Commercials for sugary cereals such as Trix, Coco Puffs, and Lucky Charms already were a familiar part of the Saturday-morning-TV landscape. Though their proliferation in the 1960s and 70s had prompted outcries from consumer groups and child advocates, attempts to regulate them had failed. (The 1970s advocacy campaign against sugared cereals and other food marketing on television is discussed in more detail in chapter 4 of this book.) See also Michael F. Jacobson and Laurie Ann Mazur, *Marketing Madness: A Survival Guide for a Consumer Society* (Boulder, CO: Westview Press, 1995.)

36. Sonia Reyes, "Into the Mouths of Babes," *Brandweek*, May 6, 2002, 26–31.

37. Ibid.

38. Schor, *Born to Buy*, 123.

39. Schor, 24.

40. As Susan Linn explains, the famous study was: "The Nag Factor," conducted by Western Media International (now Initiative Media Worldwide) and Lieberman Research Worldwide. "According to the press release . . . headlined 'The Fine Art of Whining: Why Nagging Is a Kid's Best Friend,' the study identifies which kinds of parents are most likely to give in to nagging. Not surprisingly, divorced parents and those with teenagers or very young children ranked the highest." Linn, *Consuming Kids*.

41. McNeal, *The Kids Market*, 14, 160.

42. For a description and analysis of contemporary practices by market-research firms focusing on children, see chapter 6, "Dissecting the Child Consumer," in Juliet B. Schor, *Born to Buy: The Commercialized Child and the New Consumer Culture* (New York: Scribner, 2004).

43. McNeal, *The Kids Market*, 13.

44. Russakoff, "Marketers," 16.

45. Barbara Kantrowitz and Pat Wingert, "The Truth about Tweens," *Newsweek*, October 18, 1999, 62. As trade publication *Broadcasting & Cable* later observed: "The tween demo is an invention of modern marketing. The term popped into the zeitgeist as advertisers homed in on those critical gap years between kids and teen consumers" Allison Romano, "Tween and Mean," *Broadcasting & Cable* 16 (April 19, 2004): 17.

46. Kantrowitz and Wingert, "The Truth about Tweens," 62.

47. Timothy J. Coffey, Gregory Livingston, and David L. Siegel, *The Great Tween Buying Machine: Marketing to Today's Tweens* (Ithaca, NY: Paramount Marketing Publishers, 2001), 2, 8.

48. Kay S. Hymowitz, "Tweens: Ten Going on Sixteen," *City Journal*, Autumn 1998, www.city-journal.olrg/html/8_4_al.html. See also Kathleen McDonnell, *Honey, We Lost the Kids: Rethinking Childhood in the Multimedia Age* (Toronto: Second Story Press, 2001).

49. Author's notes from "Targeting Tweens" conference, Atlanta, GA, January 20, 2000.

50. Author's notes, "Targeting Tweens" conference.

51. Linn, *Consuming Kids*, 130–131.

52. Author's notes, "Targeting Tweens" conference.

53. Catherine Schetting, "Tween Shall Meet: Kids 9–14 Are the Newest Hot Programming Demo," *Broadcasting & Cable*, March 5, 2001, 16.

54. Sonia Reyes, "Into the Mouths of Babes," *Brandweek*, May 6, 2002, 26–31.

55. Linda Haugsted, "Kidvid Nets Reach for 'Tween' Audience," *Multichannel News*, November 27, 2000, 76.

56. Ibid.

57. See Juliet Schor, *Born to Buy*, chapter 9, "Empowered or Seduced?" for a discussion of some of these efforts. See also Susan Linn, *Consuming Kids*. In addition to these books, a number other works published in recent years decry the growing commercialism in children's lives. These include: Michael Jacobson and Laurie Ann Mazur, of the Center for Science in the Public Interest (and the Center for the Study of Commercialism), *Marketing Madness*; and David Walsh, President of the nonprofit National Institute on Media and the Family, *Selling Out America's Children* (Minneapolis, MN: Deaconess Press, 1994).

58. This research was updated and the warnings renewed in a report issued by the APA Task Force on Advertising and Children in 2004, "Report of the APA Task Force

on Advertising and Children: Section: Psychological Issues in the Increasing Commercialization of Childhood," http://www.apa.org/releases/childrenads.pdf.

59. Consumers Union, 1997, *Captive Kids: A Report on Commercial Pressures on Kids at School,* http://www.consumersunion.org/other/captivekids/index.htm. CU also published *Selling America's Kids:* Consumers Union, 1998, *Commercial Pressures on Kids of the 90's.* http://www.consumersunion.org/other/sellingkids/index.htm. See also Alex Molnar, *Giving Kids the Business: The Commercialization of America's Schools* (Boulder, CO: Westview Press, 1996).

60. The Center for Commercial-Free Public Education, About Page, http://www.ibiblio.org/commercialfree/about.html. A lack of funding forced the center to close its doors in 2002. Other new groups that were created during the 1980s to fight commercialism included: the Coalition to Stop Commercial Exploitation of Children, the Center for the New American Dream, the Motherhood Project, and Commercial Alert.

61. *Roper Reports,* "Today's Kids—Especially Teens—Are Wired to the Hilt, Gender Gap Still Exists," November 12, 1998.

62. Seymour Papert, *Mindstorms: Children, Computers, and Powerful Ideas* (New York: Basic Books, 1980).

63. Joseph Turow, *The Internet and the Family: The View from Parents, The View from the Press* (Philadelphia: The Annenberg Public Policy Center of the University of Pennsylvania, May 4, 1999), 6–14. http://www.annenbergpublicpolicycenter.org/02_reports_releases/report_1999.htm.

64. Jeffrey D. Stanger and Natalia Gridina, *Media in the Home 1999* (Philadelphia: The Annenberg Public Policy Center of the University of Pennsylvania, June 28, 1999), 3. http://www.annenbergpublicpolicycenter.org/02_reports_releases/report_1999.htm.

65. John Cassidy, *Dot.con: How America Lost its Mind and Money in the Internet Era* (New York: HarperCollins, 2002), 86.

66. Cassidy, *Dot.con,* 92.

67. "100 Best Business Web Sites," *Interactive Age,* April 24, 1995, 45.

68. Edwin L. Artzt, "Artzt Enthusiastic about CASIE Gains," *Advertising Age,* March 13, 1995. See also Joseph Turow, *Breaking Up America,* 168–169.

69. Nancy Coltun Webster, "The New Whiz Kids," *Advertising Age,* February 13, 1995, S1–S2.

70. Michael Haile, "Multimedia Companies Set Their Sights on the Lucrative Cybertot Category," *Hollywood Reporter,* February 21, 1995.

71. "MSN Divulges Few Home/Family Plans," *The Digital Kids Report* (New York: Jupiter Communications, June 1995).

72. Diana Simeon and Aimee Pamintuan, "Special Report: Children and the Web," *The Digital Kids Report* (New York: Jupiter Communications, April, 1995).

73. Press release, Da Vinci Time and Space, February 13, 1995.

74. Connie Guglielmo, "Child's Play; Hollywired Start-up Da Vinci Time & Space Is Creating the First Interactive Television Environment for Kids," *Wired*, January, 1995, 86.

75. Kathryn C. Montgomery, "Digital Kids: The New Online Children's Consumer Culture," in Dorothy Singer and Jerome Singer, eds., *Handbook of Children and the Media* (Thousand Oaks, CA: Sage Publications, 2001).

76. Comments from Laura Groppe, president of GirlGames, speaking at the conference on "Ensuring a Quality Media Culture for Children in the Digital Age," Washington, DC, October 22, 1998. Quoted in Gary O. Larson, Kathryn Montgomery, and Yalda Nikoomanesh, *Ensuring a Quality Media Culture for Children in the Digital Age* (Washington, DC: Center for Media Education, 1999), 11.

77. For example, Jupiter Communications, established in 1986, quickly grew from a tiny firm in Manhattan to a large organization with branches in San Francisco and London. Excite, "Jupiter Communications," http://www1.excite.com/home/careers/company profile/0,15623,279,00.html. Forrester Research, a high-tech research firm in Cambridge, Massachusetts, began in 1986, signing on dozens of technology companies as clients. In 1993, the company's report, "The New Public Internet," predicted that the Internet would become a household tool. Forrester Research, "Timeline," http://www.forrester.com/timeline. Both firms were cited repeatedly in news coverage of trends in the dot-com market.

78. "Defining the Digital Consumer IV: Digital Kids Pre-Conference Seminar," New York, NY, October 25, 1995. I attended this event.

79. Transcript, "Defining the Digital Consumer IV Agenda: Digital Kids Pre-Conference Seminar," New York, NY, October 25, 1995.

80. Ibid.

81. Don Tapscott, *Growing Up Digital: The Rise of the Net Generation* (New York: McGraw-Hill, 1998).

82. "MaMaMedia's Idit Harel: Camp Counselor to the New Clickerati," *Min's New Media Report*, November 22, 1999.

83. Remarks by Idit Harel at a CME-sponsored conference in Washington, DC, October 22–23, 1998. *Center for Media Education*, "Ensuring a Quality Media Culture for Children in the Digital Age," May 1999.

84. Press release, MaMaMedia, May 18, 1999. MaMaMedia debuts broadband content on road runner's cable modem service.

85. MaMaMedia, http://www.mamamedia.com.

86. "MaMaMedia's Idit Harel," 1999.

87. Anya Sacharow, "General Mills, MaMaMedia Go Online, On Boxes to Reach Kids," *Adweek*, September 28, 1998. You Rule School, http://www.youruleschool.com.

88. Bob Thompson, "The Selling of the Clickerati; Today's Kids Have a Visceral Connection to Computers. Now the Shopping Gods Are Starting to Connect As Well," *Washington Post Magazine*, October 24, 1999, W11.

89. Ibid. I was one of the speakers at this conference.

90. Ibid.

91. "What is really happening [on the Web]," explained *Red Herring*, a trade publication for high-tech investors, "is what will ultimately happen on interactive television: the infomercialization of all programming. Services will deliver some content, with lots of appeals (some soft, some hard) to purchase. Requesting literature and additional information (read: volunteering for a mailing list), and actual buying, will be easily enabled. . . . This is not advertising as you and I understand it, but a more viewer-engaged, browse-and-buy genre just beginning to emerge as a form of programming unto itself." K. Davis and O. Driscoll, "Road Map for the Internet," *Red Herring* 19 (1999), http://www.redherrring.com/mag/issue19/roadmap.html.

92. Thompson, "Selling of the Clickerati," W11.

93. McNeal, *Kids as Customers*, 93.

94. E. Gruen, "Defining the Digital Consumer IV Agenda: Digital Kids Pre-Conference Seminar," New York (1995). Author's notes from the Digital Kids conference, 1999.

95. J. Roberts, "Defining the Digital Consumer IV Agenda: Digital Kids Preconference Seminar," New York (1995).

96. Author's notes, Digital Kids conference.

97. Thompson, "Selling of the Clickerati," W11.

98. Elizabeth Weise, "Parents Pay for Kids' Virtual Wallets," *USA Today*, June 9, 1999, sec. 9D.

99. Rebecca Quick, "New Web Sites Let Kids Shop, Like, Without Credit Cards," *Wall Street Journal*, June 14, 1999, B1–4.

100. Harris Interactive, http://www.harrisinteractive.com (accessed December 11, 2000).

101. "Between . . . May 2000 and July 2001, 538 Internet companies closed." Jon Swartz, "Webbys Go On Despite Dot-Bomb Threat," *USA Today*, July 18 2001, sec. B.

102. Kathleen Tracy, "Indie Days May be Numbered in Webland," *Kidscreen*, November 1, 2000, 53.

103. I made a personal visit to the MaMaMedia offices in Manhattan in 2001.

104. Zack Martin, "Perfect Teen Payment Vehicle Elusive," *Card Marketing*, May–June, 2002, 13.

105. Tracy, "Indie Days," 53.

106. Katie Hafner, "Teenage Overload, or Digital Dexterity," *New York Times*, April 12, 2001, sec. G.

107. Catherine Schetting Salfino, "Great Expectations; Younger Set Demands Strong Complementary Online Experience," *Broadcasting & Cable*, March 19, 2001, 58.

108. A. S. Berman, "Paddling Deep into 'Creek': Desktop Takes Fans Inside Dawson's Head," *USA Today*, April 13, 2000, sec. 3D; Marc Graser, "'Dawson's Creek' Site Gives Fans Peek into Series' World," *Variety*, November 16, 1998, 37–39.

109. Dawson's Desktop, http://www.dawsonsdesktop.com (accessed April 24, 2001).

110. Alex Frangos, "Between Shows: For Fans of 'Dawson's Creek,' the Angst Never Stops," *Wall Street Journal*, March 16, 2001, sec. R.

111. http://store.yahoo.com/spestore/dawrepnec.html (accessed July 21, 2000).

112. Adage.com, "A Primer on Behavioral Marketing," http://www.adage.com/MarketingIntel/pdf/ClariaBehaviorialBasics.pdf.

Chapter 3

1. Scholar Henry Jenkins has written about the significance of this compelling image: "The figure of the endangered child surfaced powerfully in campaigns for the Communications Decency Act, appearing as a hypnotized young face awash in the eerie glow of the computer terminal on the cover of *Time*." From Henry Jenkins, "Introduction," *The Children's Culture Reader* (New York: New York University Press, 1998), 9.

2. Phillip Elmer-Dewitt, Wendy Cole, and Joshua Quittner, "On a Screen Near You: Cyberporn," *Time*, July 3, 1995, 38–46.

3. Henry Jenkins, "Empowering Children in the Digital Age: Towards a Radical Media Pedagogy," *Radical Teacher*, April 30, 1997, 30.

4. See Eric Nuzum, *Parental Advisory: Music Censorship in America* (New York: Harper Collins, 2001), 13–43; Danny Goldberg, *Dispatches from the Culture Wars: How the Left Lost Teen Spirit* (New York: Miramax, 2003); Bruce Haring, "Lyrics Concerns Escalate," *Billboard*, November 11, 1989, 1–2.; John Burgess, "Video Game Industry Plans Ratings System; Move in Response to Congressional Pressure," *Washington Post*, December 8, 1993, F1.

5. Quoted from remarks prepared for delivery by Vice President Al Gore, Royce Hall, UCLA, Los Angeles, California, January 11, 1994.

6. Elmer-Dewitt, Cole, and Quittner, "On a Screen Near You," 40.

7. Steven Levy, "No Place for Kids? A Parents Guide to Sex on the Net," *Newsweek*, July 3, 1995, 46.

8. Elmer-Dewitt, Cole, and Quittner, "On a Screen Near You," 41.

9. Pew Internet & American Life Project, "Featured Reports," http://207.21.232.103/index.asp.

10. The National Information Infrastructure: Agenda for Action, Executive Summary, http://www.ibiblio.org/nii/NII-Executive-Summary.html.

11. See Patricia Aufderheide, *Communications Policy in the Public Interest* (New York: Guilford Press, 1999). See also Jeff Chester, *Digital Destiny: New Media and the Future of Democracy* (New York: The New Press, 2007).

12. Marjorie Heins, *Not in Front of the Children: "Indecency," Censorship, and the Innocence of Youth* (New York: Hill and Wang, 2001), 157–179; Jonathan Wallace and Mark Mangan, *Sex, Laws, and Cyberspace* (New York: M&T Books, 1996).

13. For a discussion of the Pacifica case and its impact, see Heins, *Not in Front of the Children*, 89–136. See also William Triplett, "Broadcast Indecency," *CQ Researcher* 14, no. 14 (2004): 323–330.

14. People for the American Way, "Right Wing Organizations: Christian Coalition of America," http://www.pfaw.org/pfaw/general/default.aspx?oid=4307. Christian Coalition of America, "About page," http://www.cc.org/about.cfm.

15. Mark J. Rozell and Clyde Wilcox, "Second Coming: The Strategies of the New Christian Right," *Political Science Quarterly* 11 (Summer 1996): 271–295.

16. Mike Godwin, *Cyber Rights: Defending Free Speech in the Digital Age* (New York: Times Books, 1998), 215.

17. See Howard Rheingold, *Smart Mobs: The Next Social Revolution* (Cambridge, MA: Perseus Publishing, 2002).

18. Electronic Frontier Foundation, "Home page," http://www.eff.org/about/.

19. Electronic Privacy Information Center, "About EPIC," http://www.epic.org/epic/about.html.

20. "Democratic Values for a Digital Age: 1995 Activities of the Center for Democracy and Technology," February 1995, http://www.cdt.org/mission/overview1995.html.

21. Comment from Marc Rotenberg of the Electronic Privacy Information Center, in Elmer-Dewitt, Cole, and Quittner, "On a Screen Near You," 41.

22. Wallace and Mangan, *Sex, Laws, and Cyberspace*, 175.

23. Cecile S. Holmes, "Christian Broadcasters Voice Their Vision; They Find More Americans Are Receptive to Their Goals," *Houston Chronicle*, February 10, 1996, Religion, 1. Note: Gingrich did not support the Communications Decency Act and was a defender of the Internet.

24. Jim Exon, "Editorial," *New York Daily News*, April 7, 1995, 31.

25. Jenkins, *The Children's Culture Reader*, 9.

26. Elmer-Dewitt, Cole, and Quittner, "On a Screen Near You," 41.

27. From Paul Andrews, "Will Censorship Muffle Internet," *Seattle Times*, December 17, 1995, A3.

28. Peter H. Lewis, "Protest, Cyberspace-Style, for New Law," *New York Times*, February 8, 1996, A16.

29. Heins, *Not in Front of the Children*, 158–159.

30. James Kim, "Straight to the 'Net: Services Give Direct Access," *USA Today*, November 1, 1995, 2B.

31. David Johnston, "Use of Computer Network for Child Sex Sets Off Raids," *New York Times*, September 14, 1995, A1.

32. "Remarks by the President in the Signing Ceremony for the Telecommunications Act Conference Report," Library of Congress, Washington, DC, February 8, 1996, http://clinton4.nara.gov/WH/EOP/OP/telecom/release.html.

33. Ibid. The Telecommunication Act of 1996 extended the provisions for "universal service" to the new technologies, by requiring subsidies for providing affordable access to the Internet for schools, libraries, and rural health-care facilities. Federal Communications Commission, "E-rate," http://www.fcc.gov/learnnet/; Federal Communications Commission, "The FCC's Universal Service Program for Schools and Libraries," October 7, 2005, http://www.fcc.gov/cgb/consumerfacts/usp_Schools.html; Federal Communications Commission, "The FCC's Universal Service Program for Rural Health Care Providers, October 11, 2005, http://www.fcc.gov/cgb/consumerfacts/usp_RuralHealthcare.html; Federal Communications

Commission, "The FCC's Universal Service Support Mechanisms," http://www.fcc.gov/cgb/consumerfacts/universalservice.html.

34. John Schwartz, "On-line Obscenity Bill Gains in Senate; Panel Backs Legislation; Critics See Threat to First Amendment," *Washington Post*, March 24, 1995, A1.

35. "So You Want to Buy a President?" *Frontline*, PBS. http://www.pbs.org/wgbh/pages/frontline/president/players/sculley.html.

36. This *USA Today* poll is cited in Holmes, "Christian Broadcasters," Religion, 1.

37. "Sex, Violence, and the Media," *The CQ Researcher* 5, no. 43 (1995): 1033.

38. "Washington Wire: Dirty Secret," *Wall Street Journal*, March 31, 1995, sec. A.

39. Ramon G. McLeod and Reynolds Holding, "Clinton OKs Telecom Overhaul; Rights Groups File Suit—Censorship Concerns," *San Francisco Chronicle*, February 9, 1996, A1.

40. "Remarks by the President," February 8, 1996, http://clinton4.nara.gov/WH/EOP/OP/telecom/release.html.

41. Alexandra Marks, "Washington Turns up the Debate on TV Violence," *Christian Science Monitor*, July 14, 1995.

42. President William Jefferson Clinton, State of the Union Address, U.S. Capitol, January 23, 1996, http://clinton2.nara.gov/WH/New/other/sotu.html.

43. Meeting with executives: "Clinton Calls for V-Chip and Telecom Bill Passage," *Communications Daily* January 25, 1996.

44. "Sex, Violence, and the Media," 1017–1040.

45. One of the pollsters was Dick Morris, who later became involved in an extramarital scandal that was about as far removed from family values as possible. Denise Lavoie, "Morris Returns Home after Resigning in Sex Scandal," Associated Press, August 30 1996; Dan Balz, "White House Has Had Its Fill of Dick Morris; Failure to Keep Low Profile after Scandal Does Not Sit Well with President's Staff, "*Washington Post*, September 8, 1996, A12.

46. As Tony Podesta recalls: "The White House policy was to find 20 issues that most of the country would agree on and make a state of the union speech on it." Author's interview with Tony Podesta, April 19, 2004.

47. See Kathryn C. Montgomery, *Target: Prime Time. Advocacy Groups and the Struggle over Entertainment Television* (New York: Oxford University Press, 1989); Geoffrey Cowan, *See No Evil: The Backstage Battle over Sex and Violence on Television* (New York: Simon and Schuster, 1979). See also James T. Hamilton, *Channeling Violence: The Economic Market for Violent Television Programming* (Princeton, NJ, Princeton University Press, 1998).

48. This is one of the arguments made by Willard Rowland in his book *The Politics of TV Violence: Policy Uses of Communications Research* (Beverly Hills, CA: Sage Publications, 1983).

49. Mark M. MacCarthy, "Broadcast Self-Regulation: The NAB Codes, Family Viewing Hour, and Television Violence," *Cordozo Arts & Entertainment Law Journal* 13, no. 5: 667–696.

50. See Montgomery, *Target: Prime Time*, 51–74.

51. See Dale Kunkle and Brian Wilcox, "Children and Media Policy," in Dorothy Singer and Jerome Singer, eds., *Handbook of Children and the Media* (Thousand Oaks, CA: Sage Publications, 2001).

52. "1990 Year in Review," *Electronic Media*, January 7, 1991, 44.

53. Doug Halonen and Thomas Tyrer, "Violence Gets Cool Reception," *Electronic Media*, July 5, 1993, 1.

54. "Congressional Skeptics Challenge Network Violence Agreement," *Communications Daily*, July 2, 1993.

55. "Simon Suggests Industry Group to Monitor Violence," *Communications Daily*, August 3, 1993, 1.

56. "Differing TV Violence Bills Introduced in House and Senate," *Communications Daily*, August 6, 1993, 1.

57. "V-Chip Technology Invented by Professor Tim Collings," www.tri-vision.ca/documents/Collings%20As%20Inventor.pdf.

58. Mike Boone, "Computer Chip Promises to Censor Children's Viewing," *The Gazette* (Montreal, Quebec), May 7, 1994, F6.

59. Ibid.

60. Mike Boone, "Winning the Battle," F6. In February 1993, advocacy groups that had been working on the TV-violence issue in the United States were invited to a special meeting at the Canadian Embassy, where Keith Spicer encouraged them to follow the "Canadian example" for dealing with TV violence. I was one of the participants in that meeting.

61. "Differing TV Violence Bills," 1.

62. Doug Halonen, "Lawmakers Aim to Limit TV Violence," *Electronic Media*, August 9, 1993, 3.

63. See James H. Snider, *Speak Softly and Carry a Big Stick: How Local TV Broadcasters Exert Political Power* (iUniverse, Inc., 2005).

64. Barry Cole and Mal Oettinger, *Reluctant Regulators* (Reading, MA: Addison Wesley, 1978), 35–49.

65. Jenny Hontz and Jon Lafayette, "NAB Upbeat about Digital Future," *Electronic Media*, March 28, 1994, NAB '94, 1. For background on the policy debates over the broadcast spectrum issue, see Joel Brinkley, *Defining Vision: How Broadcasters Lured the Government into Inciting a Revolution in Television* (Orlando: Harcourt Brace & Co., 1997); see also Snider, *Speak Softly.*

66. *Communications Daily*, February 3, 1994, 7.

67. "Programmers Must Take Responsibility for TV Violence, Surgeon General Says," *Communications Daily*, September 16, 1993.

68. "Reno Endorses Bills to Deal with TV Violence," *Communications Daily*, October 21, 1993.

69. "Hundt Seen Winning Quick Senate Approval as FCC Chairman," *Communications Daily*, September 23, 1993.

70. Ellen Edwards, "GOP Win Stirs Fears of TV Curbs; Some See Restrictions on Language, Sex," *Washington Post*, November 17, 1994, D1.

71. As Jim Snider explains in his report, Senator Bob Dole publicly attacked the provision in December 1995, labeling it as a "giveaway" and "corporate welfare," and calling for a public auction. Broadcasters responded with a massive advertising and lobbying campaign in both the print and the broadcast media, with virtually no reporting of other positions. See J. H. Snider, *Speak Softly and Carry a Big Stick: How Local TV Broadcasters Exert Political Power*. Lincoln, NE, iUniverse, 2005.

72. "The V-chip Diversion," Editorial, *Electronic Media*, June 19, 1995, 8.

73. Federal Communications Commission, "Excerpts from V-Chip Legislation," http://www.fcc.gov/vchip/legislation.html; The Library of Congress, "Telecommunications Act of 1996," http://thomas.loc.gov/cgi-bin/query/D?c104:4:./temp/~c10442gHfR::.

74. Monroe E. Price, "The V-chip and the Jurisprudence of Ratings," in Monroe E. Price, Ed., *The V-Chip Debate: Content Filtering From Television to the Internet*. Lawrence Erlbaum Associates, Mahwah, NJ, 1998), xx.

75. Public Notice FCC 97–34, "Commission Seeks Comment on Industry Proposal for Rating Video Programming," February 7, 1997.

76. Monroe Price, ed., *The V-Chip Debate: Content Filtering from Television to the Internet* (Mahwah, NJ: Lawrence Erlbaum Associates, 1998), 260.

77. See Jeff Chester, *Digital Destiny.* See also Patricia Aufderheide, *Communications Policy.*

78. Frank Rich, "The Idiot Chip," *New York Times*, February 10, 1996, sec. 1, 23.

79. Tom Shales, "Chip of Fools: Any Way You Program It, the V-chip Is a Long Stride Toward Censorship," *Washington Post*, March 10, 1996, G1.

80. John M. Broder and Jane Hall, "TV Networks to Rate Sexy, Violent Programs," *Chicago Sun-Times*, March 1, 1996, 18.

81. See Eric Nuzum, *Parental Advisory: Music Censorship in America* (New York: Harper Collins, 2001).

82. John Burgess, "Video Game Industry," F1.

83. "Mortal Kombat at Hearing: TV Violence Dispute Spills over into Videogames," *Communications Daily* December 10, 1993, 4; "The Rating Game," *U.S. News and World Report*, November 21, 1994, 91. See also "Media Ratings: A Comparative Review," in Price, *The V-chip Debate.*

84. Daniel B. Wood, "Ratings Content of TV Programs; View from Hollywood Studio 41," *Christian Science Monitor*, December 19, 1996, 1.

85. Paul Farhi, "The V-chip Blip; If History Is Any Guide, TV Violence May Actually Increase," *Washington Post*, February 18, 1996, C2. Not all parts of the creative community were opposed to the ratings. The Caucus for Producers, Writers, and Directors, which had played a very prominent role in entertainment-industry political and regulatory battles for decades, not only supported the ratings but subsequently weighed in during the FCC review process in support of the content based system as opposed to the age-based system.

86. Nina J. Easton, "The Final Battle," *Los Angeles Times Magazine*, October 20, 1996, 10. Cynthia Littleton, "Valenti Calls V-chip 'Quick Fix,'" *Broadcasting & Cable*, July 8, 1996, 25.

87. Farhi, "V-chip Blip," C2.

88. In a comparative survey of the MPAA age-based movie ratings and the HBO content ratings, conducted by the National Television Violence Study—an academic research initiative funded by the cable-television industry—parents favored the content system. However, warnings such as "parental discretion advised," had a "forbidden fruit effect" on minors making the shows even more appealing, especially among boys 10–14. See Joanne Cantor and Kristen Harrison (1996). "Ratings and Advisories for Television Programming," in *National Television Violence Study* (Thousand Oaks, CA: Sage Publications, 1996), vol. 1: 361–410.

89. Littleton, "Valenti," 25.

90. This private conversation took place between the FCC official and me.

91. Kathryn C. Montgomery, "Advocating Children's Television," in J. Alison Bryant and Jennings Bryant, eds., *The Children's Television Community: Institutional, Critical, Social Systems, and Network Analyses* (Mahwah, NJ: Lawrence Erlbaum Associates, 2006). Like Action for Children's Television, CME was reluctant to endorse any policy proposals that might restrict speech. At the same time, however, the group shared the concerns of advocacy colleagues who were alarmed by the rise in popular

children's programs such as *The Mighty Morphin' Power Rangers*, in which violence had become a dominant theme. See Nancy Carlsson-Page and Diane Levin, *Who's Calling the Shots?: How to Respond Effectively to Children's Fascination with War Play and War Toys* (Philadelphia: New Society Publishers, 1990).

92. Paul Frahi, "TV Ratings to Have 6 Vague Levels; Shows' Sex, Violence Won't Be Specified," *Washington Post*, December 10, 1996, A1.

93. From author's notes during meeting, National Association of Broadcasters, December 10, 1996.

94. Ibid.

95. Statement of Rep. Edward J. Markey concerning public support for TV ratings system, December 18, 1996.

96. Editorial, "Kids and Television," *Washington Post*, December 26, 1996, A26; Editorial, "TV Ratings Ignore Core Problem," *Rocky Mountain News* (Denver, Colo.), December 15, 1996, 74A; Rosslein, "TV Industry Needs to Do Better Than This," *Milwaukee Journal-Sentinel* (Wisconsin), December 16, 1996, 10; Editorial, "Ratings Should Tell More," *St. Petersburg Times* (Florida), December 17, 1996, sec. A; Mona Charon, "Insufficient TV Ratings," St. Louis Post-Dispatch (Missouri), December 16, 1996, sec. B; Editorial, "Ratings More Hype Than Help," *Boston Herald*, December 8, 1996, 32.

97. "TV Program Ratings Assailed at Senate Hearing," *Communications Daily*, February 28, 1997.

98. "Coalition Urges FCC to Reject TV Industry's Rating Proposal," *Communications Daily*, April 9, 1997.

99. Paul Farhi, "Chorus of Boos Greets TV Rating System," *Washington Post*, April 25, 1997, G2.

100. "'Safe Harbor' Bill Clears Senate Commerce Panel," *Communications Daily*, May 2, 1997.

101. I was a participant in that meeting.

102. The industry leaders were joined by Anthony Podesta, a consultant who represented the broadcasters, several studios, and networks, and also was an advisor to the White House, where his brother, John, was Clinton's chief of staff.

103. As the process came close to resolution, Congress and the White House engaged in an intense political tug-of-war over who would take public credit for the final victory. At one point during the last few weeks of negotiations, several senators summoned the press to Capitol Hill to announce that the process finally was over. In fact, negotiations were still underway; participants learned of the premature press conference as they were haggling over major issues at MPAA

headquarters. When administration representatives insisted on announcing the agreement at the White House, the advocacy groups successfully argued that congressional lawmakers from both parties be invited.

104. Montgomery, "Advocating Children's Television."

105. "Canadians Propose 7–Point TV Ratings System," *Communications Daily*, May 6, 1997. "Canadian entertainment industry recommended Monday that government regulators adopt 7–point rating system for TV programming that bears very close resemblance to the controversial 6–point system proposed by U.S. entertainment industry."

106. Kyle Pope, "Canadians Pull TV Rating Plan Hailed as Model," *Wall Street Journal*, December 18, 1996, C1.

107. John Aloysius Farrell, 'Decency' Law for the Internet Is Struck Down," *Boston Globe*, June 27, 1997, A1. Heins, *Not in Front of the Children*, 157–179.

108. Electronic Privacy Information Center, EPIC press release, July 16, 1997. From the author's personal files.

109. John M. Broder, "The Supreme Court: The Policy: Clinton Readies New Approach on Smut," *New York Times*, June 27, 1997, A21.

110. *Reno v. American Civil Liberties Union*, 521 U.S. 844 (1997).

111. Rebecca Vesely, "White House Unveils E-Commerce Policy," *Wired News*, July 1, 1997. http://www.wired.com/news/print/0,1294,4885,00.html. There were some inconsistencies in the government's free-market approach. Both industry and privacy advocates had been alarmed by the U.S. policy on encryption, a contradiction that had not gone unnoticed by the press. "The Clinton Administration's track record on cyberspace is suspect," commented the *Orange County Register* in an editorial. "The president signed the Communications Decency Act just invalidated by the Supreme Court. The administration pushed the 'clipper chip' to give government agencies instant access to private communications, and has shown a preference for government action to provide Internet services to schools and libraries." "Editorial: Clinton Sees Strong Role for the Government in Cyberspace," *Orange County Register*, July 8, 1997.

112. Vesely, "White House," http://www.wired.com/news/print/0,1294,4885,00.html.

113. Rebecca Vesely, "The Clinton-Gore Porn-Filtering 'Toolbox,'" *Wired News*, July 16, 1997, http://www.wired.com/news/politics/0,1283,5241,00.html.

114. Steven Levy, "No Place for Kids," *Newsweek*, July 3, 1995, 47.

115. *Business Wire*, "Telecommunications Bill endorses ban on distribution of obscenities over the Internet; Congressman Markey cites Cyber Patrol as the V-chip

for the Internet," February 2, 1996. See Heins, *Not in Front of the Children*, 180–200.

116. For example, the Gay and Lesbian Alliance Against Defamation (GLAAD) released the report "Access Denied: An Impact of Internet Filtering Software on the Gay and Lesbian Community" in December 1997; it documented how "the majority of software products on the market, as well as new products in development, place information sites serving the lesbian, gay, bisexual and transgender community in the same categories as sexually explicit sites." http://www.glaad.org/publications/archive_detail.php?id=941&.

117. K. K. Campbell, "Who's Watching the 'Watchers'?" *The Toronto Star*, January 30, 1997, J3.

118. Bennett Haselton, personal interview, February 25, 2003.

119. As chapter 7 shows, Hazelton was only one of the first of a new generation of cyberactivists.

120. Rebecca Vesely, "Cybersitter Goes After Teen," *Wired.com*, December 9, 1996, http://www.wired.com/news/politics/0,1283,901,00.html. Haselton interview, February 25. 2003. Campbell, "Watchers," J3.

121. Daniel J. Weitzner, "Yelling 'Filter' on the Crowded Net: The Implications of User Control Technologies," in Price, *The V-Chip Debate*, 207. As Weitzner explains, many of the same groups that embraced new filtering technologies in their challenges to CDA later were to become critics of such devices.

122. Heins, *Not in Front of the Children*, 164.

123. Electronic Privacy Information Center, "Faulty Filters: How Content Filters Block Access to Kid-Friendly Information on the Internet," December 1997, http://www.epic.org/reports/filter_report.html.

124. This quote from the Web site Family.netshepherd.com was included in Electronic Privacy Information Center, "Faulty Filters: How Content Filters Block Access to Kid-Friendly Information on the Internet," December 1997, http://www.epic.org/reports/filter_report.html. The report comments that "Family Search is the first product to incorporate two of the goals identified at the July White House meeting—content rating and filtered search engines."

125. To make this inherently subjective process more systematic, RSAC offered its own "objective" online tool, described as "a fully automated paperless system that relies on a quick, easy-to-use questionnaire that the Web master completes at RSAC's home page." The questionnaire contained a number of questions about the "level, nature and intensity" of the "sex, nudity, violence or offensive language" found on the site. After the online questionnaire was completed, the RSACi Web Server would get to work on it, tabulating the results and then spewing out the appropriate "html

advisory tags that the Web master then places on the Web site/page." Press Release, W3C, "Recreational Software Advisory Council Launches Objective, Content-Labeling Advisory System for the Internet," February 28, 1996, http://www.w3.org/PICS/960228/RSACi.html. See also Jonathan Weinberg, "Rating the Net," in Price, *The V-Chip Debate*, 221–242.

126. In 1999, RSAC was folded into a new organization, the Internet Content Rating Association (ICRA), "an international, non-profit organization of internet leaders working to develop a safer internet." "About ICRA," http://www.icra.org/about/.

127. See Heins, *Not in Front of the Children*, 180–200; see also Daniel J. Weitzner, "Yelling 'Filter,'" in Price, *The V-chip Debate*, 207–220.

128. From Statement of Barry Steinhardt, ACLU Associate Director, December 2, 1997.

129. See *Internet Family Empowerment White Paper: How Filtering Tools Enable Responsible Parents to Protect Their Children Online.* Center for Democracy and Technology, July 16, 1997. See also Jerry Berman and Daniel Weitzner, "Abundance and User Control: Renewing the Democratic Heart of the First Amendment in the Age of Interactive Media," *Yale Law Journal* (104), 1995, 1619, 1634–1635. The ALA ultimately decided not to support filtering software policies, and it became a strong opponent of such subsequent legislation to mandate the systems in schools and libraries.

130. Dan Brekke, "Lobbying on Eve of Clinton Net Summit," *Wired News*, December 1, 1997. Sen. Coats introduced this legislation (S1482) November 8, 1997. http://thomas.loc.gov/cgi-bin/query/D?c105:1:./temp/~c105wntOzU. Senator Patty Murray (D-WA), who described herself as "one of the few senators who actually understands the Internet," tried to develop legislation that would address parental concerns about Internet safety without running afoul of the First Amendment. The CDA was "not the right approach," she told the press, "but now that it's gone, I can tell you there is a huge vacuum out there. Everybody wants to use the Internet, but first they need to know their kids are not going to learn how to make bombs or how to perform some ritual from devil worship." Murray introduced a bill in July 1997 designed to exempt those who voluntarily rated their Web sites from legal liability if children found objectionable material there. The ALA and other civil-liberty groups immediately attacked the proposal, saying that it would amount to privatized censorship. Even filtering-software company NetNanny objected to the bill, arguing that it would only "pollute" cyberspace. Danny Westneat, "Sen. Murray's Internet Controls Draw Fire—Senator's Plan to Rate, Sift Content Put on Hold; Critics Say It Amounts to Privatized Censorship," *Seattle Times*, July 15, 1997, A12.

131. The official mission statement of the first summit was revised to include a laundry list of issues: "We recognize that examining children's relationships with

the Internet is complex," the statement reads, "involving issues of equitable access, marketing and advertising practices, quality content, privacy, and safety from harmful content, and illegal activity, among others." "Internet Online Summit: Focus on Children, Mission Statement," http://www.kidsonline.org/mission. I was one of the speakers at the summit.

132. Dan Brekke, "Lobbying on Eve of Clinton Net Summit," *Wired News*, December 1, 1997.

133. Electronic Privacy Information Center, "Faulty Filters: How Content Filters Block Access to Kid-Friendly Information on the Internet," December 1997, http://www.epic.org/reports/filter_report.html.

134. From Statement of Barry Steinhardt, ACLU Associate Director, December 2, 1997.

135. Brekke, "Lobbying."

136. As Amy Harmon reported in the *New York Times*, under the proposed Internet ratings system, "parents can also choose to block unrated sites entirely, a feature that critics say renders most of the Web invisible, since only 50,000 of the million-plus sites on the Web have rated themselves so far." Though about 5,000 new sites were rating themselves every month, enthusiasm for the RSACi self-rating device already had begun to wane, with most of the major news-related sites refusing to rate themselves. "And although three of the major Internet search companies agreed to encourage new-listings applicants to rate themselves, none currently do so on a regular basis," she observed. Christine Varney, who recently had been a Clinton-appointed Federal Trade Commissioner, and now was chairing the summit, admitted to the *New York Times* that the administration had been forced to revisit its original goals. "What's happened is people realized this is a far more complex issue than anyone ever imagined," she explained. Amy Harmon, "Ideological Foes Meet on Web Decency, *New York Times*, December 1, 1997, D1.

137. In addition to these initiatives, the Interactive Services Association, the Commercial Internet eXchange Association, and other online trade groups came forward with a "Zero Tolerance Policy," agreeing to "aggressively cooperate with law enforcement authorities in investigating child pornography incidences on the network." The Department of Justice and private industry provided a "computer training initiative" to train "local law enforcement officials on catching and prosecuting cyber predators and child pornographers." The ALA unveiled its new "cybercollection" of links to more than seven hundred "great sites" for kids. And researchers from AT&T Labs and the University of Michigan issued "A Catalog of Tools That Support Parents' Ability to Choose Online Content Appropriate for their Children," http://www.kidsonline.org/.

138. "Participants in Internet/Online Summit Announce Individual Initiatives to Help Children and Families," Kids Online Press Release, December 1, 1997, http://www.kidsonline.org/news/advisory_971201.html.

139. "Internet/Online Summit Highlights Cooperation and Action to Enhance the Safety and Benefits of Cyberspace for Children and Families," Kids Online Press Release, December 2, 1997, http://www.kidsonline.org/news/advisory_971202a.html.

140. Magid, "The Starr Report," http://www.safekids.com/starr.htm.

141. "Youth, Pornography, and the Internet," Computer Science and Telecommunications Board, The National Academies of Science, 2002, http://books.nap.edu/catalog/10261.html.

142. GetNetWise, "Tools for Families," http://kids.getnetwise.org/tools/.

143. J. Cory Allen and Kimberly Duyck Woolf, "The 5th Annual APPC Conference on Children and Media: A Summary" (Philadelphia: Annenberg Public Policy Center of the University of Pennsylvania, 2001); D. Davis, "The Second Annual Annenberg Public Policy Center's Conference on Children and Television: A Summary" (Philadelphia: Annenberg Public Policy Center of the University of Pennsylvania, 1997); "Major New Study of the V-chip Rating System," Kaiser Family Foundation, 1998, http://www.kff.org/content/archive/1434rating.html; "Few Parents Use V-chip to Block TV Sex and Violence, but More Use TV Rating to Pick What Kids Can Watch," 2001, http://kff.org/content/2001/3158/V-Chip%20releasehtm.

144. I served on the MPAA's TV Parental Guidelines Monitoring Board from 1997 to 2004.

Chapter 4

1. Joe Kilsheimer, "Web Has Sites Just for Kids," *Times-Picayune* (New Orleans), August 12, 1995, C8.

2. Federal Trade Commission, *Complaint and Request for Investigation in the Matter of Deceptive Internet Sites Directed toward the Young Child Audience*, May 13, 1996. From the author's personal files, 1. http://www.ftc.gov/os/1997/07/cenmed.htm#1) CME.

3. Sam Vincent Meddis, "Net: New and Notable," *USA Today*, January 4, 1996, 3D.

4. Federal Trade Commission, *Complaint*, 5.

5. Lynn Van Dine, "Computers: Cyber Playground a Winner with Kids Worldwide," *Detroit News*, January 31, 1996.

6. Federal Trade Commission, *Complaint*, 5.

7. SpectraCom, "Company Description," 1996, http://www.spectracom/description.html (accessed January 26, 1996).

8. SpectraCom, "Research Services," 1996, http://www.spectracom/research.html (accessed January 26, 1996). As cited in Federal Trade Commission, *Complaint*, 3.

9. Federal Trade Commission, *Complaint*, 4.

10. Ibid.

11. As president of the Center for Media Education, I was directly involved in many of the events described in this chapter.

12. Patti M. Valkenburg, "Media and Youth Consumerism," *Journal of Adolescent Health* 27S (2000): 52–56.

13. Inger Stole, "Consumer Protection in Historical Perspective: The Five-Year Battle over Federal Regulations of Advertising 1933–1938," *Mass Communication & Society* 3 (2000): 351–372. Established in 1910, the Association of National Advertisers (ANA) represented hundreds of major U.S. brands, dedicating itself exclusively to marketing and brand building for these companies. The ANA had played a prominent role over the years in the many controversies surrounding television advertising. Responding to the pressures about sex and violence in entertainment programming, ANA helped create the Family Friendly Programming Forum, "a group of over 40 major national advertisers . . . who are taking positive steps to increase family friendly programming choices on television." http://www.ana.net/about/about.htm. The American Association of Advertising Agencies (AAAA)—known in the business as the "4 As"—had been around since 1917, representing the big ad agencies that accounted for 75 percent of total advertising volume in the United States. One of its primary goals was "to resist unwise or unfair legislation and regulation." http://www.ana.net/family/default.htm. Representing the grassroots arm of the ad lobby, the American Advertising Federation could mobilize its hundreds of local ad clubs and college chapters against regulatory threats at any level of government. http://www.aaf.org/about/index.html. For a history of earlier consumer group efforts, and the counter strategies of the advertising industry, see Inger L. Stole, *Advertising on Trial: Consumer Activism and Corporate Public Relations in the 1930s* (Champaign: University of Illinois Press, 2006).

14. Established at the early part of the twentieth century, the Federal Trade Commission had jurisdiction over trade matters, ensuring fair competition. Over the years, it took on more and more responsibility as the government watchdog of the advertising industry. In the 1970s, amid a burgeoning consumer movement, the FTC became more aggressive in regulating advertising under the new leadership of Democrat Michael Pertschuk, an appointee of President Jimmy Carter. "Mike Pertschuk and the Federal Trade Commission," Case Study, Kennedy School of Government, Harvard University, 1981. C16-81-387.0.

15. Dale Kunkel, "The Role of Research in the Regulation of Children's Television Advertising," *Knowledge: Creation, Diffusion, Utilization* 12, no. 1 (1990): 101–119.

16. Kunkel, Dale, "The Implementation Gap: Policy Battles about Defining Children's Educational Programming," *Annals of the American Academy of Political and Social Science*, 557 (1998) 39–53; Dale Kunkel, "Crafting Media Policy: The Genesis and Implications of the Children's Television Act of 1990," *American Behavioral Scientist* 35 (1991): 181–202. Barry Cole and Mal Oettinger, *Reluctant Regulators* (Reading, MA: Addison Wesley, 1978), 243–288. See also Michael Pertschuk, *Revolt against Regulation: The Rise and Pause of the Consumer Movement* (Berkeley: University of California Press, 1982).

17. "Mike Pertschuk and the Federal Trade Commission," Case Study.

18. Dale Kunkel, "Children and Television Advertising" and Dale Kunkel and Brian Wilcox, "Children and Media Policy," in *Handbook of Children and the Media* (Thousand Oaks, California: Sage Publications, 2001), 375–394, 385–604.

19. Dale Kunkel and Brian Wilcox, "Children and Media Policy," 385–604. Dale Kunkel, "The Role of Research," 101–119.

20. The law stipulates that commercial time in children's programming be limited to twelve minutes per hour on weekdays and ten and a half minutes per hour on weekends. Dale Kunkel, "Children and Television Advertising," 386. Dale Kunkel, "Crafting Media Policy: The Genesis and Implications of the Children's Television Act of 1990," *American Behavioral Scientist* 35 (1991): 181–202.

21. Ibid. Angela Campbell, "Self Regulation and the Media," *Federal Communications Law Journal* 51 (1999): 711–771. As critic Susan Linn later pointed out, while CARU's principles "look great on paper," many of them run "exactly counter to what advertisers themselves say that they are doing." Susan Linn, *Consuming Kids: The Hostile Takeover of Childhood* (New York: New Press, 2004), 197.

22. Stuart Elliott, "The Media Business: Advertising; Watch What the Internet Asks Children, Sponsors Are Warned, or See the Government Step In," *New York Times*, September 14, 1998, 11.

23. Kunkel, "Crafting Media Policy."

24. Edmund Andrews, "Broadcasters, to Satisfy Law, Define Cartoons As Education," *New York Times*, September 30, 1992, A1. See also Kathryn C. Montgomery, "Advocating Children's Television," in J. Alison Bryant and Jennings Bryant, eds. *The Children's Television Community: Institutional, Critical, Social Systems, and Network Analyses* (Mahwah, NJ: Lawrence Erlbaum Associates, 2006).

25. Kathryn C. Montgomery and Shelley Pasnik, "Web of Deception: Threats to Children from Online Marketing," *Center for Media Education* (1996).

26. Stuart Elliott, "Talks End on Show Using Ad Character," *New York Times*, March 16, 1992, D9.

27. Montgomery and Pasnik, "Web of Deception," 8.

28. Ibid.

29. Federal Trade Commission, *Complaint*.

30. Montgomery and Pasnik, "Web of Deception," 8–9.

31. Don Peppers and Martha Rogers, *The One to One Future: Building Relationships One Customer at a Time* (New York: Doubleday, 1993). See Kathryn Montgomery, "Digital Kids: The New Online Children's Consumer Culture," in Dorothy Singer and Jerome Singer, eds., *Handbook of Children and the Media* (Thousand Oaks, CA: Sage Publications, 2001), 635–650.

32. Don Peppers, "Foreword," in Seth Godin, *Permission Marketing: Turning Strangers into Friends, and Friends into Customers* (New York: Simon and Schuster, 1999), 12.

33. M. Perkins, "Mining the Internet" in *The Red Herring* (March 1996)

34. See Kathryn Montgomery, "Relationship Marketing," *Encyclopedia of Children, Adolescents, and the Media* (Thousand Oaks, CA: Sage Publications, 2006).

35. Author's personal account of the meeting.

36. Lawrie Mifflin, "Advertisers Chase Young People in Cyberspace," *New York Times*, March 29, 1996, A16; Leslie Miller, "Kids Snared in Web of Ads, Group Says," *USA Today*, March 29, 1996, 1D; Jeannine Aversa, "Watchdog Group Sounds Web Warning," *Chicago Sun-Times*, March 28, 1996, 4. Denise Gellene, "KIDS: Group Calls Internet Marketing a Web of Deceit," *Los Angeles Times*, March 29, 1996, A1.

37. Miller, "Kids," 1D.

38. Alan Bunce, "TV Activist: Curb Internet Ads for Kids," *Christian Science Monitor*, April 11, 1996, 12.

39. "Roadkill; Bits & Bytes; Of Trix and Charms and Kids on the Web; Luxury Cars Caught Sleeping at the Wheel," *Advertising Age*, April 8, 1996.

40. Desda Moss, "A Push to Get Kids off Mail Lists," *USA Today*, April 1, 1996, 6D.

41. Federal Trade Commission, *Complaint*; Montgomery and Pasnik, "Web of Deception," 7.

42. Gwen Moran, "You Are Your Customer's Keeper," 2002, http://www.gwenmoran.com/youkeep.html.

43. Miller, "Kids," 1D.

44. Speaking from the floor of the FTC hearing, June 1996, Clarke explained:

> It's important to involve and educate web site developers about what needs to be done to resolve privacy issues and security issues, because I think you will find that in this industry the developers that are in it often tend to be in their twenties and have not run into these issues in other areas. . . . It is important to make sure that we are included because a lot of the things that you will find that have been done have not been done maliciously, but have been done out of ignorance. And as we get educated, we will make sure that the work that we are doing includes privacy concerns.

Federal Trade Commission, "Public Workshop on Consumer Privacy on the Global Information Infrastructure" Washington, DC, June 4–5, 1996, http://www.ftc.gov/bcp/privacy/wkshp96/pw960604.pdf and http://www.ftc.gov/bcp/privacy/wkshp96/pw960605.pdf, 429.

45. Author's personal account.

46. The following are only a handful of the many articles in the press about privacy concerns online: "Stall Cyber-criminals Cruising Superhighway, "*Times Picayune* (New Orleans), July 26, 1995, E3; Clinton Wilder and Bob Violino, "Online Theft," *InformationWeek*, August 28, 1995, 30–35; Robert Rossney, "Privacy Is the Price We Pay for Convenience," *San Francisco Chronicle*, February 23, 1995, E4; Gary H. Anthes, "Security Risks Lurk on the Web," *Computer World*, February 27, 1995, 20. Richard Karpinsky, "Security Breaches Plague the Internet," *Interactive Age*, February 27, 1995, 4. Surveys also showed increasing public concern over privacy threats. For example, a survey by Equifax in 1995 found that

> nearly one out of every two people in the U.S. is "very" concerned about threats to their personal privacy today (47 percent) and another 35 percent are "somewhat" concerned. Although the proportion of people expressing this level of concern has remained fairly stable since 1983, 1995 is the second consecutive year that this proportion has topped 80 percent. Notably, the percentage of people who say they are very concerned declined from 51 percent in 1994 to 47 percent in 1995. Substantial proportions of the American public are concerned about how businesses are handling their personal information with many consumers saying they have refused to provide information. The vast majority of Americans (80%) agree that "consumers have lost all control over how personal information about them is circulated and used by companies." The number of people who feel this way has been climbing steadily since 1990 when this question was first asked.

1995 Equifax/Harris Consumer Privacy Survey, "Executive Summary," http://www.mindspring.com/~mdeeb/equifax/cc/parchive/svry95/docs/summary.html.

47. Simon G. Davies, "Re-engineering the Right to Privacy: How Privacy Has Been Transformed from a Right to a Commodity," in Philip E. Agre and Marc Rotenberg, eds., *Technology and Privacy: The New Landscape* (Cambridge, MA: MIT Press, 1997), 164.

48. Colin J. Bennett, "Convergence Revisited: Toward a Global Policy for the Protection of Personal Data?" in *Technology and Privacy*, 113. According to Xiaomei Cai and Walter Gantz, laws were enacted to regulate the government and corporations from collecting personal information from citizens (e.g., The Privacy Act of 1974;

Computer Matching and Privacy Protection Act of 1988; Electronic Communication Privacy Act of 1986). "However, these laws regulate the potential use and disclosure of the personal information already gathered. Prior disclosure of information collection practices, as well as obtaining prior consent from citizens, fall outside the scope of these laws." "Online Privacy Issues Associated with Web Sites for Children," *Journal of Broadcasting and Electronic Media* 44, no. 2 (Spring 2000): 197–215.

49. David Masci, "Internet Privacy," *CQ Researcher* 8, no. 41 (November 1998): 953.

50. Bennett, "Convergence," 111.

51. Cai and Gantz, "Online Privacy Issues," 346.

52. "Clinton Administration Seeks Preemptive Strike to Block Limits on the Internet," *Communications Daily*, September 13, 1996, 5.

53. Federal Trade Commission, "Public Workshop on Consumer Privacy on the Global Information Infrastructure" Washington, DC, June 4–5, 1996, http://www.ftc.gov/bcp/privacy/wkshp96/pw960604.pdf and http://www.ftc.gov/bcp/privacy/wkshp96/pw960605.pdf.

54. The legislation included three key principles:

> (1) Consumers should know what information is being collected about them, often through technologies that operate in background. (2) There should be "adequate and conspicuous" notice when information collected by telecom companies will be reused or sold. (3) Consumers should have right to prevent reuse or resale of information collected.

Markey labeled the principles: "Knowledge, Notice and No." "Markey Plans Bill for On-Line Privacy," *Communications Daily* June 7, 1996, 4.

55. Ibid.

56. Electronic Privacy Information Center, "EPIC Bill Track," http://www.epic.org/privacy/bill_track.html. There were seventy-five house bills in the 105th Congress (1997–1998), and fifty-three in the Senate.

57. Federal Trade Commission, "FTC: Consumer Privacy Comments Concerning the Coalition for Advertising Supported Information and Entertainment–P954807," http://www.ftc.gov/bcp/privacy/wkshp97/comments3/casie.038.htm.

58. Federal Trade Commission, "Public Workshop on Consumer Privacy on the Global Information Infrastructure" Washington, DC, June 4–5, 1996, http://www.ftc.gov/bcp/privacy/wkshp96/pw960604.pdf.

59. Author's recollection of private meeting at the FTC.

60. Direct Marketing Association, "What Is the Direct Marketing Association?" 2002, http://www.the-dma.org/aboutdma/whatisthedma.shtml.

61. "Direct Marketing Association Develops Privacy Protections for Online Media; Participates at FTC Online Privacy Workshop," *PR Newswire*, June 4, 1996.

62. Leslie Miller, "A Call to Protect Kids' Privacy On Line," *USA Today*, June 13, 1997, 1D.

63. Federal Trade Commission, "Public Workshop on Consumer Privacy," http://www.ftc.gov/bcp/privacy/wkshp96/pw960604.pdf.

64. As CFA's general council, Mary Ellen Fise, explained, information collected from children offline

> contemplates the use of an envelope and a stamp, but it also in almost every case required some type of small payment for whatever the free thing that the child is getting. And so that involves a parent. If not a formal consent, it's at least implied consent. You can't allow children to be saying, "Yeah, mom said it was okay."

The groups also called for online companies to post privacy policies that children and their parents could read and understand, instead of the fine-print legalese that often was tucked away in the obscure corners of some Web sites. Companies would be required to allow parents to control any information their children gave out online. Federal Trade Commission, "Public Workshop on Consumer Privacy," http://www.ftc.gov/bcp/privacy/wkshp96/pw960605.pdf, 329–330.

65. Federal Trade Commission, "Public Workshop on Consumer Privacy," http://www.ftc.gov/bcp/privacy/wkshp96/pw960605.pdf, 381.

66. "Consumers Union Experts to Provide New Research about Kids' Online Privacy at FTC Workshop," *PR Newswire*, June 12, 1997.

67. Bill Pietrucha, "Online Blocking Software Lacking, FTC Told," *Post-Newsweek Business Information Newsbytes*, June 13, 1997.

68. CME and CFA pointed out in their FTC filing that blocking children from giving out any information online also might keep them from valuable noncommercial activities on the Web. "For example, a father certainly may wish to prevent his son from releasing his name and street address to an online stranger or to an online salesperson, but he may want his son to give the same information to his baseball coach or scout leader." Federal Trade Commission, "Public Workshop on Consumer Privacy," http://www.ftc.gov/bcp/privacy/wkshp96/pw960605.pdf.

69. I visited the CARU offices in 1995.

70. Federal Trade Commission, "Public Workshop on Consumer Privacy," http://www.ftc.gov/bcp/privacy/wkshp96/pw960605.pdf.

71. Mifflin, "Guidelines," 5.

72. Dean Anason, "Capital Briefs: Self-Regulation Urged for Online Privacy," *American Banker*, February 11, 1997, 4.

73. In the United States, it is currently "not a good time to legislate," Varney told national and international regulators at the Sixth Conference on Computers, Freedom, and Privacy in April 1996, adding that the FTC is the U.S. federal agency that is best equipped to provide the public with "guidance" on the protection of privacy. "Computers and Privacy—World Govts Tackle Web," *Post-Newsweek Business Information Newsbytes*, April 1, 1996.

74. Bill Pietrucha, "Web Consortium Unveils Platform for Privacy Project," *Post-Newsweek Business Information Newsbytes*, June 11, 1997.

75. Ira Teinowitz, "Net Privacy Debate Spurs Self-Regulation," *Advertising Age*, June 9, 1997, 36.

76. Bill Pietrucha, "Direct Marketers Offer Proposals for Online Privacy," *Post-Newsweek Business Information Newsbytes*, June 10, 1997.

77. "The McGraw-Hill Companies Unveil Privacy Policy at FTC Conference on Electronic Consumer Privacy; Groundbreaking Policy Protects Cyberspace Privacy of Consumers," *PR Newswire*, June 11, 1997.

78. Patrick Flanagan, "Web Privacy Issues Getting Rapid Resolution," *Telecommunications Americas* 31, no. 8 (August 1997): 10–12.

79. Hiawatha Bray, "Firms Launch Plan to Police Own Web Sites: TRUSTe Group Looks to Avoid FTC Action by Addressing Privacy Concerns Voluntarily," *Boston Globe*, June 10, 1997, C6.

80. Tom Abate, "Nonprofit Wants Web Sites to Protect Your Privacy," *San Francisco Chronicle*, December 24, 1997, D3.

81. Rajiv Chandrasekaran, "Database Firms Set Privacy Plan; Personal Information Curbs Intended to Preempt Regulation," *Washington Post*, June 10, 1997, A1.

82. Electronic Privacy Information Center, "Surfer Beware: Electronic Privacy and the Internet," June 1997, http://www.epic.org/reports/surfer-beware.html.

83. Seth Scheisel, "America Online Backs off Plan to Give out Phone Numbers," *New York Times*, July 25, 1997, D1; Rajiv Chandrasekaran, "AOL Cancels Plan for Telemarketing; Disclosure of Members' Numbers Protested," *Washington Post*, G1.

84. Rajiv Chandrasekaran, "Protecting Children's Privacy Online; Administration Wants Firms to Get Parental Consent to Gather Data," *Washington Post*, June 14, 1997, D1. Center for Media Education, Consumer Federation of America, "An Update of Children's Web Site's Information Collection Practices," submitted to the Federal Trade commission, June 12, 1997.

85. Bill Pietrucha, "Regulations Needed for Children's Web Sites," *Post-Newsweek Business Information Newsbytes*, June 13, 1997.

86. Leslie Miller, "A Call to Protect Kids' Privacy On Line," *USA Today*, June 13, 1997, 1D.

87. The White House, "A Framework for Global Electronic Commerce," July 1 1997, http://www.technology.gov/digeconomy/framewrk.htm.

88. Ibid.

89. Steve Lohr, "U.S. and European Union Agree Not to Have Internet Tariffs," *New York Times*, December 15, 1997, 14.

90. "Internet Executive Summits Move to One Global Summit in Washington DC," *Business Wire*, January 9, 1998.

91. Heather Green, "A Little Privacy, Please," *Business Week*, March 16, 1998, 98.

92. Denise Caruso, "An On-line Tug-of-War over Consumers' Personal Information," *New York Times*, April 13, 1998, 5.

93. Ibid.

94. Ibid.; Rory J. O'Connor, "Internet Safety Big Worry for Parents," *Pittsburgh Post-Gazette*, June 14, 1997, A2. Joseph Turow, "Privacy Policies on Children's Websites: Do They Play by the Rules?" Annenberg Public Policy Center, March 2001, http://www.annenbergpublicpolicycenter.org/04_info_society/family/2001_privacyreport.pdf.

95. O'Connor, "Internet Safety," A2. Turow, "Privacy Policies," http://www.annenbergpublicpolicycenter.org/04_info_society/family/2001_privacyreport.pdf.

96. U.S. Federal Trade Commission, "Public Workshop on Consumer Information Privacy," http://www.ftc.gov/bcp/privacy/wkshp97/volume3.pdf.

97. As the *Washington Post* reported,

> The administration will ask Web site operators to obtain parental consent before taking children's personal information, such as their name and address, said presidential adviser Ira Magaziner, who heads an administration task force on electronic commerce. President Clinton is expected to present the task force's recommendations, which include the request for parental consent, on July 1.

Rajiv Chandrasekaran, "Protecting Children's Privacy Online; Administration Wants Firms to Get Parental Consent to Gather Data," *Washington Post*, June 14, 1997, D1.

98. Ira Teinowitz, "FTC Applies More Pressure to Sites Marketing to Kids: Government Demands Tougher Guidelines for Online Privacy," *Advertising Age*, June 23, 1997, 16.

99. Joseph Reagle, Jr., "W3C Activities Related to the US Framework for Global Electronic Commerce," July 6 1997, http://www.w3.org/TR/NOTE-framework-970706.html#privacy.

100. Ibid.

101. Mike Snider, "Dual Efforts Aim at Making Net Safe for Kids," *USA Today*, July 17, 1997, 1A.

102. Letter from Jodie Bernstein, Director, Bureau of Consumer Protection, Federal Trade Commission, to Kathryn C. Montgomery, President, and Jeffrey A. Chester, Executive Director, Center for Media Education, July 15, 1997. As with many FTC procedures, regulators had been in communication with the company throughout the confidential investigation process. By the time the final ruling was issued, KidsCom had curtailed many of its questionable practices. "KidsCom now sends an e-mail to parents when children register at the site, providing notice of its collection practices," the FTC letter explained. "Parents are provided with the option to object to release of information to third parties on an aggregate, anonymous basis. Most important, KidsCom does not release personally identifiable information (in the form of Key Pal information) to third parties without *prior* parental approval."

The letter also found that "there is no evidence that KidsCom at any time released any personally identifiable information to third parties for commercial marketing or any other purposes (other than for the Key Pal program). Such practices would have been of particular concern in light of the absence of adequate disclosure and prior parental consent."

103. Rajiv Chandrasekaran, "FTC Rules on Online Data Collection; Parental Notice Must Be Posted to Get Information from Children," *Washington Post*, July 17, 1997, C3.

104. Stuart Elliott, "A Clinton Adviser Argues the Economic Case for Self-Regulation of Sales Pitches in Cyberspace," *New York Times*, November 4, 1997, D13.

105. Stuart Elliott, "Self-Regulation in Cyberspace: The Web Site for Beanie Babies Undergoes Several Changes," *New York Times*, December 8, 1997, D16.

CARU guidelines state: "Advertisers to children who collect identifiable information online should make reasonable efforts, in light of the latest available technology, to ensure that parental permission is obtained." Children's Advertising Review Unit, Council of Better Business Bureaus, Inc., "Self-Regulatory Guidelines for Children's Advertising Guidelines for Interactive Electronic Media," 1997, http://www.ftc.gov/reports/privacy3/comments/002-caru.htm.

106. The Center for Media Education had faulted the Beanie Babies Web site for failing to post a privacy policy. Stuart Elliott, "Self-Regulation in Cyberspace: The Web Site for Beanie Babies Undergoes Several Changes," *New York Times*, December 8, 1997, D16.

107. Brad Edmondson, "Washington's Wild about Privacy," *American Demographics, Inc. Forecast*, August, 1997.

108. Eric Sorensen, "'Net Privacy Issues a Concern of U.S." *Denver Post*, January 12, 1998, D11.

109. Green, "Privacy," 98.

110. Norm Alster, "Web-Loving White House May Toughen on Privacy," *Investor's Business Daily*, June 2, 1998, A8.

111. Steve Lewis and Grant Butler, "Australia May Sign Up for E-Commerce Coup," *Australian Financial Review*, April 17, 1998, 3.

112. "Industry Plans to Avert Govt. Internet Privacy Regulation," *Communications Daily*, April 20, 1998.

113. *PR Newswire*, May 11, 1998. "BBBOnLine Privacy Program Created," *Electronic Buyers News*, July 6, 1998, 60. "BBBOnLine Privacy Program Created to Enhance User Trust on the Internet," *PR Newswire*, June 22, 1998. Kim M. Bayne, "Privacy Still a Burning Web Issue: Marketers Scramble to Come Up with Self-Regulation Methods," *Advertising Age*, June 29, 1998, 37. "BBBOnLine Privacy Program will be Tough, Industry-Driven and Consumer-Friendly, BBB Tells House Subcommittee; "Authentication, Verification, Recourse and Consequences Are Activities the Better Business Bureau System Does Best," *PR Newswire*, June 21, 1998.

114. "Industry Reacts in Advance to FTC Privacy Concerns," *Communications Daily*, June 4, 1998.

115. "High Tech Industry Leaders Announce Self-Regulatory Plan to Ensure Online Privacy," *PR Newswire*, June 3, 1998.

116. Elizabeth Weise, "Industry Groups Address On-line Privacy Issues," *USA Today*, June 23, 1998, 1D.

117. The CME proposals included:

1) As a general rule, marketers should not collect personally identifiable information from children 13 and under. Children and their parents have a right to anonymity and autonomy. 2) If such information is necessary, then marketers must bear the burden of responsibility for justifying the collection and use of personally identifiable information from children. 3) Marketers must clearly and prominently display a privacy statement which describes what information (personally identifiable as well as aggregate and anonymous) they collect and how it is used. 4) If a marketer can justify the collection of personally identifiable information about children, that information may be collected only if prior valid parental consent has been obtained. 5) The collected information may only be used for the purposes that have been disclosed. 6) Marketers must not use forms of enticement such as spokescharacters or the lure of free merchandise to solicit personally identifiable information from children. 7) Parents must have rights to access, to correction and the right to prevent the future use of the collected information.

CME press backgrounder, "Protecting Children's Online Privacy," May 28, 1998. From author's personal files.

118. Leslie Miller, "FTC Seeks Laws for Kids' Privacy On Line," *USA Today*, June 4, 1998, 1D. "Internet Groups Suggest Privacy Self-Regulation," *Houston Chronicle* June 23, 1998, 3.

119. Louise Kehoe, "Appeal for Internet Law to Limit Child Data," *The Financial Times of London*, June 5, 1998, 4; Federal Trade Commission, "Privacy Online: A Report to Congress," June 1998, http://www.ftc.gov/reports/privacy3/priv-23a.pdf. Federal Trade Commission Media Advisory, "FTC Releases Report to Congress on Protecting Consumers Online Privacy," June 1 1998, http://www.ftc.gov/opa/1998/06/privacyonline.htm.

120. Kehoe, "Appeal," 1998.

121. Federal Trade Commission, "Privacy Online: A Report to Congress," June 1998 http://www.ftc.gov/reports/privacy3/priv-23a.pdf (accessed 3 January, 2006).

122. Federal Trade Commission Media Advisory, "FTC Releases Report to Congress on Protecting Consumers Online Privacy," 1 June 1998, http://www.ftc.gov/opa/1998/06/privacyonline.htm (accessed January 3, 2006).

123. Miller, "FTC Seeks Laws," 1D. Federal Trade Commission Media Advisory, June 1, 1998, "FTC Releases Report to Congress on Protecting Consumers Online Privacy," http://www.ftc.gov/opa/1998/06/privacyonline.htm (accessed January 3, 2006).

124. This move, observed the *Financial Times of London*, "signals a significant reversal," noting that the commission previously had "backed industry demands for 'self regulation' of the Internet." Kehoe, "Appeal," 4.

125. Jonathan Gaw, "Report Says Web Sites Far Short on Privacy; FTC Upset with Questioning of Children, Urges Laws to Protect Consumers," *Star Tribune*, June 5, 1998, 14A.

126. Miller, "FTC Seeks Laws," 1D.

127. Kehoe, "Appeal," 4.

128. Robert O'Harrow Jr., "Firms Prepare Plan for Protecting Privacy on Internet," *Washington Post*, June 20, 1998, D03. Secretary Daley's response was less than enthusiastic. Speaking to the summit gathering, he remarked that the announcement was a positive step but expressed disappointment that the group still needed more time to finalize its program. "Protect Privacy or Feds Will—Daley," *Post-Newsweek Business Information Newsbytes*, June 23, 1998.

129. Bill McConnell, "Private Sector Won't Hit FTC's Target for On-line Privacy Plan," *American Banker* 163, no. 138 (July 1998): 3.

130. "Industry Presses for Online Privacy Self-Regulation," *Post-Newsweek Business Information Newsbytes*, July 21, 1998.

131. O'Harrow Jr., "Firms Prepare Plan," 1998.

132. "Industry Internet Privacy Practices Criticized at Hearing," *Communications Daily*, July 22, 1998. Bill Tauzin was chair. "Meanwhile, Sen Bryan (D-NV) and Senate Commerce Committee Chairman McCain (R-AZ) introduced legislation requiring FTC to come up with privacy rules.

133. Ira Teinowitz, "FTC Delays Seeking Adult Web Site Curbs," *Advertising Age*, July 27, 1998, 30.

134. At New York University in May Vice President Gore had delivered a speech in which he explained that the administration was approaching the privacy issue in stages, concentrating initially on "four key areas, including protecting sensitive personal information, stopping identity theft, protecting children's privacy online, and urging the private sector to voluntarily protect privacy." Bill Pietrucha, "Gore Seeks Electronic Bill of Rights," *Post-Newsweek Newsbytes*, July 31, 1998. This position would be articulated at a White House event in July, where the vice president called for an "electronic bill of rights." See Ted Bridis, "White House Starts Online Privacy Push," *Chicago Sun-Times*, July 31, 1998, 31.

135. Two separate bills were introduced in Congress, both of which complied with the general framework first articulated by CME and CFA. The first (S. 2326) was introduced by Senator Bryan (D-NV) and cosponsored by Senator John McCain (R-AZ). In the House of Representatives, Congressman Ed Markey (D-MA) introduced HR 4667, the "Electronic Privacy Bill of Rights Act of 1998." Author's personal files.

136. Center for Media Education, "Legislative Alert: S.2326 Children's Online Privacy Protection Act of 1998." ND. Author's personal files.

137. "Statement of American Library Association to Senate Communications Subcommittee Regarding S.2326, The Children's Online Privacy Protection Act of 1998," September 23, 1998. American Library Association, Washington, DC. From author's personal files.

138. The CME/CFA proposal, which had been amended in April 1998, was designed to protect children age 13 and under. "Guidelines and Policy Principles for the Collection and Tracking of Information from Children on the Global Information Infrastructure and Interactive Media," The Center for Media Education/Consumer Federation of America, April 1998. Personal files of author.

139. The FCC's children's-advertising guidelines were based on social-science research supporting the need for special protections for younger children—because developmental differences made them more susceptible to manipulation by certain advertising techniques. See Dale Kunkel, "Children and Host-Selling Television Commercials," *Communication Research* 15 (1988): 71–92.

140. In fact, the research applied to children eight years and under. See Dale Kunkel, "Children and Television Advertising" and Dale Kunkel and Brian Wilcox, "Children and Media Policy," in *Handbook of Children and the Media*, 375–394, 385–604.

See also Ellen Wartella and James Ettema, "A Cognitive Developmental Study of Children's Attention to Television Commercials," *Communication Research* 1 (1974): 69–88.

141. See Angela J. Campbell, "Ads2Kids.com: Should Government Regulate Advertising to Children on the World Wide Web?" *Gonzaga Law Review* 33 (1998): 311.

142. CME also called for requirements that would give teenagers "access to information already collected, and opportunity to correct, prevent or curtail the use of personal information." U.S. Senate, "Testimony of Kathryn Montgomery, Ph.D., President, Center for Media Education, Before the Senate Commerce, Science, and Transportation Committee Communications Subcommittee. Subject—S. 2326, The Children's Online Privacy Protection Act of 1998," 2000 CIS S 26136, Washington, DC, Federal News Service, September 23, 1998.

143. One of the industry lobbyists was representing both America Online and the American Library Association.

144. Jeri Clausing, "Senate Panel Debates Children's Online Privacy," *New York Times*, September 24, 1998, http://partners.nytimes.com/library/tech/98/09/cyber/articles/24privacy.html.

145. "Statement of Managers for S. 2326" ND. From the author's personal files. The document was faxed from Senator Bryan's office to the Center for Media Education on October 8, 1998.

146. Jamie Beckett, "Web Sites Use Treats to Pry Data from Youngsters," *San Francisco Chronicle*, September 22, 1998, C1.

147. Will Rodger, "Industry Finally Gives In to Privacy Bill," *Interactive Week*, September 28, 1998.

148. Jeri Clausing, "Technology Bills Languish as Congress Races for Exit," *New York Times*, October 12, 1998.

149. Will Rodger, "Senate Passes Bill Banning New Internet Taxes," *Interactive Week*, October 12, 1998.

150. Arik Hesseldahl, "Adult Sites' Tax Break Curbed," *Wired News*, October 6, 1998.

151. "Daley Says Europe Likely to Accept U.S. Online Privacy Rules," *Washington Telecommunications Newswire*, July 29, 1998.

152. Louise Kehoe, "U.S., EU at Odds over Cyber-Privacy," *Financial Times*, August 29, 1998, 15.

153. Robert O'Harrow Jr., "Companies Plan Ad Blitz," *Washington Post*, October 7, 1998, C12.

Electronic Privacy Information Center, "Critical Infrastructure Protection and the Endangerment of Civil Liberties: An Assessment of the President's Commission on Critical Infrastructure Protection (PCCIP)," 1998, http://www.epic.org/reports/epic-cip.html.

154. Reuters, "Anti-porn Bill Clears House," *Wired News*, October 8, 1998, http://www.wired.com/news/politics/0,1283,15491,00.html.

From the EPIC website: "The 'Child ONLINE Protection Act' makes it a federal crime to 'knowingly' communicate 'for commercial purposes' material considered 'harmful to minors.' Penalties include fines of up to $50,000 for each day of violation, and up to six months in prison if convicted of a crime." Electronic Privacy Information Center, "EPIC Alert," October 28 1998, http://www.epic.org/alert/EPIC_Alert_5.15.html.

155. "Congress Closing in on Internet Issues," *Communications Daily*, October 8, 1998.

156. Reuters, "Anti-porn," http://www.wired.com/news/politics/0,1283,15491,00.html.

157. Robert MacMillan, "Tax Freedom Act, CDAII Sail Smoothly," *Post-Newsweek Business Information Newsbytes*, October 15, 1998.

158. Will Rodger, "Net Bills Virtually Law," *Interactive Week*, October 19, 1998.

159. Ibid.

160. Jeri Clausing, "Controversial Internet Proposal Is Attached to Budget Bill," *New York Times* online, Cybertimes, October 15, 1998.

161. Electronic Privacy Information Center, "EPIC Alert," October 28, 1998, http://www.epic.org/alert/EPIC_Alert_5.15.html.

162. John Borland, "Activists: Kids' Privacy Law Is Only First Step," *TechWeb*, October 23, 1998, http://www.techweb.com/wire/story/TWB19981023S0017.

163. *Global News Wire*, October 23, 1998.

164. David L. Aaron, U.S. Department of Commerce, "Letter to Industry Representatives," November 4, 1998, http://www.epic.org/privacy/intl/doc-safeharbor-1198.html (accessed October 2, 2005).

165. See Marc Rotenberg, *Privacy Law Sourcebook 2004: United States Law, International Law, and Recent Developments* (Washington, DC: Electronic Privacy Information Center, 2005).

166. Mark Sandalow, "Homeland Law Evokes Fears of 'Big Brother.'" *San Francisco Chronicle*, November 28, 2002, A24; Rob Lever, "Fears of Police State," *The Advertiser*, November 25, 2002, 23; Carrie Kirby, "Personal Privacy Takes Alarming Hit, Critics

Say; Trolling the Web for Terror Plots," *San Francisco Chronicle*, November 20, 2002, A13; Bruce Alpert, "Terror Information System Facing Copious Criticism; Many Worry Plan Is Like 'Big Brother.'" *Times Picayune* (New Orleans), July 24, 2002, 6.

167. Jupiter communications analyst Michele Slack in Jennifer Gilbert and Ira Teinowitz, "Compliance Deadline Looms for Kids' Sites," *Advertising Age* 71, no.17 (April 2000): 54–56.

168. "COPPA the First Year: A Survey of Sites," Center for Media Education (April, 2001), author's personal files; William Glanz, "FTC Fines Web Sites over Privacy of Children," *Washington Times*, April 21, 2001.

169. The FTC's COPPA rule applies to:

> Operators of commercial web sites and online services directed to children under 13 that collect personal information from them; operators of general audience sites that knowingly collect personal information from children under 13; and operators of general audience sites that have a separate children's area and that collect personal information from children under 13.

Children's Online Privacy Protection Act, Federal Trade Commission. http://www.ftc.gov/privacy/privacyinitiatives/childrens.html.

170. Mike Shields, "Surfing Lessons," *Mediaweek*, July 25, 2005, 4–6.

171. As the FTC's Web site explained the rules:

> If the website is sharing the personal information with a company or person whose only role is to provide support for the internal operations of the website—like a fulfillment house or a shipping company—the disclosure of the personal information is not to a "third party" and does not have to be spelled out in the privacy policy. The Rule specifically defines "third party" to exclude people who provide internal support. These providers are obligated to use the personal information only to carry out their specific obligations. They cannot use the information for any other purpose.

Federal Trade Commission, "Introduction to COPPA." http://www.ftc.gov/bcp/conline/edcams/coppa/intro.htm.

172. E-Consultancy, "Email Marketing to Tweens (Kids Ages 8–12): COPPA Loopholes, Demographics, Creative Samples," News, PR & Features, http://www.e-consultancy.com/newsfeatures/152058/email-marketing-to-tweens-kids-ages-8-12-coppa-loopholes-demographics-creative-samples.html?keywords=demographics.

173. Nonetheless, the Web site Bolt.com decided to take its own precautions to ensure compliance with COPPA by prohibiting children younger than fifteen from becoming registered users, a loss of about 50,000 of the site's 2.25 million customers. (See full profile of this Web site in chapter 5.) William Glanz, "New Privacy Act Spurs Web Sites to Oust Children," *Washington Times*, April 20, 2000.

174. John Fetto, "Candy for Cookies," *American Demographics* 22, no. 8 (August 2000): 10; Annenberg Public Policy Center, "The Internet and the Family 2000," May 16, 2000, http://www.annenbergpublicpolicycenter.org/04_info_society/family/finalrepor_fam.pdf.

175. Glanz, "FTC."

176. Ibid.

177. Letter from Mary K. Engle, Associate Director, Division of Advertising Practices, Federal Trade Commission to Marc Rotenberg, Executive Director, Electronic Privacy Information Center, November 24, 2004.

178. Hiawatha Bray, "Firms Launch Plan to Police Own Web Sites: TRUSTe Group Looks to Avoid FTC Action by Addressing Privacy Concerns Voluntarily," *Boston Globe*, June 10, 1997, C6; "TRUSTe Introduces Children's Privacy Seal Program, Yahooligans! First Web Site to Participate; TRUSTe to Apply for Safe Harbor Status with the Federal Trade Commission when Children's Online Legislation Passes," *PR Newswire*, October 13, 1998.

179. Mike Shields, "Surfing Lessons," 4–6.

180. "KidsCom.com Privacy Policy," 1995–2006, http://www.kidscom.com/info/privacy_policy.html.

181. "Jorian Clarke and Circle 1 Network, "Experts Advice: Articles, ParentsTalk," http://www.parentstalk.com/expertsadvice/ea_fs_clarke.html, http://www.spectracom.com/site/a01_about/a01b04_team/bios.jsp?pg=jori.

Chapter 5

1. Quoted in Susannah R. Stern, "Sexual Selves on the World Wide Web: Adolescent Girls' Home Pages as Sites for Sexual Self-Expression," in Jane D. Brown, Jeanne R. Steele, and Kim Walsh-Childers, *Sexual Teens, Sexual Media: Investigating Media's Influence on Adolescent Sexuality* (Mahwah, NJ: Lawrence Erlbaum Associates, 2002).

2. Susannah R. Stern, "Adolescent Girls' Expression on Web Home Pages: Spirited, Sombre, and Self-Conscious Sites," *Convergence* 5, no. 4 (1999): 22–41. As Stern explained,

> The home pages ultimately suggested that the girls disclosed online to reach an audience, and subsequently, to make themselves known. Their home pages allowed the girls to envision multiple audiences and to disclose more freely than in more conventional disclosure vehicles. Girls' willingness to disclose and explore their developing identities online demonstrated their agency as resilient media producers.

Susannah R. Stern, "Making Themselves Known: Girls' WWW Homepages as Virtual Vehicles for Self-Disclosure" (PhD diss., University of North Carolina, Chapel Hill. 2000).

3. Stern, "Adolescent Girls' Expression," 22–41.

4. Amanda Lenhart, Lee Rainie, and Oliver Lewis, *Teenage Life Online: The Rise of the Instant-Message Generation and the Internet's Impact on Friendship and Family Relationships* (Washington, DC: Pew Internet and American Life Project, 2001), 37,

http://www.pewinternet.org/pdfs/PIP_Teens_Report.pdf. The study showed that 45 percent of all American children under the age of 18 go online. Almost three quarters (73 percent) of those between 12 and 17 go online. In contrast, 29 percent of children 11 or younger go online.

5. Some of the research for this section was published in CME's 2001 report: *TeenSites.com: A Field Guide to the New Digital Landscape* (Washington, DC: Center for Media Education, 2001). See Lee Rainie and Dan Packel, *More Online, Doing More* (Washington, DC: Pew Internet & American Life Project, Feb. 2001). Lenhart, Rainie, and Lewis, *Teenage Life Online*, 37.

6. Malcolm Gladwell "The Coolhunt," *New Yorker*, March 17, 1997, 78–89.

7. Cheskin Research, "Online Teens Give Clues to the Future of the Web," August 23, 1999, www.cheskin.com/who/press/release_netteens.html (accessed March 24, 2001).

8. Susan Harter, "Processes Underlying the Construction, Maintenance, and Enhancement of the Self-Concept in Children," *Psychological Perspective on the Self* 3 (1990).

9. Hazel Marcus and Paula Nurius, "Possible Selves," *American Psychologist* 41, no. 9 (1986): 954–969.

10. Jane Brown, "Teenage Room Culture: Where Media and Identities Intersect," *Communication Research* 21, no. 6 (1994): 813–827.

11. Jeanne Roqqe Steele and Jane D. Brown, "Adolescent Room Culture: Studying Media in the Context of Everyday Life," *Journal of Youth and Adolescence* 24, no. 5 (1995): 551–576.

12. Jeffrey Jensen Arnett, Reed Larson, and Daniel Offer, "Beyond Effects: Adolescents as Active Media Users," *Journal of Youth and Adolescence* 24, no. 5 (1995): 511–518; Reed Larson, "Secrets in the Bedroom: Adolescents' Private Use of Media," *Journal of Youth and Adolescence* 24, no. 5 (1995): 535–551; Steele and Browne, "Adolescent Room Culture," 551–576. See also Sonia Livingstone and Moira Bovill, *Young People and the Media,* Report of the Research Project, "Children, Young People and the Changing Media Environment," (London: 1999).

13. Larson, "Secrets in the Bedroom," 535–551; See also Keith Roe, "Adolescents' Use of Socially Disvalued Media: Towards a Theory of Media Delinquency," *Journal of Youth and Adolescence* 24, no. 5 (1995): 617–632.

14. A number of scholars have written about the role of the Internet in adolescent identity development. See, for example: Kaveri Subrahmanyam, Patricia M. Greenfield, and Brendesha Tynes, "Constructing Sexuality and Identity in an Online Teen Chat Room," *Journal of Applied Development Psychology* 25 (2004): 651–666; Sandra Calvert, "Identity Construction on the Internet," in Sandra L. Calvert, Amy B.

Jordan, and Rodney R. Cocking, *Children in the Digital Age: Influences of Electronic Media on Development* (Westport, CT: Praeger, 2002), 57–70; Crispin Thurlow and Susan McKay, "Profiling 'New Communication' Technologies in Adolescence," *Journal of Language and Social Psychology* 22, 1 (March 2003): 94–104; Elisheva F. Gross, Jaana Juvonen, and Shelly L Gable, "Internet Use and Wellbeing in Adolescence," *Journal of Social Issues* 58, no. 1 (Spring 2002): 75–91; Gill Valentine and Sarah L. Holloway "Cyberkids? Exploring Children's Identities and Social Networks in On-line and Off-line Worlds," *Annuls of the Association of American Geographers* 92, no. 2 (2002): 302–320.

15. In her landmark book *Life on the Screen*, Sherry Turkle argues that the Internet and other digital media have created a new "culture of simulation," with profound implications for the development of identity.

> From scientists trying to create artificial life to children "morphing" through a series of virtual personae, we shall see evidence of fundamental shifts in the way we create and experience human identity. But it is on the Internet that our confrontations with technology as it collides with our sense of human identity are fresh, even raw. In the real-time communities of cyberspace, we are dwellers on the threshold between the real and virtual, unsure of our footing, inventing ourselves as we go along.

Sherry Turkle, *Life On the Screen: Identity in the Age of the Internet* (New York: Touchstone, 1995).

16. At the time that Stern conducted her study, only 8.4 million teenagers in the United States had Internet access, and they tended to be in the higher socioeconomic brackets. And, as she points out, most of the girls who created home pages required some knowledge of HTML and familiarity with Web-site construction. As she explained, "altogether, we can assume that the girls who create home pages are at least currently an elite group." She also only looked at a handful of Web sites. Nonetheless, what she found provided one of the first glimpses of the world of online personal expression by teens. Stern, "Making Themselves Known," 5–6.

17. Stern, "Making Themselves Known," 54–55.

18. Angelfire, "Angelfire Building Guides: Build a Great Celebrity Fan Site!" http://angelfire.lycos.com/doc/guides/celebrity1.html (accessed May 9, 2001). Although the Angelfire service was free, users were compelled to carry ads, which were inserted by Angelfire programmers, on their sites.

19. Daniel Chandler and Dilwyn Roberts-Young, "The Construction of Identity in the Personal Homepages of Adolescents," University of Wales, Aberystwyth, 1998, http://www.aber.ac.uk/media/Documents/short/strasbourg.html.

20. Lenhart, Rainie, and Lewis, *Teenage Life Online*, 43; Ariana Eunjung Cha, "Dear Web Diary, SO Much to Tell! Teens Tell Secrets, Get Gifts, Attract Nuts, Scare Mom," *Washington Post*, September 2, 2001, A1.

21. Stern, "Making Themselves Known," 94.

22. Daniel Chandler, "Personal Home Pages and the Construction of Identities on the Web," http://www.aber/ac/uk/~dgc/Webident.html (accessed December 10, 1999).

23. Chandler and Roberts-Young, "The Construction of Identity."

24. Sandy Coleman, "Battling the Web's Dark Side; Schools Balance Student Rights, Rules in Incidents on Net," *Boston Globe*, March 27, 2000, B1.

25. Tamar Lewin, "Schools Challenge Students' Internet Talk," *New York Times*, March 8, 1998, sec.1, 16.

26. See William Wresch, *A Teacher's Guide to the Information Highway* (Upper Saddle River, NJ: Merrill, 1997); "FCC to Seek Comment on Commerce Plan for Filtering in Schools," *Washington Telecom Newswire*, May 4, 1999; Gordon Keith Dahlby, "Identification of Emerging Information Technology Issues of the 21st Century Affecting Public School Board Policies" (PhD diss., Iowa State University, 2004); Hilary Cowan, "Hillside Elementary School Acceptable Use Policies," *Electronic Learning* 16, no. 1 (1996): 21; Kevin Johnson and Nancy Groneman, "Legal and Illegal Use of the Internet: Implications for Educators," *Journal of Education for Business*. 78, no. 3 (2003): 147.

27. Henry J. Kaiser Family Foundation, "Key Facts: Teens Online," Fall 2002, http://www.kff.org/entmedia/upload/Key-Facts-Teens-Online.pdf.

28. Lenhart, Rainie, and Lewis, *Teenage Life Online*, 42. The Children's Online Privacy Protection Act required that children under the age of 13 be prevented from entering a chat room on a commercial Web site unless they had prior parental approval, which may have inhibited some younger teens from entry, though the Pew Internet & American Life Report did not ask this question in its survey.

29. Lenhart, Rainie, and Lewis, *Teenage Life Online*, 4.

30. Patricia M. Greenfield, Elisheva F. Gross, Kaveri Subrahmanyam, Lalita K. Suzuki, and Brendesha Tynes, "Teens on the Internet: Interpersonal Connection, Identity, and Information," in Robert Kraut, ed., *Information Technology at Home* (New York: Oxford University Press, forthcoming). Accessed at Children's Digital Media Center, University of California, Los Angeles, http://www.cdmc.ucla.edu/papers.html.

31. Kaveri Subrahmanyam, Patricia M. Greenfield and Brendesha Tynes, "Constructing Sexuality and Identity in an Online Teen Chat Room," *Journal of Applied Development Psychology* 25 (2004): 651–666.

32. From Robert Roy Britt, "Report: Teens Routinely Exposed to Pornography and Sexual Advances on Internet," LiveScience.com, March 2005, http://www.livescience.com/othernews/050302_teen_sex.html. See also Patricia M. Greenfield, "Developmental Considerations for Determining Appropriate Internet Use Guidelines for Children and Adolescents," *Journal of Applied Developmental Psychology* 25, no. 6 (2004): 751–762.

33. Greenfield, Gross, Subrahmanyam, Suzuki, and Tynes, "Teens on the Internet," 24.

34. See Elisheva Gross, "Adolescent Internet Use: What We Expect, What Teens Report," *Journal of Applied Developmental Psychology* 25, no. 6 (2004): 633–649.

35. Elisheva F. Gross, Jaana Juvonen, and Shelly L. Gable, "Internet Use and Well-Being in Adolescence," *Journal of Social Issues* 58, no. 1 (2002): 75–90.

36. Michael Pastore, "Internet Key to Communication among Youth," CyberAtlas, http://cyberatlas.internet.com/big_picture/demographics/article/0,,5901_961881,00 html.

37. Lenhart, Rainie, and Lewis, *Teenage Life Online*, 3.

38. The study found that social interaction was the primary reason most teens used the Web. The four most popular activities among online teens were sending or reading e-mail (with 92 percent of teens participating), surfing the Web for fun (84 percent), visiting an entertainment site (83 percent), and sending an instant message (74 percent). Lenhart, Rainie, and Lewis, *Teenage Life Online*, 6.

39. Rebecca E. Grinter and Leysia Palen, "Instant Messaging in Teen Life," 2002, http://www.cs.colorado.edu/~palen/Papers/grinter-palen-IM.pdf.

40. The AIM service offered the insertion of sixteen distinct graphical faces into messages, each with an associated emotion or thought: "smile, frown, wink, sticking out tongue, oh oh, kissy face, yelling, big grin, put your money where your mouth is, foot in mouth, embarrassed. angel, hmmm, crying, my lips are sealed, and Joe Cool." AOL Instant Messenger (SM) for Macintosh, Version 4.1.1068 (2000).

41. See David Elkind, *The Hurried Child: Growing Up Too Fast Too Soon* (Reading, MA: Addison-Wesley, 1988).

42. Grinter and Palen, "Instant Messaging in Teen Life."

43. Lenhart, Rainie, and Lewis, *Teenage Life Online*, 22.

44. Grinter and Palen, "Instant Messaging in Teen Life."

45. Ibid. A number of other researchers have examined teen IM use. For example, see Gross, Juvonen, and Gable, "Internet Use and Well-Being," 75–90; Naomi S. Baron, "Instant Messaging and the Future of Language," *Communications of the ACM* 48, no. 7 (July 2005); Gloria E. Jacobs, " 'ur part of it': Portfolio People and Adolescent Use of Instant Messaging" (PhD diss., University of Rochester, 2005).

46. Louise Rosen, "Why IM Matters So Much," *Upside Today*, September 19, 2000, http://www.upside.com/texis/mvm/print-it?id=39c289380&t=1.

47. "Sticky content" was a popular catch phrase in the industry trade literature. For example, see "Uproar and Zing Announce Strategic Partnership and Launch

Co-Branded Site Featuring Games for Prizes, Online Photo Albums and Greeting Cards," *Business Wire* (June 27, 2000); "Giving Consumers What They Want," *New Media Age* (November 30, 2000).

48. For example, AOL Time Warner was far and away the largest commercial presence on the Internet, with over 29 million members worldwide. According to Jupiter Media Metrix, AOL accounted for nearly a third of all time spent online in the United States in January 2001. Half of the company's member households had children between the ages of two and seventeen. MTV, the popular teen cable network, had created an interactive media experience for viewers of their network and users of their Web sites (which included MTV.com, VH1.com, and Sonicnet.com). Reaching nearly 5 million users a month, the sites allowed visitors to watch music videos online, to vote for the videos that subsequently would be aired on various cable-TV programs, and to link to a variety of MTV branded Web "radio" stations that played popular music featured on MTV, TRL, and MTV2. See *TeenSites.com*, Center for Media Education, 2001, 15–57.

49. Matthew Mirapaul, "Music Videos Enter the Digital Age," *New York Times*, Aug 21, 2000, C3.

50. Lee Gomes, "Free Tunes for Everyone! MP3 Music Moves into High-School Mainstream," *Wall Street Journal*, June 15 1999, B1.

51. Ibid.

52. Dan Lippe, "It's All in Creative Delivery; Gurus in the Teen Universe Build a Track Record by Gauging Where the Market Is Going," *Advertising Age*, June 25, 2001, S8.

53. "Building the Web Site That Every Kid Wants," *Selling to Kids*, April 5, 2000, 3.

54. Joe Nickell, "Teen Scene Maker," *Business 2.0*, March 2000, 108.

55. Interview with Dan Pelson, Founder, Chairman and CEO. Interviewed by Nathan Kaiser, Founder and CEO, nPost.com, Friday, October 19, 2001. http://wsww.npost.com/interpelson.html.

56. Jane Mount, personal interview, May 4, 2000.

57. VentureReporter.net, "Bolt CEO on Surviving the Downturn and Changing Realities of the Teen Market" July 30, 2002.

58. Philip Connors ". . . Entertain the Teens," *Wall Street Journal Interactive Edition*, 1999, December 6, 1999, http://www.wsj.com.

59. Bolt.com, http://www.bolt.com (accessed June 11, 2001), cited in *TeenSites.com*, 110.

60. Bolt.com press kit. ND.

61. From Pelson interview in nPost.com.

62. Advertisement in the *Wall Street Journal,* January 20, 2000.

63. VentureReporter.net, "Bolt CEO on Surviving the Downturn."

64. Beth Vivieros, "Bolt Inc.: Teenage Lightning," *Direct Magazine,* May 1, 2001. This also was confirmed in my interviews with Jane Mount.

65. Connors, ". . . Entertain the Teens."

66. VentureReporter.net, "Bolt CEO on Surviving the Downturn."

67. Bolt.com press kit.

68. Erin Kelly, "This Is One Virus You Want to Spread," *Fortune,* November 27, 2000, 297–300.

69. For example, the highly popular e-mail service Hotmail embedded Web-site information in all e-mail messages its users sent, enabling friends to click the link and sign up in an instant. Steve Jurvetson, "What Exactly Is Viral Marketing?" *Red Herring,* May 2000.

70. VentureReporter.net "Bolt CEO on Surviving the Downturn."

71. Ryan Naraine, "Bolt.com Slashes Staff, Unveils Credit Card for Teens," Atnewyork.com, March 1, 2001, http://www.atnewyork.com/news/article/0,1471,8471_702411,00.html; Beth Cox, "Internet Profits—Not an Oxymoron," *E-Commerce—Trends,* June 5, 2001, http://ecommerce.internet.com/opinions/print/0,,3551_778841,00.html.

72. VentureReporter.net, "Bolt CEO on Surviving the Downturn."

73. Vivieros, "Bolt Inc." In a March 2006 press release, the company announced it was "relaunching" its site, summarized its successes in the ten years since it was founded, and announced its most recent features:

> Bolt Media's 10-year-old web site, bolt.com, relaunched early today. The upgraded site enables its 10.8 million monthly unique users to create, store and market their personal content—including videos, photos and music. Combining creative networking with social interaction, Bolt.com motivates people to build a personal audience. Bolt caters particularly to the millions of burgeoning auteurs looking to distribute their content to their own personal audience. . . . Bolt Media's traffic has been on the rise over the past few months, jumping 300% in 2005 to 10.8 million unique users per month, compared with the prior year, according to Jupiter's Media Metrix. This increase, due primarily to increased functionality through product development and acquisitions, has attracted more advertisers, including Verizon Wireless, Cingular, Coca Cola, Fox Television and Sony Pictures among its top 50 advertisers. . . . To provide value to its original audience, Bolt Media today also launched Bolt2.com, enabling kids, tweens, and younger teens to create content, meet people, and play games in a safe, no pressure, and age-appropriate environment. Bolt2.com is organized around games, pop culture and self-discovery.

"Bolt Media Re-Launches bolt.com; Company Focuses on 18–34 Year-Old Creation Generation, Connecting People via Their Creativity; Bolt's Original Core Membership Grows Up," *Business Wire*, March 6, 2006.

74. According to Alloy, Inc.'s Form 10-K, filed April 18, 2005, and accessed through EdgarOnline, the company was "incorporated in January 1996, launched our Alloy website in August 1996 and began generating meaningful revenues in August 1997 following the distribution of our first Alloy catalog."

75. Brian Morrissey, "Alloy Returns to Red, Vows Full-Year Profits," *Silicon Alley Daily*, May 31, 2001.

76. Brian Morrissey, "Do Not Abandon Hope, All Ye Who Enter the Alley: Alloy Posts Profit," *Silicon Alley Daily*, March, 16, 2001.

77. From the 10-K. As of 2005, the company owned and operated fifty-five of the stores, including six outlets in twenty-two states.

78. Ibid.

79. Alloy Media + marketing, "Elementary School Out-of-Home Media," 2006, http://www.alloymarketing.com/media/tweens/outofhome.htm.

80. 360 Youth, "About 360 Youth," http://www.360youth.com/aboutus/index.html (accessed July 21, 2005). Its 2005 form 10-K says the company has acquired a database of "31 million generation Y consumers." The 10-K also includes a listing of clients: "During fiscal year ended January 31, 2005, we had over 1800 advertising clients, including AT&T Wireless (Cingular), Citibank, Geico, Paramount Pictures, Procter & Gamble, Qwest Communications, Simon Brand Ventures, and Verizon Wireless."

81. 360 Youth, "About 360 Youth."

82. 360 Youth, "Teen Online Advertising and Promotion," http://www.360youth.com/teens/online_advertising.

83. Alloy Media + marketing, "Teen Online Advertising and Promotion," http://www.alloymarketing.com/media/teens/onlineadvertising.htm.

84. Alloy, Inc.'s Form 10-K, filed April 18, 2005.

85. For example, a visitor to Alloy.com who clicked the "win prizes" link would be serenaded by pop singer Brie Larson and offered a chance to win a DVD player by filling out an elaborate questionnaire that required name, address, e-mail address, and telephone number (cell phone number was "optional"). After giving out this personal information, the visitor then would be asked a series of market research questions and given an invitation to receive e-mail newsletters with "the latest news and info via email about Brie Larson, including contests, tour dates, contests, new music, and more." Alloy.com, http://www.brielarsonshesaid.com/survey.asp?referrerCode=alloy (accessed July 22, 2005).

86. Securities and Exchange Commission, *Alloy, Inc. Annual Report 10-K for 2004*, filed April 18, 2005.

87. eCRUSH was not created by Alloy, but was acquired after it had been online for several years. Alloy Media + marketing, "Teen Online Advertising and Promotion."

88. eCRUSH, http://www.ecrush.com/?sess_sid=&cobrand (accessed July 21, 2005).

89. Ibid.

90. Ibid. Accessed July 24, 2005.

91. "Staying Safe Online," eSpintheBottle, http://www.espinthebottle.com/safety.phtml?sid= (accessed May 1, 2006).

92. The privacy policy states:

> During the registration process, users are given the option to opt-in or out of our mailing (postal and email) lists as well third party promotions. Our own newsletters and promotions are sent by us and contain offers on behalf of advertisers. By not opting out, the user is agreeing to receive current and future offers. For our cobranded offers at registration we obtain your consent to sell or rent the email and/or postal addresses of registered users to bring you great offers directly from our sponsors, partners or advertisers. By checking the boxes the user is agreeing to such use.

eCRUSH, "Privacy Policy," April 29, 2004, http://www.ecrush.com/privacy.phtml?sess_sid=&cobrand= (accessed April 20, 2006).

93. Even for the information that is not "personally identifiable," this Web site enables Alloy to collect very valuable research that could become part of its database of "aggregate" information about the teen market. Though the site explains that this information is needed in order to make the social-networking functions work, it is also exactly the kind of information that marketers are seeking and that Alloy trades in. "In order to bring people together," the site explains:

> we must gather personal information, which may or may not include: customer specified username, gender, age, ethnic background, appearance, religion, income range, occupation, preference/lifestyle information and general geographic location. Much of this information is optional, and meant to be seen by members and visitors to the site but cannot be used to identify a specific member by name, address, etc.

eCRUSH, "About Us," http://www.ecrush.com/aboutus.phtml?sess_sid=&cobrand (accessed July 24, 2005).

94. In fact the privacy policy on e-SPIN-the-Bottle, which is part of the eCRUSH network, explains that the company is a member of TRUSTe, one of the safe-harbor programs established during the privacy debates of the 1990s. http://www.ecrush.com/privacy.phtml?sess_sid=&cobrand= (accessed April 20, 2006).

95. Comments by Greg Livingston, speaking at the "Born to Be Wired" conference, Yahoo! headquarters, Sunnyvale, California, July 24, 2003. I was a presenter at this conference.

96. "A New Media Landscape Comes of Age," Executive Summary, *Born to Be Wired: The Role of New Media for a Digital Generation*, commissioned by Yahoo! and Carat Interactive. Research conducted by Harris Interactive and Teenage Research Unlimited, July 23, 2003. From the author's personal files.

97. Cliff Peale, "P & G Targets Teenage Buyers," *Cincinnati Enquirer*, Sunday, October 27, 2002, http://enquirer.com/editions/2002/10/27/biz_tremor27.html.

98. Malcolm Gladwell discusses the notion of "connectors" as a key strategy in peer-to-peer marketing at length in his popular book *The Tipping Point.* Malcolm Gladwell, *The Tipping Point: How Little Things Can Make a Big Difference* (Boston: Little, Brown, 2000). For a discussion of this and other similar strategies used to market to teens, see Alissa Quart, *Branded* (New York: Basic Books, 2003).

99. Peale, "P & G."

100. PR Newswire, "Blue Dingo Puts the Shakes in Tremor; Online Teen Social Network Goes to Work for Proctor & Gamble," March 22, 2004.

101. Peale, "P & G."

102. "Blue Dingo Puts the Shakes in Tremor; Online Teen Social Network Goes to Work for Proctor & Gamble," *PR Newswire*, March 22, 2004; http://www.tremor.com/login/about_us.aspx.

103. See Martin Lindstrom and individual contributors, *Brandchild* (London: Kogan Page, 2003), 137–156. See also Mark Hughes, *Buzzmarketing: Get People to Talk about Your Stuff* (New York: Portfolio, 2005); Buzz Marketing, http://www.buzzmarketing.com/index.html; Emanuel Rosen, *The Anatomy of Buzz: How to Create Word of Mouth Marketing* (New York: Doubleday, 2000).

104. Burston Marseller, "Unconventional Marketing Becomes the Convention," e-Fluentials Blog, September 26, 2005, http://blog.efluentials.com/index.php?title=unconventional_marketing_becomes_the_con&more=1&c=1&tb=1&pb=1.

105. Christine Frey, "Web Friend or Foe?: Digital Buddies Are Elaborate Marketing Tools, But Their Lifelike Responses in Online Instant Messages Can Be Misleading," *Ottawa Citizen*, August 15, 2002, G4.

106. Ryan Naraine, "ELLEgirl Targets Teen Girls with IM Buddy," Atnewyork.com, February 25, 2002, http://www.instantmessagingplanet.com/public/article.php/980271.

107. Frey, "Web Friend or Foe?"

108. Ibid.

109. Ibid.

110. Catherine Arnold, "The ONE Technology Marketers need NOW," *Marketing News*, February 17, 2003, 1.

111. New Line Cinema also used a bot campaign to promote some of its other films, including the *Lord of the Rings* series. Arnold, "The ONE Technology," 1.

112. The campaign included spots on MTV and a sweepstakes offering a prize of attending the MTV Video Awards. "Acuvue Ads Clearly a Dry-Eye Problem," *Brandweek.com,* August 12, 2005.

113. "America Online, Inc. Debuts New AIM.com and AIM(R) Today," *Business Wire,* August 17, 2005.

114. AIM, "IM Robots," http://aimtoday.aim.com/aimbots/index.adp.

115. "America Online," *Business Wire.*

116. Wikipedia, "Avatar (Virtual Reality)," http://en.wikipedia.org/wiki/Avatar_(virtual_reality).

117. Charlotte Goddard, "Capri-Sun Uses Web Game to Draw Kids," *Revolution* (October 9, 2003), 7.

118. PR Newswire, "Sulake Corporation and Alloy Media + marketing Announce Strategic Partnership and Joint Marketing Campaign," February 9, 2005.

119. Ibid.

120. Amanda Lenhart, Mary Madden, and Paul Hitlin, *Teens and Technology* (Washington, DC: Pew Internet and American Life Project), July 27, 2005, http://www.pewinternet.org/PPF/r/162/report_display.asp.

121. In 2004, there were 100 million gaming consoles in households, 60 million hand-held games and a growing number of game-enabled cellphones. T. L. Stanley, "Video Games: The New Reality of Youth Marketing," AdAge.com, March 22, 2004.

122. Stanley, "Video Games."

123. Massive Incorporated, http://www.massiveincorporated.com/site_advert/advert_home.htm (accessed February 2, 2005).

124. Stanley, "Video Games."

125. Stanley, "Video Games."

126. Michael McCarthy, "Disney Plans to Mix Ads, Video Games to Target Kids, Teens," *USA Today,* January 18, 2005, 6B.

127. Marketers in the category spent $414.1 million in advertising in the first 11 months of 2003." Stanley, "Video Games."

128. Stanley, "Video Games"; Beth Snyder Bulik, "Advergaming Grows in Reach and Power," AdAge.com, May 24, 2004.

129. Annette Bourdeau, "The Kids Are Online," *Strategy,* May 1, 2005.

130. Bulik, "Advergaming."

131. Massive Entertainment specializes in advergaming and other immersive advertising strategies, promising its clients customized reports on consumer interaction with brand messaging that would allow for a level of accountability and evaluation of the effectiveness of the ad placements that is impossible in other traditional broadcast and print outlets. The internet link "allows Massive's software to modify the ads as players progress through a game." In 2005, the company predicted it would put its software into "40 games by the end of the year," and reported it had signed agreements with advertisers such as "Dunkin' Donuts, Intel, Paramount Pictures, Coca-Cola, Honda and Universal Music Group." Massive Incorporated, http://www.massiveincorporated.com/site_advert/advert_home.htm (accessed February 2, 2005).

132. Dart Motif, http://www.dartmotif.com (accessed September 27, 2005).

133. Michael Saxon, Vice President, Media Products—NetRatings, Inc. "Moving Towards Marketing's Holy Grail: Connecting Online and Offline User Behavior," Nielsen/NetRatings, http://thearf.org/downloads/Councils/Online/2004-05-17_ARF_NetRatings.pdf (accessed September 25, 2005).

134. Bourdeau, "The Kids Are Online."

135. Matt Richtel, "A New Reality in Video Games: Advertisements," *New York Times*, April 11, 2005, C1.

136. Mike Shields, "Surfing Lessons," *Mediaweek*, July 25, 2005; Joe Pereira, "Junk-Food Games," *Wall Street Journal*, May 3, 2004, B1–B4. The Walt Disney Company launched its own "branded entertainment" site in 2005, with the interactive multi-player game *Virtual Magic Kingdom*. With a target market of tweens and young teens, the site beckoned visitors on a fun-filled virtual tour of Disney's five global resorts and eleven theme parks, where they could engage in free online games based on real rides, including the Haunted Mansion and Jungle Cruise. They could also chat, create their own avatars, and earn virtual points that could be redeemed for prizes when they visited the actual theme parks. McCarthy, "Disney," 6B.

137. Lenhart, Madden, and Hitlin, *Teens and Technology*, http://www.pewinternet.org/PPF/r/162/report_display.asp.

138. Elisa Batista, "She's Gotta Have It: Cell Phone," Wired.com, May 16, 2003, http://www.wired.com/news/print/0,1294,58861,00.html (accessed September 27, 2005).

139. Press release, "The [R]evolution Is Now: 'The Mobiles' Defines an Emerging Wireless Lifestyle," Context-Based Research Group, Baltimore, MD, February 21, 2003.

140. Ibid.

141. Robbie Blinkoff speaking at the Born to Be Wired Summit. Author's personal notes.

142. Elisa Batista, "She's Gotta Have It." According to one female focus-group participant in the Yahoo! study: "Without having a cell phone, I would just feel completely disconnected. . . . I'd be worried about people trying to contact me. . . . It would just be a disaster." "From 'My Generation' to 'My Media Generation': Yahoo! and OMD Global Study Finds Youth Love Personalized Media," Press release from Yahoo and OMD, *Business Wire,* September 27, 2005.

143. Alice Z. Cuneo, "Brands Seek Higher Calling As 'Virtual' Cellphone Providers," *Advertising Age* (Midwest region edition), Chicago: April 11, 2005, 32.

144. "Mobile Advertising to Take the Place of TV Advertising," *PR Newswire,* April 11, 2005.

145. Virgin Mobile, http://www.virginmobileusa.com/ringmore/home.do (accessed September 27, 2005).

146. Ibid.

147. Virgin Mobile, "Rescue Rings," http://www.virginmobileusa.com/xl/dailydose/rescueRings.do (accessed September 29, 2005).

148. Virgin Mobile, "Wake Up Calls," http://www.virginmobileusa.com/xl/dailydose/wakeupCalls.do (accessed September 29, 2005).

149. Virgin Mobile, "Daily Dose," http://www.virginmobileusa.com/xl/dailydose/virginGopher.do (accessed September 29, 2005).

150. Virgin Mobile, "VirginXtras," http://www.virginmobileusa.com/xtras/home.do (accessed September 29, 2005).

151. Virgin Mobile, "Ringtones and More," http://www.virginmobileusa.com/choosePhone.do (accessed September 29, 2005).

152. Lenhart, Madden, and Hitlin, *Teens and Technology,* http://www.pewinternet.org/PPF/r/162/report_display.asp.

153. Virgin Mobile, "Great Rates," http://www.virginmobileusa.com/greatrates/greatrates.do (accessed September 29, 2005).

154. According to Lenhart, Madden, and Hitlin, in their report *Teens and Technology,* 50 percent of IM-using teens have included a link to an interesting or funny article or Web site in an instant message, 45 percent have used IM to send photos or documents, and 31 percent have sent music or video files via IM.

155. According to the Privacy Rights Clearinghouse,

> Commercial messages [on cell phones] can only be sent to individuals who have consented. This ruling does not apply to Short Message Service (SMS) messages transmitted solely to phone

numbers (as opposed to those sent to standard e-mail addresses). But, according to the FCC, any such messages that are autodialed, which is likely for spam messages, are already covered by the Telephone Consumer Protection Act.

Rules on location-based cell-phone advertising, however, appear to be less clear. "The wireless industry is aware of consumers' privacy concerns and has been working to develop consent-based guidelines for the development of wireless advertising," though the Federal Communications Commission so far has decided not to enact such rules. "Because of the federal government's reluctance to regulate location-based wireless services, consumers must research carefully the privacy implications of these services before subscribing." "Wireless Communications: Voice and Data Privacy," Privacy Rights Clearinghouse, posted October 1994, revised January 2006. http://www.privacyrights.org/fs/fs2-wire.htm (accessed May 1, 2006).

156. In 2002, the music industry won a court injunction against Napster under the Digital Millennium Copyright Act, forcing the company to cease its online file-sharing business. Brad King, "The Day the Napster Died," *Wired News,* May 15, 2002, http://www.wired.com/news/mp3/1,52540-0.html. According to the Pew Internet & American Life Project: "In 2004, 17% of music downloaders said they were actively using paid services while 7% reported using them in the past. In 2005, 34% of current music downloaders say they now use paid services and 9% say they have tried them in the past." Pew Internet Project Data Memo, "Music and Video Downloading Moves beyond P2P," March 2005. http://www.pewinternet.org/PPF/r/153/report_display.asp (accessed March 14, 2006).

157. Mya Frazier, "iTunes Drums Up Marketing Ties," *Advertising Age,* July 25, 2005.

158. Friendster, explained *Wired News,* "helps users find dates and new friends by referring people to friends, or friends of friends, or friends of friends of friends, and so on." Leander Kahney, "Making Friendsters in High Places," *Wired News,* July 17, 2003.

159. Andrew Trotter, "Social-Networking Web Sites Pose Growing Challenge for Educators," *Education Week,* February 15, 2006, 8.

160. Kris Oser, "MySpace: Big Audience, Big Risks," *Advertising Age (*Midwest region edition), February 20, 2006, 3. As columnist Nat Ives observed, "Marketers hope these sites will make it easier to start and track communication about brands among friends and contacts," Nat Ives, "A New Type of Pitch to the Online Crowd Mixes Pop Stars and Personals," *New York Times,* December 3, 2004, C6.

161. Jessi Hempel and Paula Lehman, "The MySpace Generation," *Business Week,* December 12, 2005, 86.

162. Ives, "A New Type of Pitch," C6.

163. Michelle Halpern, *Marketing* 111, no. 3 (2006): 5

164. Hempel and Lehman, "The MySpace Generation," 86.

165. The quote in Bourdeau, "The Kids Are Online" is from Rob Davy, commercial manager at Nexopia, which is based in Edmonton, Canada.

166. Blogs are designed to be updated regularly, with new entries made daily or hourly. They also include many links to other sources. See chapter 7 for a discussion of blogs in the 2004 election cycle.

167. Pew Internet & American Life Project, "Data Memo: The State of Blogging," January 2005, http://www.pewinternet.org/pdfs/PIP_blogging_data.pdf.

168. "Web Image Hosting Sites Show Explosive Growth, Corresponding to Sharp Rise in Blogging Activity, According to Nielsen/Net5Ratings; Teens Index Higher Than Other Demographic Groups for Image Hosting Web Sites," *PR Newswire*, September 13, 2005.

169. Ibid.

170. Shruti L. Mathur, "Teens Clicking into New Cliques," *Minneapolis Star Tribune,* August 17, 2005, 12W.

171. David Huffaker, "Gender in Online Identity and Language Use among Teenage Bloggers," (masters thesis, Communication, Culture, and Technology Program, Georgetown University, 2004), http://www.soc.northwestern.edu/gradstudents/huffaker/.

172. Nat Ives, "Nike Tried a New Medium for Advertising: The Blog," *New York Times,* June 7, 2004, C14. Diane Anderson, "Blogs: Fad or Marketing Medium of the Future?, Nike, Dr. Pepper, Mazda, SBC Try Marketing in the Blogosphere," *Brandweek*, November 29, 2004.

173. "Weblogs: Windows of Marketing Opportunity," *PR Week,* June 2, 2003. Mazda was attacked for creating "a fake blog with viral videos by a fictional 22-year-old blogger." Brian Morrissey, "Blogs Growing into the Ultimate Focus Group," *Ad Week,* June 20, 2005, 12.

174. For example, Google's AdSense and blog-advertisement networks like Blogads have been set up to facilitate the process. "In a Forrester Research survey last month, 64 percent of marketers expressed interest in advertising on blogs." Morrissey, "Blogs" 12.

175. Morrissey, "Blogs" 12.

176. Ibid.

177. Quote from Howard Kaushansky, CEO of Umbria, which was founded in March 2004 and counts Sprint and Electronic Arts among its clients. Morrissey, "Blogs" 12.

178. John Cate, Vice President and National Media Director, who heads Carat Interactive's blog practice. Quoted in: Morrissey, "Blogs," 12.

179. Look-Look, "Who We Are," http://www.look-look.com/looklook/html/Test_Drive_Who_We_Are.html (accessed April 20, 2006).

180. Ibid.

181. Look-Look, Inc., PowerPoint Presentation, 2005, http://www.thearf.org/downloads/Councils/Youth/2005-08-09_ARF_Youth_Look-Look.pdf.

182. E-mail sent to author from mycoke.com.

183. Lenhart, Madden, and Hitlin, *Teens and Technology*.

184. Ibid.

185. Ibid.

186. Ibid.

187. The report also noted that almost all teenagers in households with income levels greater than $75,000 per year were online, most of them with high-speed connections. Lenhart, Madden, Hitlin, *Teens and Technology*.

188. "From 'My Generation' to 'My Media Generation.'"

Chapter 6

1. Ginny Holbert, "MTV's N.Y. Loft Sleeps 7, Free Rent, No Privacy," *Chicago Sun-Times*, May 20, 1992, 49.

2. See Su Holmes and Deborah Jermyn, eds. *Reality Television* (New York: Routledge, 2004).

3. MTV, *The Real World: Las Vegas*, Episode 14, http://www.mtv.com/onair/dyn/realworld-season12/episode.jhtml?episodeID=60486.

4. Episode Guide, *The Real World: Las Vegas*, Season 12. http://www.mtv.com/onair/realworld/season12/episodeguide/index.jhtml?intNum=11.

5. MTV, "MTV Fight for Your Rights Protect Yourself Online Talk Show," December 10, 2002, http://www.mtv.com/onair/realworld/season12/chat.jhtml.

6. Social marketing is a widely used strategy in health communication. It is based on the principle that the same techniques for selling products can be used to sell ideas and influence health behaviors. See Philip Kotler, Ned Roberto, and Nancy Lee, *Social Marketing: Improving the Quality of Life*, 2d ed. (Beverly Hills, CA: Sage Publications, 2002).

7. Ad Council, "Campaigns," http://www.adcouncil.org/default.aspx?id=15; Kaiser Family Foundation, "Shouting to Be Heard: Public Service Advertising in a New Media Age, February 21, 2002, http://www.kaisernetwork.org/health_cast/hcast_index.cfm?display=detail&hc=464.

8. Personal Responsibility and Work Opportunity Reconciliation Act of 1996. Public Law 104–193. http://www.acf.dhhs.gov/programs/ofa/prwora96.htm.

9. While the Rose Garden signing ceremony was taking place, about a hundred protesters marched across the street, chanting "Shame, shame, shame!" The Republican challenger, Senator Bob Dole, who had lobbied behind the scenes to prevent the bill from getting to the president before the election, told the press that by signing the welfare-reform legislation into law, Clinton had "done everything but change parties." Larry Lipman, "Clinton Signs Welfare Bill Despite Objections of Allies," *Atlanta Journal-Constitution*, August 23, 1996, 12A.

10. Priscilla Pardini, "Federal Law Mandates 'Abstinence-Only' Sex Ed," Rethinking Schools Online 12, no. 4 (1998), http://www.rethinkingschools.org/archive/12_04/sexmain.shtml.

11. Nicholas D. Kristof, "Shaming Young Mothers," *New York Times*, August 23, 2002, A17.

12. See Jane D. Brown, Jeanne R. Steele, and Kim Walsh-Childers, eds., *Sexual Teens, Sexual Media: Investigating Media's Influence on Adolescent Sexuality* (Mahwah, NJ: Lawrence Erlbaum Associates, 2002). See L. Monique Ward and Rocio Rivadenyra, "Contributions of Entertainment Television to Adolescents' Sexual Attitudes and Expectations: The Role of Viewing Amount versus Viewer Involvement," *Journal of Sex Research* 36, no. 3 (1999): 237–250.

13. In the 1970s, communications professors George Gerbner and Larry Gross developed what became known as "cultivation theory," which argues that television programming shapes viewers' understanding of social reality. George Gerbner and Larry Gross, "Living with Television: The Violence Profile," *Journal of Communication* 26 (1976a): 172–199. This concept has influenced the work of many media scholars attempting to understand the influence of media content on behaviors and attitudes.

14. See Arvind Singhal and Everett M. Rogers, *Entertainment-Education and Social Change: History, Research, and Practice* (Lawrence Erlbaum Associates: 2004); See also "Entertainment Education and Health in the United States—Issue Brief" (Henry J. Kaiser Family Foundation: April 2004), http://www.kff.org/entmedia/7047.cfm. For additional background on the origins of entertainment-education efforts in the United States see Kathryn C. Montgomery, *Target: Prime Time: Advocacy Groups and the Struggle over Entertainment Television* (New York: Oxford University Press, 1989), 174–193.

15. David O. Poindexter, "A History of Entertainment-Education, 1958–2000," in Singhal and Rogers, eds., *Entertainment-Education and Social Change: History, Research, and Practice*.

16. The Media Project employed many of the standard strategies of other entertainment-education efforts, providing a constant stream of factual information about reproductive health, teen pregnancy, and other trends to producers and writers; working with individual shows to provide technical assistance, storyline ideas, and help for them to craft episodes in ways that are consistent with the organization's mission. The project set up a helpline, as an "on-call resource offering prompt assistance to TV and film writers on sexual health issues, including: contraception, unintended pregnancy, abortion, sexually transmitted diseases, including AIDS, adolescent sexuality, parent/child communication, peer pressure, healthy decision making, sexual orientation." The group also held regular briefings in Hollywood on the latest "hot topics" in teen sexual health. Through its SHINE awards, the Media Project also showcased and highlighted TV programs that had done a good job of depicting teen sexual-health issues. http://www.themediaproject.com/about/mission.htm. Another group working with entertainment television on teen pregnancy is the National Campaign to Prevent Teen Pregnancy. This private, nonprofit organization was established in 1996 and officially endorsed by the Clinton Administration prior to the passage of the welfare-reform bill. The group's goal was to reduce teen pregnancies by one-third by 2005. M. A. J. McKenna, "Teen Pregnancy; New Campaign Seeks to Determine Why U.S. Has Highest Rate and to Reduce 1 Million Yearly Figure by One-third by 2005," *Atlanta Journal and Constitution*, March 26, 1996. 5C. The National Campaign to Prevent Teen Pregnancy, "About Us," http://www.teenpregnancy.org/about/accomplishments.asp.

17. Advocates for Youth, "Who We Are, What We Believe, How We Are Different," http://www.advocatesforyouth.org/about/vision.htm.

18. Personal interview with Robin Smalley, Advocates for Youth, Media Project, June 6, 2002.

19. Sarah N. Keller and Jane D. Brown, "Media Interventions to Promote Responsible Sexual Behavior," *Journal of Sex Research* 39, no. 1 (February 2002): 67–73.

20. "Making a Difference One Scene at a Time," The Media Project, http://www.themediaproject.com/news/shows/index.htm (accessed June 17, 2005).

21. The Media Project, "The SHINE Awards," http://www.themediaproject.com/shine/index.htm.

22. As TV critic Lisa de Moraes of the *Washington Post* described the plot, "a female student who is running for class president performs oral sex on a male opponent in exchange for his support, a student works as a stripper, and a teacher has an affair with a student, among others." Lisa de Moraes, "Family Values Groups Accuse Fox of 'Boston Public' Indecency," *Washington Post*, February 6, 2002, C7.

23. Advocates For Youth, "Youth Action Center," http://www.advocatesforyouth.org/youth/advocacy/yan/index.htm.

24. A 2001 survey by the Kaiser Family Foundation found that two-thirds of young people had used the Internet to search for information about health. Many were seeking information on sexual issues as well, finding the Internet to be a particularly valuable tool because they could look up topics without anybody knowing about it. Forty-four percent of online teens had turned to the Internet for information on sexual health, including pregnancy, birth control, HIV/AIDS, and other sexually transmitted diseases. Kaiser Family Foundation, "Generation Rx.com: How Young People Use the Internet for Health Information," http://www.kff.org/entmedia/20011211a-index.cfm (accessed October 8, 2005).

25. American Social Health Association, IWannaKnow, http://www.iwannaknow.org/expert/index.html (accessed October 8, 2005).

26. Nancy L. Parello, Elizabeth M. Casparian, *The Roadmap: A Teen Guide to Changing Your School's Sex Ed,* Network for Family Life Education, Center for Applied Psychology, Rutgers University, 2000, http://www.sexetc.org/file/roadmap/roadmap_all.pdf.

27. Jennifer Egan, "Lonely Gay Teen Seeking Same," *New York Times Magazine,* December 11, 2000.

28. OutProud, http://www.outproud.org/.

29. Researchers studying the impact of media on adolescent sexual behavior have created a special "sexual media diet" measure to assess "the relative exposure to sexual content" in various media outlets. See Jane D. Brown, Kelly Ladin L'Engle, Carol J. Pardun, Guang Guo, Kristin Kenneavy, and Christine Jackson, "Sexy Media Matter: Exposure to Sexual Content in Music, Movies, Television and Magazines Predicts Black and White Adolescents' Sexual Behavior," *Pediatrics* 117, no. 4 (April 2006): 1018–1027; See also Jane D. Brown, Jeanne R. Steele, and Kim Walsh-Childers, "Introduction and Overview," in *Sexual Teens, Sexual Media: Investigating Media's Influence on Adolescent Sexuality,* 1–24.

30. Kaiser Family Foundation and MTV, "Reaching the MTV Generation," December 2003, http://www.kff.org/entmedia/Reaching-the-MTV-Generation-Report.cfm.

31. According to the Poynter Institute, the Kaiser Family Foundation had $550 million in assets and income in the $30–40 million range in 2001. Rick Edmonds, "A Kaiser Prescription for Healthcare News," *Poynter Report,* Spring 2001.

32. The organization's mission statement, as it appears on the Web site: "The Henry J. Kaiser Family Foundation is a nonprofit, private operating foundation focusing on the major health care issues facing the nation. The Foundation is an independent voice and source of facts and analysis for policymakers, the media, the health care community, and the general public." Kaiser Family Foundation, "About Us," http://www.kff.org/about/index.cfm.

33. Mollyann Brodie, Ursula Foehr, Vicky Rideout, Neal Baer, et al. "Communicating Health Information Through Entertainment Media," *Health Affairs* 20, no. 1 (2001).

34. Vince Stehle, "Sex, Truth, and Video"; The Chronicle of Philanthropy, October 8, 1998; telephone interview with Vicky Rideout, January 13, 2003; personal interview with Tina Hoff, Kaiser Foundation, March 24, 2005; personal interview with Robin Smalley, June 6, 2002.

35. Its 2003 study revealed that while overall sexual activity continued to be high, with 64 percent of all shows featuring some sexual content, at least more of them were including some reference to "safer sex" issues, such as "waiting to have sex, using protection, or the possible consequences of unprotected sex." The study showed that about a quarter of the shows that included either "talk about" sex or depictions of sexual intercourse also included some reference to safer sex, which was nearly double the rate the foundation had found four years earlier. A third of the shows with teen characters involved with sexual activity included a safe-sex message. And of the top twenty shows among teenaged viewers, "nearly half (45%) of the episodes that included a reference to sexual intercourse also included a reference to a safer sex topic." "TV Sex Getting 'Safer,' Kaiser Family Foundation Study Finds," Press Release, February 4, 2003. Kaiser Family Foundation, http://www.kff.org/entmedia/20030204a-index.cfm (accessed June 20, 2005).

36. Interview with Tina Hoff; Kaiser Family Foundation and MTV, "Reaching the MTV Generation," December 2003, http://www.kff.org/entmedia/Reaching-the-MTV-Generation-Report.cfm.

37. Kaiser Family Foundation, "Reaching the MTV Generation," December 2003, http://www.kff.org/entmedia/Reaching-the-MTV-Generation-Report.cfm.

38. "Viacom Looks to Web for Worldwide Growth," *Domains Magazine*, April 15, 2006, http://domainsmagazine.com/domain/Domains_19/Domain_151.shtml.

39. Kaiser Family Foundation, "Fact Sheet: Teenage Sexual and Reproductive Behavior in the United States," February 1997, http://www.kff.org/entmedia/loader.cfm?url=/commonspot/security/getfile.cfm&PageID=14459.

40. National Campaign to Prevent Teen Pregnancy, "United States Birth Rates for Teens 15–19," http://www.teenpregnancy.org/resources/data/brates.asp.

41. Kaiser Family Foundation, "Reaching the MTV Generation," December 2003, http://www.kff.org/entmedia/Reaching-the-MTV-Generation-Report.cfm.

42. Personal interview with Jaime Uzeta, Senior Director of Strategic Partnerships and Public Affairs, MTV Music Television, New York, December 9, 2002.

43. See R. Serge Denisoff, *Inside MTV* (New Brunswick, NJ: Transaction Books, 1988).

44. Kaiser Family Foundation, "Shouting to Be Heard"; Walter Gantz and Nancy Schwartz, *A Report on Television Content* (Indiana University, 2002), http://www.kaisernetwork.org/health_cast/uploaded_files/ContentStudy.KaiserPSAs.pdf.

45. Kaiser Family Foundation, "Shouting to Be Heard: Public Service Advertising: An Overview," "PSA's in a New Media Age," February 2002, http://www.kff.org/entmedia/20020221a-index.cfm (accessed October 12, 2005).

46. Even the groups that have been successful at working with Hollywood producers and writers often have had to frame their issues carefully in order to integrate them successfully into the storylines of television programs, which can require a careful balancing to avoid offending program advertisers. See Kathryn C. Montgomery, *Target: Prime Time*, 194–215.

47. Matt James, Tina Hoff, Julia Davis, and Robert Graham, "Leveraging the Power of the Media to Combat HIV/AIDS," *Health Affairs* 24, no. 3 (2005): 854–857.

48. Vince Stehle, "Sex, Truth, and Video."

49. Ibid.

50. Kaiser Family Foundation, "Reaching the MTV Generation," December 2003, http://www.kff.org/entmedia/Reaching-the-MTV-Generation-Report.cfm.

51. Vince Stehle, "Sex, Truth, and Video."

52. Press Release: "Rock for Life Slams MTV for Promoting Sexual Behavior among Teens," PR Newswire, October 28, 1998.

53. Kaiser Family Foundation, "Reaching the MTV Generation," 4.

54. MTV, "Fight For Your Rights: Protect Yourself, Videos," http://www.mtv.com/onair/ffyr/protect/videos.jhtml.

55. Kaiser Family Foundation, "Reaching the MTV Generation," 4.

56. MTV's other partners in the Fight for Your Rights: Protect Yourself initiative included: Planned Parenthood, Advocates for Youth, Rock the Vote, and Sex, etc., all of which took much more of an advocacy approach than that of the Kaiser Family Foundation. http://www.mtv.com/onair/ffyr/protect/takeaction.jhtml.

57. Kaiser Family Foundation, "Reaching the MTV Generation," 19.

58. Ibid.

59. Interview with Jaime Uzeta.

60. MTV, "Fight for Your Rights: Protect Yourself," "Sex Quiz," http:/www.mtv.com/onair/ffyr/protect/sexquiz.jhtml.

61. Interview with Jaime Uzeta.

62. Kaiser Family Foundation, "Reaching the MTV Generation," 5.

63. The tent also offered a sexual-health game quiz and a lounge area where PSAs and related shows were broadcast. According to the foundation, between September 1999 and April 2003, the campaign's TV specials ran a total of 84 times, reaching an audience of more than 96 million. Kaiser Family Foundation, "Reaching the MTV Generation," 5.

64. "Ramping It Up; Pro-Social Activity Heats Up, As Two New Campaigns and More Ambitious Programs Head for the Air," *Multichannel News*, May 2, 2005.

65. Interview with Tina Hoff.

66. MTV "Fight for Your Rights: Protect Yourself," "Online Talk Show," December 10, 2002, http://www.mtv.com/onair/realworld/season12/chat.jhtml (accessed August 10, 2004).

67. Mike Antonucci, "MTV to Carry Safe-Sex Message to Youths," *San Jose Mercury News*, March 22, 2000.

68. Kaiser Family Foundation, "Reaching the MTV Generation," 7–12.

69. "Kaiser Family Foundation, "Reaching the MTV Generation," 20. "List of the 2004 George Foster Peabody Award Winners," *The Associated Press*, March 31, 2004.

70. When the campaign began, the CBS network was part of Viacom. In 2006, CBS and Viacom split into two companies. "CBS, Viacom Formally Split," CBS News, January 3, 2006. http://www.cbsnews.com/stories/2006/01/03/business/main1176111.shtml. Kaiser partnered separately with CBS after the split. KNOW HIV/AIDS, "Learn about the Campaign," http://www.knowhivaids.org/learn.html.

71. BET, "About Rap-It-Up," http://www.bet.com/Health/aboutrapitup.htm.

72. Among the campaign's primary targets in the United States were "youth, people of color, women, and men who had sex with men." The choice of words was purposeful, designed to include not only gay men but also a particular subgroup of African-American men who had sexual relations with other males but who did not think of themselves as homosexual, a practice known as being "on the downlow." Interview with Tina Hoff.

73. Viacom, Press Release, http://www.viacom.com/view_brand.jhtml?inID=2§ionid=1.

74. E-mail to author from Vicky Rideout, Kaiser Foundation, May 21, 2002.

75. The episode aired May 12, 2003. I received a copy of the e-mail a few days before. It was distributed online through various listservs, including this one: http://rome.dartmouth.edu/scripts/wa.exe?A2=ind0305&L=wrac-l&T=0&P=4377.

76. Interview with Tina Hoff.

77. "KNOW HIV/AIDS Launches Groundbreaking Series of Messages on HIV Testing; Viacom and Kaiser Team Up with Crispin Porter + Bogusky to Create 13 New PSAs," *PR Newswire*, June 9, 2004.

78. Ric Kahn, "Knowing Is Beautiful—Or Is It? To Encourage Young People to Get Tested, AIDS Campaign Puts a Glossy Spin on Fighting a Deadly Disease—But Some Fear the Consequences," *Boston Globe*, March 20, 2005, 1.

79. Kaiser Family Foundation, "HIV/AIDS Public Education Campaign Wins Emmy," October 22, 2004, http://www.kff.org/entpartnerships/phip102204nr.cfm; The Peabody Awards, "Winners," http://www.peabody.uga.edu/archives/search.asp.

80. Kaiser Family Foundation, "Assessing Public Education Programming on HIV/AIDS: A National Survey of African Americans," October 2004, http://www.kff.org/entmedia/upload/Assessing-Public-Education-Programming-on-HIV-AIDS-A-National-Survey-of-African-Americans-Topline-Results.pdf.

81. Kaiser was by no means the first nonprofit to launch these kinds of media efforts in other countries. The U.S.-based Population Institute had been working with India and with other developing countries for decades on various entertainment-education initiatives aimed at education audiences about pro-social and public-health issues. See Singhal and Rogers, eds., *Entertainment-Education and Social Change*.

82. Interview with Tina Hoff; James, Hoff, Davis, and Graham, "Leveraging the Power of the Media to Combat HIV/AIDS."

83. Ibid.

The Heroes Project, launched in July 2004, is a national initiative in India that uses media and societal leaders to address HIV/AIDS. It was launched by Richard Gere and Parmeshwar Godrej and is conducted in partnership with the Kaiser Family Foundation and the Gates Foundation's Avahan Initiative. Heroes Project seeks to harness India's communication power and potential to address the spread of HIV/AIDS and reduce stigma and discrimination, by influencing public perception and policy through two platforms: mass media and advocacy. It correspondingly works through two avenues in its initiative: a mass media campaign and a societal leaders program.

Kaiser Family Foundation, "Heroes Project," http://www.kff.org/entpartnerships/india/index.cfm.

84. Wendy Melillo, "PSAs Fight to Be Heard," Adweek.com, October 31, 2005.

85. For a case history of the negotiations that led to the Master Settlement Agreement, see Michael Pertschuk, *Smoke in Their Eyes: Lessons in Movement and Leadership from the Tobacco Wars* (Nashville, TN: Vanderbilt Press, 2001).

86. Steven Schroeder, "Tobacco Control in the Wake of the 1998 Master Settlement Agreement," *New England Journal of Medicine* 350, no. 3 (2004): 293. Stuart Elliott, "'Youth' Rhymes with 'Truth,' and Both Are Central to a Big New National Campaign against Smoking," *New York Times*, February 4, 2000, C6.

87. "Tobacco Settlement Proceeds to Be Released to States, Tobacco Sales Down during First Year Since Settlement," *PR Newswire*, November 12, 1999; "FTC Drops Its Complaint against Cartoon Camel," *Associated Press*, March 1, 1999.

88. "Tobacco Settlement Proceeds to Be Released to States."

89. Its ambitious charge included: reducing the public's exposure to secondhand smoke, encouraging more smokers to quit, and ensuring that vulnerable populations could receive equitable access to smoking prevention and cessation programs. Cheryl Healton, Commentary, "Who's Afraid of the Truth," *American Journal of Public Health*, 91, no. 4 (2001): 554.

90. "Tobacco Use among Middle and High School Students—United States, 1999," Centers for Disease Control and Prevention, *Journal of the American Medical Association* 283, no. 9 (2000): 1134.

91. As scholar Constance Nathanson explained, organizations such as Group Against Smokers' Pollution (GASP) argued against the long-promoted notion that cigarette smoking is an individual decision that only affects the smoker.

> The hazards of smoking were relocated from the individual's risky behavior to the behavior of his smoking neighbor; exposure was no longer a matter of choice but of involuntary victimization; and finally, the responsibility for risk reduction was shifted away from the individual at risk to the 'polluting' smoker and to the regulatory agencies of government.

Constance A. Nathanson, "Social Movements as Catalysts for Policy Change: The Case of Smoking and Guns," *Journal of Health Politics, Policy and Law* 24, no. 3 (1999): 421–489.

92. For example, see "Special Report: Higher Cigarette Taxes Reduce Smoking, Save Lives, Save Money," and other similar reports at Campaign for Tobacco-Free Kids, http://www.tobaccofreekids.org/reports/prices.

93. K. Michael Cummings and Hillary Clarke, Department of Cancer Control and Epidemiology, Roswell Park Cancer Institute for the Advocacy Institute's Health Science Analysis Project, April 14, 1998. http://www.advocacy.org/publications/mtc/counterads.htm (accessed June 22, 2005).

94. California Department of Health Services, "Tobacco Control Section," http://www.dhs.ca.gov/tobacco/ (accessed June 22, 2005).

95. Stanton A. Glantz and Edith D. Balbach, *Tobacco War: Inside the California Battles* (Berkeley: University of California Press, 2000), 126–130.

96. Bob Moseley, "Smoke Signals: Can Magazines Handle the Truth.com?" *Folio*, August 2000.

97. Glantz and Balbach, *Tobacco War*, 330–343.

98. Campaign for Tobacco-Free Kids, "Big Tobacco Still Targeting Kids," September 19, 2005, http://www.tobaccofreekids.org/reports/targeting; Jack O'Dwyer,

"Tobacco-Free Kids Set to Kick Butts," *Jack O'Dwyer's Newsletter*, February 25, 1998, 7.

99. Ira Teinowitz, "Anti-Smoking Campaign Delayed," *Advertising Age* (Midwest Region Edition), January 31, 2000, 3–5.

100. Cheryl Healton, Commentary, "Who's Afraid of the Truth," 554.

101. "Multistate Settlement Agreement: The 'Vilification Clause' Does Not Restrict State Tobacco Prevention Spending," Campaign for Tobacco-Free Kids, April 18, 2000, http://www.tobaccofreekids.org/research/factsheets/index.php?CategoryID=8. Complete MSA available at http://www.naag.org/backpages/naag/tobacco/msa.

102. Healton, "Who's Afraid of the Truth," 554.

103. Adam Bryant, "In Tobacco's Face: The Latest Antitobacco Ads Are Hyperaggressive, and That Has Giant Philip Morris Smoking Mad," *Newsweek*, March 20, 2000, 40; "Estimated Exposure of Adolescents to State-Funded Anti-Tobacco Television Advertisements—37 States and the District of Columbia, 1999–2003," *Journal of the American Medical Association* 295, no. 7 (2006): 751–752.

104. Healton, "Who's Afraid of the Truth," 554.

105. Personal interview with Cheryl Healton, President, Chief Executive Officer, American Legacy Foundation, September 24, 2004.

106. Teressa Iezzi, "Truth," *Clientology*, June 1, 2001, http://www.boardsmag.com/articles/magazine/20010601/truth.html.

107. Some of the funds from the MSA were used to create an online archive of tobacco industry documents from the court cases. See Stanton A. Glantz, John Slade, Lisa A. Bero, Peter Hanauer, and Deborah E. Barnes, eds., *The Cigarette Papers* (Berkeley, CA: University California Press, 1996). http://ark.cdlib.org/ark:/13030/ft8489p25j.

108. See *The CQ Researcher* 4, no. 36 (September 1994): 36, 842–860.

109. Interview with Thornton. Teressa Iezzi, "Truth," *Clientology*, June 1, 2001, http://www.boardsmag.com/articles/magazine/20010601/truth.html.

110. Iezzi, "Truth," http://www.boardsmag.com/articles/magazine/20010601/truth.html.

111. "Truth (SM) Outbreak Tour Hits the Road," *PR Newswire*, June 5, 2001.

112. American Legacy Foundation, "Progress Report," 2002–2003.

113. Betsy Spethmann, "Responsibility Marketing," *Promo*, October 2001.

114. Interview with Cheryl Healton.

115. See Montgomery, *Target: Prime Time.*

116. Stuart Elliott, "'Youth' Rhymes with 'Truth.'"

117. Bruce Butterfield, "Networks Balk at Government's Anti-Smoking Ads," *Boston Globe*, January 25, 2000.

118. Interview with Cheryl Healton.

119. Chris Reidy, "Anti-Smoking Spots Embroil Boston Advertising Agency in Controversy."

120. Ira Teinowitz and David Goetzl, "Anti-Tobacco Ads Hit Big Roadblock; Foundation Pushing $150 Million Effort Rebuffed by ABC and CBS," *Advertising Age*, February 7, 2000, 59; Adam Bryant, "In Tobacco's Face," 40.

121. Adam Bryant, "In Tobacco's Face," 40; "TV Networks Rebuff New Anti-Smoking Ads," *Boston Globe*, February 8, 2000, D10.

122. Adam Bryant, "In Tobacco's Face," 40.

123. Skip Wollenberg, "Antismoking Group Pulls Two of Its First Four Commercials Amid Complaints," *Associated Press*, February 15, 2000.

124. Marc Kaufman, "Fuming over Smoking Ads; Advocates Outraged as Anti-Tobacco Foundation, Pressured by Industry, Pulls Disputed Spots from Airwaves," *Washington Post*, February 20, 2000, A3.

125. Skip Wollenberg, "Once-Rejected Ad against Smoking Makes American Network TV Debut on Olympics," *The Associated Press*, September 19, 2000.

126. Cheryl Healton, "Big Tobacco's Broken Vows," *Advertising Age*, February 5, 2001, 18.

127. Ogilvy PR, "Industry Awards," http://www.ogilvypr.com/about-ogilvy-pr/awards.cfm; Alicia Griswold, "'Truth' Shares Grand Effie Award," AdWeek.com, June 5, 2003.

128. Matthew C. Farrelly, Kevin C. Davis, M. Lyndon Haviland, Peter Messeri, and Cheryl G. Healton, "Evidence of a Dose-Response Relationship between "truth" Antismoking Ads and Youth Smoking Prevalence," *American Journal of Public Health* 95, no. 3 (2005): 425–232.

129. Bob Moseley, "Smoke Signals: Can Magazines Handle the Truth.com?" *Folio*, August 2000.

130. Personal interview with Amber Thornton, Executive Vice-president, American Legacy Foundation, November 5, 2004.

131. Street Theory, http://www.streettheory.org (accessed June 10, 2004).

132. See Kathryn Montgomery, Barbara Gottlieb-Robles, and Gary O. Larson, *Youth as e-Citizens: Engaging the Digital Generation* (Washington, DC: American University, 2004), 21. http://www.centerforsocialmedia.org/ecitizens/index2.htm.

133. Interview with Cheryl Healton.

134. Interview with Amber Thornton.

135. "Truth Debuts Three New Ads during MTV's Video Music Awards; Campaign also Sponsors a Brand New MTV2 Award on mtv.com," *PR Newswire*, September 5, 2001.

136. Lisa Bertagnoli, "A Little Close to Home," *Marketing News*, August 27, 2001, 3.

137. Bill McConnell, "Bad Spot; Tobacco Company Asks FCC to Ban Dog-Urine Ads," *Broadcasting & Cable*, November 12, 2001.

138. "Hopefully this is a one-time incident in which the tobacco companies demonstrate that they haven't really changed by threatening the foundation whenever it produces hard-hitting ads that put the tobacco industry in a bad light," Matthew Myers, head of the Campaign for Tobacco-Free Kids, told the press. Skip Wollenberg, "Antismoking Group Pulls Two of Its First Four Commercials." Officials at the American Legacy Foundation explained their decision as a necessary move to prevent the campaign from being undermined and derailed by the powerful tobacco lobby. As Cheryl Healton commented to reporters: "There are very strong incentives for the political forces dependent on the larger streams of money to weigh in." Explaining that the board believed it had been on sound legal ground when it decided to approve the ads, there was "another political interpretation that might be different." Marc Kaufman, "Fuming over Smoking Ads," A3.

139. "Truth Debuts Three New Ads."

140. Other measures also were identified as important factors, including the banning of tobacco marketing targeted at youth, another outcome of the settlement with the tobacco companies. "Cigarette Smoking among American Teens Declines Sharply in 2001," *Ascribe Newswire*, December 19, 2001.

141. "The Week," *Advertising Age*, January 28, 2002, 10.

142. Greg Winter, "Antismoking Group Sues to Preserve an Ad Campaign's Tone," *New York Times*, February 14, 2002, A23.

143. Andrea Weigl, "Greensboro, N.C.-Based Tobacco Company Sues over 'Truth' Ads," *News & Observer*, February 20, 2002.

144. In addition to the urine ad, Lorillard lawyers cited the body-bag spot, as well as another ad that had run on April Fools Day in which an actor playing a tobacco-company spokesman announced that the industry was going to recall all cigarettes until it could develop one that would not cause health problems. The ad ended with

a voice over, saying "April Fool." Rita K. Farrell, "Tobacco Maker Says It Was 'Vilified' in Ads," *New York Times*, May 11, 2005, C5.

145. Weigl, "Greensboro, N.C.-Based Tobacco Company Sues over 'Truth' Ads."

146. Interview with Healton.

147. Farrell, "Tobacco Maker," C5.

148. "Tobacco Firm Loses Bid to Cut Group's Funding," *Los Angeles Times*, August 23, 2005. C6.

149. In a national sample of approximately 50,000 students in grades 8, 10, and 12 surveyed each spring between the years 1997 and 2002, researchers at the Research Triangle Institute in North Carolina found prevalence among all students declined from 25.3 percent to 18 percent between 1999 and 2002, and that the truth® campaign had accounted for 22 percent of that decline. The numbers translated into 300,000 fewer smokers as a result of the campaign. The study was the first to examine behavioral outcomes of the campaign. It found that while smoking among teens already was declining when the campaign began, rates had accelerated during its first two years, more than doubling from 3 percent to 7 percent. Farrelly, Davis, Haviland, Messeri, and Healton, "Evidence of a Dose-Response Relationship between "truth" Antismoking Ads and Youth Smoking Prevalence."

150. Kim Krisberg, "Anti-Smoking Campaign Lowers Youth Smoking Rates with 'truth'," *The Nation's Health*, April 2005.

151. Some advocates believed such a precise figure had been carefully calculated by the tobacco companies to trigger a stop in the payments when the slightest decline in market share occurred. Interview with Phillip Wilbur, November 1, 2006.

152. Ira Teinowitz, "Question Is Whether 'Truth' Will Win Out," *Advertising Age*, July 16, 2001, 3.

153. In 2003, ALF launched a public-service campaign with the Advertising Council. Its message was much tamer than the controversial countercommercials that continued to be the foundation's trademark. The Ad Council, which the advertising industry created during World War II, had run countless PSA campaigns for decades. The images from these PSAs made memorable impressions on the public mind—from Smokey the Bear, to the Crash Dummies of seatbelt-use fame, to the crying Indian who urged people to protect the environment. But these mainstream, prosocial messages all played it safe, focusing on individual responsibility and avoiding any content that might ruffle the feathers of the corporations that were part of the large advertising and media industry infrastructure. The ALF campaign with the Ad Council was no exception to this long-standing tradition. Wendy Melillo, "American Legacy Adds Ad Council to Address Issues," *Brandweek*, June 2, 2003, 8. The Ad Council "Public Service Advertising That Changed a Nation," http://www.adcouncil.org/research/adweek_report (accessed October 24, 2005).

154. Wendy Melillo, "Legacy Seeks Partner as Ad Budget Dwindles," *Ad Week*, May 23, 2005, 11.

155. Kim Krisberg, "Anti-Smoking Campaign Lowers Youth Smoking Rates with 'Truth,'" 1–2.

156. Robert Putnam, *Bowling Alone: The Collapse and Revival of American Community* (New York: Simon & Schuster, 2000); Michael X. Delli Carpini, "The Youth Engagement Initiative Strategy Paper," Pew Charitable Trusts, September 2000, http://www.pewtrusts.com/misc_html/pp_youth_strategy_paper.cfm.

157. Ad Council, "Youth Civic Engagement," http://www.adcouncil.org/default.aspx?id=26.

158. "2004 Youth Vote: A Comprehensive Guide," Center for Information and Research on Civic Learning and Engagement. http://www.civicyouth.org/PopUps/2004_votereport_final.pdf.

159. The Center for Information and Research in Civic Learning and Engagement (CIRCLE) and the Carnegie Corporation of New York, *The Civic Mission of Schools* (New York: Carnegie Corporation, 2003), http://www.civicmissionofschools.org (accessed December 4, 2003), 4.

160. Dana Markow, "Editorial: Our Take on It," *Trends and Tudes* July 2003, http://www.harrisinteractive.com/news/newsletters/k12news/HI_Trends&Tudes News2003_V2_iss7.pdf.

161. Elizabeth Hollander, "Disengaged Youth, What's a Politician to Do?" Brown University News Service, August 18, 2000, http://www.brown.edu/Administration/News_Bureau/2000-01/00-011.html.

162. Michael X. Delli Carpini, *Gen.com: Youth, Civic Engagement, and the New Information Environment* Working Paper, http://depts.washington.edu/ccce/events/carpini.htm (accessed August 8, 2001).

163. Harvard University's Institute of Politics, for example, published a guide to reaching young voters. Harvard University Institute of Politics, "Polling," http://www.iop.harvard.edu/research_polling.html. The Pew Charitable Trusts and the Carnegie Corporation of New York funded the Center for Information & Research on Civic Learning & Engagement at the University of Maryland. Center for Information & Research on Civic Learning & Engagement, http://www.civicyouth.org/. See also University of Washington's Center for Communication and Civic Engagement, http://www.youth04.org/, http://depts.washington.edu/ccce/Home.htm.

164. For example, in 1993 the National Community Service Act was passed, establishing the Corporation for National and Community Service, and launching AmeriCorps.

AmeriCorps is a network of local, state, and national service programs that connects more than 70,000 Americans each year in intensive service to meet critical needs in education, public safety, health, and the environment. AmeriCorps incorporated two existing national service programs: the longstanding VISTA (Volunteers in Service to America) program, created by President Lyndon Johnson in 1964 and the National Civilian Community Corps (NCCC).

From AmeriCorps, "History," http://www.americorps.org/about/ac/history.asp.

165. America's Promise, http://www.americaspromise.org.

166. "Save the Children is the leading independent organization creating real and lasting change for children in need in the United States and around the world. It is a member of the International Save the Children Alliance, comprising 27 national Save the Children organizations working in more than 110 countries to ensure the well-being of children." From Save the Children, "About Save the Children," http://www.savethechildren.org/about/index.asp?stationpub=i_hpdda1_au1&ArticleID=&NewsID=.

167. Personal interview with Diane Ty, Washington, DC, January 14, 2002.

168. Montgomery, Gottlieb-Robles, and Larson, *Youth as e-Citizens*, 13–53.

169. Interview with Diane Ty, January 14, 2002.

170. Ibid. See also "Quick Facts: Volunteering/Community Service." Center for Research & Information on Civic Learning & Engagement. http://www.civicyouth.org/quick/volunteer.htm.

171. Interview with Diane Ty, January 14, 2002.

172. Emphasis in original. "Summary of KIDS4KIDS Focus Group Research," unpublished document provided to author by Diane Ty.

173. "Summary of KIDS4KIDS Focus Group Research."

174. When YN staff tried to promote the issue of land mines, for example, teens in the United States expressed no interest. Interview with Diane Ty, September 26, 2002.

175. Youth Noise, "Just 1 Click," http://www.youthnoise.com/site/CDA/CDA_Page/0,1004,79,00.html (accessed July 2, 2003).

176. Since 1985, the Working Assets nonprofit group has offered Visa or Mastercard credit cards with ten cents of every purchase donated "to nonprofit groups working for peace, human rights, equality, education and the environment—at no extra cost to you." Cause Marketing Forum, "Working Assets, http://www.causemarketingforum.com/framemain.asp?ID=132.

177. Zimmerman, Ann. "Promotional Ties to Charitable Causes Help Stores Lure Customers," *Wall Street Journal*, December 4, 2000, B1.

178. Cone Incorporated, *2000 Cone/Roper Teen Survey: Influence of Cause Marketing on GenY*, August 2000.

179. Interview with Diane Ty, January 14, 2002.

180. Youth Noise, "About Youth Noise: Meet Our Partners," http://www.youthnoise.com/site/CDA/CDA_Page/0,1004,151,00.html.

181. Interview with Diane Ty, January 14, 2002.

182. Between May 2000 and July 2001, 538 Internet companies closed. Jon Swartz, "Webbys Go On Despite Dot-Bomb Threat," *USA Today* 18 July 2001, 5B; http://www.newsbytes.com/news/01/168081.html (accessed 20 July 2001).

183. After filing for bankruptcy under Chapter 11 in January 2001, the U.S. Interactive reorganization plan won court approval in September 2001. Form 8-K, "Current Report Pursuant to Section 13 or 15(d) of the Securities Exchange Act of 1934, U.S. Interactive, Inc.," February 12, 2002.

184. Gayle Forman, "Girls in Exile" *Seventeen*, September 2001, 194–197. 203 Phil Noble, "Analysis of the Role of the Internet," http://www.cyberacties.nl/politicsonline.html (accessed September 17, 2001).

185. Phil Noble, "Analysis of the Role of the Internet," http://www.cyberacties.nl/politicsonline.html (accessed September 17, 2001).

186. Ibid.

187. Alternet, "Where Are We Now?" September 25, 2001, http://www.alternet.org/wiretapmag/story.html?StoryID=11585 (accessed October 1, 2001).

188. Interview with Diane Ty, June 24, 2003.

189. Internal YouthNOISE memo dated October 1, 2001, provided to author by Diane Ty. YouthNOISE maintained its visible profile in the online civic landscape. By 2003, it had become independent from Save the Children and had moved its headquarters from Washington, D.C., to San Francisco, where its new executive director was Ginger Thompson, cofounder of the teen online financial site, DoughNET, which had gone out of business during the dot-com crash. There were 500,000 unique visitors to the YouthNOISE site every month, with 90,000 registered users from 170 countries and all 50 states, plus Washington, D.C. Teen girls still constituted the majority of visitors, though now their numbers were only at 70 percent, down considerably from the 85 percent figure of a few years before. More than $38,000 had been raised for various nonprofit causes through the site's "Just-1-Click" campaigns. In an independent poll conducted on behalf of the project, 86 percent of survey respondents said their participation on the YouthNOISE site had "driven them to be more socially and politically active." The YouthNOISE Impact Report, 2004. Telephone interview with Ginger Thompson, November 9, 2005.

190. For an account of youth response to 9/11 on the Web, see Montgomery, Gottlieb-Robles, and Larson, *Youth as e-Citizens*, 75–85.

191. Youth Radio, http://www.youthradio.org (accessed September 15, 2001).

192. In one poem, posted on September 14, a teenager from the Latin American Youth Center's YouthBuild program offered an expression of grief and determination:

New York
They took a bite out of The Apple
They tried the White House, the Pentagon, and the Capitol
Millions startled, the core was taken
They took the wings from an angel
We all shed tears but we will not live in fear
They took my heart but I still have my soul
Danaron mi corazon pero dejaron mi alma.

Latin American Youth Center's YouthBuild Program, September 14, 2001, http://www.youthradio.org/wtc/poetry.shtml (accessed October 1, 2001).

193. DoSomething, http://www.dosomething.org/newspub/story.cfm?id=999&sid=230&cid=33 (accessed September 13, 2001).

194. Montgomery, Gottlieb-Robles, and Larson, *Youth as e-Citizens*, 78.

195. Carpini, *Gen.com*.

196. "Short-Term Impacts, Long-Term Opportunities: The Political and Engagement of Young Adults in America," Analysis and Report for the Center for Information & Research in Civic Learning & Engagement (CIRCLE), the Center for Democracy and Citizenship, and the Partnership for Trust in Government at the Council for Excellence in Government, March 2002, http://www.civicyouth.org/research/products/national_youth_survey.htm (accessed November 15, 2005).

197. Anna Greenberg, "New Generation, New Politics," *The American Prospect* 14, no. 9 (2003): A3.

Chapter 7

1. See Howard Rheingold, *Smart Mobs: The Next Social Revolution* (Cambridge, MA: Perseus Publishing, 2002).

2. David Cho, "Voting Machines on Trial in Fairfax; Ill-Fated Fall Vote Prompts Scrutiny," *Washington Post*, February 9, 2004, B1; Sarah Sennet and Adam Piore, "The Age of E-Voting," *Newsweek*, April 5, 2004, 15.

3. Joe Trippi, *The Revolution Will Not Be Televised: Democracy, the Internet, and the Overthrow of Everything* (New York: Regan Books, 2004).

4. See Martha McCaughey and Michael D. Ayers, eds., *Cyberactivism: Online Activism in Theory and Practice* (New York: Routledge, 2003).

5. Rock the Vote, "Rock the Vote 2004 Election Campaign," ND. Summary of 2004 election accomplishments of Rock the Vote. Author's personal files.

6. For detailed case histories of the PMRC campaign see Eric Nuzum, *Parental Advisory: Music Censorship in America* (New York: Harper Collins, 2001), 13–43; see also Danny Goldberg, *Dispatches from the Culture Wars: How the Left Lost Teen Spirit* (New York: Miramax, 2003).

7. As Eric Nuzum points out in his book *Parental Advisory: Music Censorship in America,* Tipper Gore and Susan Baker's connections to congressional leaders served as leverage that encouraged industry cooperation, especially because they were married to lawmakers who were involved in legislation that could benefit the industry, such as the Home Audio Recording Act. This put PMRC into a particularly powerful position and helped promote cooperation from RIAA. Nuzum, *Parental Advisory,* 20.

8. Bruce Haring, "Lyrics Concerns Escalate," *Billboard,* November 11, 1989, 1–2.

9. Bill Holland and Chris Morris, "Industry Ready to Fight La. [Louisiana] Labeling Bill," *Billboard* (July 21, 1990) 5, 84; Dave DiMartino, "Rock the Vote Calls for Support," *Billboard,* April 6, 1991, 6, 89.

10. Holland and Morris, "Industry Ready to Fight La. [Louisiana] Labeling Bill" 5.

11. Dave DiMartino, "Music Biz Hopes to 'Rock' Voting Booths," *Billboard,* September 15, 1990, 104.

12. Holland and Morris, "Industry Ready to Fight La. Labeling Bill."

13. "Madonna Joins MTV in 'Rock' the Vote' Campaign; MTV to Air 'Rock the Vote' Special Oct. 22," *Business Wire,* October 18, 1990; As *New York Times* described the show: "Madonna, the crucifix-and-lingerie pop idol who has rapped for people with AIDS, the environment and other causes, waves the American flag for freedom of speech in her latest public-service video and wiggles in red panties, bra and combat boots to get out the vote next month." Robert McFadden, "Wrapped in Flag, Madonna Raps for Vote," *New York Times*, October 20, 1990, 7.

14. DiMartino, "Rock the Vote Calls for Support," 6–8; The National Voter Registration Act, tagged the "Motor Voter Bill," required all state motor-vehicle agencies to provide voter applications to people seeking licenses and registrations. Jonathan Landay, "Voter Registration Hits High, but Will People Cast Ballots?" *Christian Science Monitor,* August 20, 1996, 1. As *Billboard Magazine* reported: "Senate staffers say the music industry's Rock the Vote campaign played an instrumental part in grassroots lobbying to convince senators to pass a pending federal voter registration bill May 20 by a vote of 61–38." Bill Holland, "Rock the Vote Assists Passage of Motor Voter Bill," *Billboard,* June 6, 1992, 8.

15. Thom Duffy, "The Spotlight Turns to Freedom in the Arts: Wife's Crusade Has Music Biz Wary of Gore," *Billboard,* July 25, 1992, 1. Anthony De Curtis, "Tipper:

Dems Send Wrong Message," *Rolling Stone,* September 3, 1992. An assessment of Al Gore's voting record in *Billboard* magazine concluded that the candidate's overall stance had been pro-industry, only "interrupted by his awkward, freshman-senator involvement in the 'information-only,' industry-bashing hearing on objectionable records lyrics by the Senate Commerce Committee." Though Tipper Gore still served on the PMRC board, it looked as if that organization was not the threat it had been a few years before, the article noted, adding that "since the agreement between the RIAA and the PMRC/PTA for a standardized sticker in 1990, the group has no longer been in the forefront of antiporn activities, but has become a clearinghouse for parents and institutions interested in the lyrics controversy." Bill Holland, "Al Gore Gets (Mostly) High Marks for His Music-Industry Record," *Billboard,* July 25, 1992, 70. In an interview with NPR, Susan Estrich, Dukakis's campaign manager, said that Gore had been less successful at raising money from Hollywood and reaching out to some liberal groups because of his wife's involvement with the issue. Morning Edition, July 10, 1992.

16. As Danny Goldberg pointed out, Bill Clinton already had criticized some popular music during his presidential campaign, chastising Jesse Jackson for including rapper Sister Souljah on a panel at the Rainbow Coalition, and joining the chorus with others on both tickets against Ice-T for the controversial "Cop Killer" album. But according to Goldberg, these were strategic moves by Clinton "to differentiate himself from Jackson and from previous Democratic nominees," pointing out that focus groups of swing voters in Michigan "had revealed that many white working-class Democratic voters had switched to Reagan and Bush because they felt that the Democratic party was too deferential toward blacks, as evidenced by the visibility at the 1988 convention of the Reverend Jesse Jackson." Goldberg also notes that Clinton's subsequent appearance playing the saxophone on Arsenio Hall and being interviewed by MTV correspondent Tabitha Soren on the network's Choose or Lose voter-registration campaign (in partnership with Rock the Vote) were a demonstration of Clinton's "political skills that made him a legend." Goldberg, *Dispatches,* 178–180.

17. "Rock the Vote," *Time,* June 15, 1992, 66–68.

18. Dierdre A. Depke, "Talk-Show Campaigning Helps Candidates Connect," *Business Week,* October 26, 1992, 34.

19. "Rock the Vote," 66. According to an article in the *Economist,* the presidential candidate had a rather uneasy experience with the youth channel. "Mr. Clinton . . . was expertly grilled by a 'representative' audience in the studios of MTV. . . . On June 15th . . . Mr. Clinton collided with the youth culture again on June 13th, at a convention of Jesse Jackson's Rainbow Coalition," where he "criticized Sister Souljah, a rap singer, . . . over an interview she did with the *Washington Post,* in which she seemed to suggest that blacks should take a week off from killing blacks and kill whites instead. Mr. Jackson, incandescent with rage, accused Mr. Clinton of

trying to use the rap singer's remarks as an opportunity to appeal to white voters." Anonymous, "The Campaign: Rockers and Rappers," *Economist,* June 20, 1992, 25.

20. Goldberg, *Dispatches,* 180.

21. Rock the Vote, "Mission & Timeline," http://www.rockthevote.org/mission.html (accessed June 20, 2001).

22. As Danny Goldberg explained:

> For the first time in history, the percentage of 18–24 year olds who voted increased—the turnout was 20 percent higher than in 1988, bringing the total youth vote to 11 million. Clinton defeated Bush by sixteen points among this group, for whom freedom of expression was a major issue. Rock the Vote, the largest registration group aimed at young voters, was explicitly formed to combat censorship in entertainment, and MTV's voter-registration slogan, "Choose or Lose," also spoke to young voters' concern with free expression, a topic that has been a mainstay of MTV Music News since Gore's wife, Tipper, embarked on her campaign for warning stickers on albums in the mid-1980s. A continuation of the Reagan/Bush FCC's policies against "indecency" would be a betrayal of those younger voters and many older ones as well.

Danny Goldberg, "Are You 'Decent'?" *The Nation,* December 21, 1992, 760–761.

23. Anne Gowen, "Motor Voter: Goin' Mobile," *Rolling Stone,* July 8, 1993, 18.

24. Cyndee Miller, "Promoting Voting: It's Goodwill—And Good Business, Too," *Marketing News,* October 26, 1992, 1–4.

25. "Smackdown Your Vote!" World Wrestling Entertainment, http://vote.wwe.com.

26. Bruce Haring and Alex Pappademas, "Onward Backstreet Soldiers! Boys Draft Army of Besotted Teens to Cyberhype Upcoming CD," FortuneCity, http://tinpan.fortunecity.com/foottap/24/articles/a224.htm (accessed October 27, 2005).

27. Carla Hay, "Rock the Vote Expands Its Efforts beyond Elections," *Billboard,* February 13, 1999, 5–7.

28. Rock the Vote, http://www.rockthevote.org (accessed 21 June 2001).

29. Hay, "Rock the Vote," 6.

30. For example, Robert Wood Johnson provided support for distribution of a booklet to U.S. youth on health-care reform, "Rock the System." Margaret Litvin, "Youth Advocates Ready to Rock; 'Pocketbook' Ruckus Ready to Roll," *USA Today,* July 20, 1994, 2A.

31. Pew Charitable Trusts, "Grant Detail: Public Policy," http://www.pewtrusts.com/search/search_item.cfm?grant_id=3626. According to the Pew Web site, the trusts gave RTV another $100,000 in 2001 "to better understand young American's issue concerns in light of the events of September 11, and their aftermath."

32. Derrick Depledge, "'Rock the Vote' Missed the Boat; Only About 30 Percent of 18– to 24–Year-Olds Cast Ballots," *Houston Chronicle,* November 23, 1996, A10. The

group did take credit for registering "a combined 1.4 million new voters" during the 1996 presidential, 1998 congressional, and 2000 presidential elections. Rock the Vote, "Mission & Timeline," http://www.rockthevote.org/mission.html (accessed June 20, 2001).

33. Rock the Vote, "Street Teams," http://rockthevote.org/streetteams.html (accessed June 6, 2003).

34. Center for Information & Research on Civic Learning & Engagement (CIRCLE), http://www.civicyouth.org/index.shtml.

35. Harvard Institute of Politics, "National Campaign," http://www.iop.harvard.edu/events_national_campaign.html.

36. Center for Information & Research on Civic Learning & Engagement, "Research: National Youth Survey 2004," http://www.civicyouth.org/research/products/national_youth_survey2004.htm.

37. Winning with Young Voters, http://www.winningwithyoungvoters.org.

38. Kathryn Montgomery, Barbara Gottlieb-Robles, and Gary O. Larson, *Youth as e-Citizens: Engaging the Digital Generation* (Washington, DC: American University, 2004) 21, http://www.centerforsocialmedia.org/ecitizens/index2.htm.

39. Republican Youth Majority, http://www.rym.org.

40. W. Lance Bennett and Michael Xenos, "Young Voters and the Web of Politics: Pathways to Participation in the Youth Engagement and Electoral Campaign Web Spheres," Working Paper 20, August 2004, Center for Information & Research on Civic Learning & Engagement. http://depts.washington.edu/bennett/about-cv.html.

41. Garance Franke-Ruta, "Virtual Politics: How the Internet Is Transforming Democracy," *The American Prospect* 14, no. 9 (2003): A6.

42. Don Hazen, "Moving On: A New Kind of Peace Activism," AlternetNet.org, February 11, 2003, http://www.alternet.org/print.html?StoryID=15163; MoveOn Documentation, http://www.moveon.org/documentation.html (accessed March 13 2003).

43. Author's notes from presentation by Wes Boyd, President, MoveOn.org, Take Back America Conference, Campaign for America's Future, Washington, DC, June 2, 2004.

44. Grant Williams, "Advocacy Group's Online Savvy Nets More Than Donations," *Chronicle of Philanthropy* 17 (April 2003): 26.

45. Bill Werde, "Friends and Foes of 'Fahrenheit' Lobby Everyone," *New York Times*, June 30, 2004, E1; Joe Garofoli, "MoveOn, a Political Force Online, Receives $5 Million Matching Gift," *San Francisco Chronicle*, November 23, 2003, A4.

46. Michael B. Cornfield, *Politics Moves Online: Campaigning and the Internet* (New York: The Century Foundation Press, 2004).

47. Gary Wolf, "Weapons of Mass Mobilization," *Wired*, September 2004, http://wired-vig.wired.com/wired/archive/12.09/moveon.html?pg=4&topic=moveon&topic_set=.

48. John Tierney, "Anti-War Demonstration That Does Not Take to the Streets," *New York Times Online*, February 26, 2003, http://www.nytimes.com/2003/02/26/national/26CND-MARCH.html?th (accessed March 1, 2003); Juliet Eilperin, "'Virtual March' Floods Senate with Calls Against an Iraq War," *Washington Post*, February 27, 2003, A22.

49. Wolf, "Weapons."

50. Michael Janofsky, "The 2004 Campaign: Advertising; Bush-Hitler Ads Draws Criticism," *New York Times*, January 6, 2004, A18.

51. Gary Wolf, "Weapons."

52. Robert Putnam, *Bowling Alone: The Collapse and Revival of American Community* (New York: Simon & Schuster, 2000).

53. Franke-Ruta, "Virtual Politics," A6.

54. About Meetup, http://www.meetup.com/about.

55. Michael X. Delli Carpini, "Gen.com: Youth, Civic Engagement, and the New Information Environment," *Political Communication* 17, no. 4 (October–December 2000): 341–349. See also The Henry J. Kaiser Foundation, *Media, Youth, and Civic Engagement—Fact Sheet*, Fall 2004, http://www.kff.org/entmedia/7168.cfm.

56. Bruce A. Williams and Michael X. Delli Carpini, "Heeeeeeere's Democracy," *Chronicle of Higher Education*, April 19, 2002, B14.

57. *Cable and Internet Loom Large in Fragmented Political News Universe*, Pew Research Center for People and the Press, January 11, 2004. http://people-press.org/reports/display.php3?ReportID=200.

58. Joseph Graf and Carol Darr, "Political Influentials Online in the 2004 Presidential Campaign," Washington, DC: Institute for Politics, Democracy, and the Internet, February 5, 2004, http://www.ipdi.org/UploadedFiles/political%20influentials.pdf.

59. Graf and Darr, "Political Influentials," 5.

60. As Katherine Murray wrote in her 2004 master's thesis, *New Media Campaign Techniques and Young Voter Engagement*, Joe Trippi, campaign manager for the Dean for America campaign embraced supporters' use of Meetup

> by adding a link to the Dean home page directing individuals to sign up for a MeetUp in their hometown. Other candidates had MeetUp followings as well, but it was the Dean MeetUps that

inspired the company to pay special attention to political MeetUps, hiring a seasoned Washington insider to serve as their director of political affairs.

Katherine Murray, "New Media Campaign Techniques and Young Voter Engagement" (masters thesis, Georgetown University, 2004).

61. "The Open Source Campaign," in Trippi, *The Revolution Will Not Be Televised,* 135–156.

62. Graf and Darr, "Political Influentials," http://www.ipdi.org/UploadedFiles/political%20influentials.pdf.

63. According to CIRCLE, "While there are more young people eligible to vote in 2004 than in the previous four election years, young people represent a declining proportion of the voting eligible population. Since the 1970s, the percentage of eligible voters who are between the ages of 18 and 29 has fallen from 30 percent in 1972 to 21 percent in 2004." Center for Information & Research on Civic Learning & Engagement, "The 2004 Presidential Election and Young Voters," Fact Sheet, October 28, 2004.

64. Memorandum, "Finding from a Recent National Survey Among 15–25 Year Olds," Lake, Snell and Perry and The Tarrance Group, commissioned by the Center for Democracy and Citizenship at the Council for Excellence in Government and the Center for Information & Research on Civil Learning & Engagement, January 15, 2004, http://www.civicyouth.org/research/products/national_youth_survey2004.htm.

65. Thea Singer, "Having Their Say; Mobilization Efforts Driving Hard to Bring Youths to Ballot in November, *Boston Herald*, October 5, 2004, 47.

66. Nat Ives, "Putting Out the Message That Registering and Then Voting Should Have a Place in the Youth Culture," *New York Times*, July 13, 2004, 15.

67. Tim Dickinson, "The Youth Vote," *Rolling Stone*, November 11, 2004, 50, 54–57.

68. Ives, "Putting out the Message," 15.

69. Singer, "Having Their Say," 47; New Voters Project, "About the New Voters Project," http://www.newvotersproject.org/about_the_new_voters_project.

70. Singer, "Having Their Say," 47.

71. Ibid; New Voters Project, http://www.newvotersproject.org/about_the_new_voters_project.

72. Declare Yourself, http://www.declareyourself.com/aboutus/aboutus.htm; Singer, "Having Their Say," 47.

73. Declare Yourself, "About Us," http://www.declareyourself.com/aboutus/aboutus.htm.

74. Ives, "Putting Out the Message," 15.

75. Ibid; Declare Yourself, "Christina Aguilera Kicks Off 'Only You Can Silence Yourself' Voter Registration Campaign," September 17, 2004, http://www.declareyourself.com/events/events.htm.

76. Music for America, "About Music for America," http://www.musicforamerica.org/about.

77. Music for America, "Music for America: Fusion of Youth Movement and Politics Ignites a Movement," http://www.musicforamerica.org/node/16673.

78. Scott Thill, "Beatboxing at the Ballot Box," AlterNet, September 17, 2004, http://www.alternet.org/mediaculture/19911 (accessed November 25, 2005).

79. Ibid.

80. Felicia R. Lee, "Hip-Hop Is Enlisted in Social Causes," *New York Times*, June 22, 2002, 7.

81. Betsy Spethmann, "Playing Politics," *Promo Magazine*, July 1, 2004, 8.

82. "Sean 'P. Diddy' Combs and His Army of Politically Conscious Celebrities Launch Massive Citizen Change Vote or Die Outdoor Campaign," PR Newswire, October 6, 2004.

83. "Influence of Popular Media on Youth: Politics and Voting," PowerPoint Presentation (ND), Lake, Snell and Perry. Presentation cited ACT African-American Battleground Survey conducted by Brilliant Corners research, March 2004, as source of information. Author's personal files.

84. As Lyor Cohen, Chairman/CEO of U.S. Recorded Music at Warner Music Group, explained:

> "We're not so naïve as to believe that putting voting-awareness stickers on millions of CDs is going to change things overnight. But the sticker is just one element in an array of initiatives that we believe can help stimulate voter turnout among young people. Following the logic that a single advertisement does not necessarily translate into a CD sale, we hope to have a cumulative impact on voter involvement by taking advantage of multiple daily touch points with music consumers."

Lyor Cohen, "Count Me In," *Billboard*, October 16, 2004, 10.

85. Aaron Baar, "Ad Council, WestWayne Call Young Voters to Action," *Ad Week*, March 29, 2004, 15. Ad Council, http://www.adcouncil.org/campaigns/register_vote (accessed November 25, 2005).

86. NALEO Educational Fund, "Voces del Pueblo," http://www.naleo.org/voces_del_pueblo.htm (accessed November 25, 2005).

87. National Coalition on Black Civic Participation, "Programs and Initiatives: Black Youth Vote!" http://www.bigvote.org. Singer, "Having Their Way," 47.

88. Redeem the Vote, http://www.redeemthevote.com (accessed November 25, 2005).

89. Lisa Neff, "Young Voters Come Out," *The Advocate* (Los Angeles: January 20, 2004) Iss. 906, 46–49.

90. Kristin V. Jones, "Who Let the Punks Out?" *The Nation*, June 7, 2004, 11.

91. Ibid.

92. Montgomery, Gottlieb-Robles, and Larson, *Youth as e-Citizens*, 22.

93. "New Web Site Encourages Student Participation in Politics," *The Daily Cardinal* (University of Wisconsin), March 23, 2004.

94. "College Campaign Warns Students That Bush Policy Is Leading to Post-Election Military Draft," *US NewsWire*, October 18, 2004.

95. Author's notes, presentation by Ben Brandzel to Principles of Strategic Communication Class, American University, Washington, DC, November 17, 2004.

96. Ariana Eunjung Cha, "Grass-roots Politics with Click of a Mouse; In Silicon Valley, Tech-Driven Support Groups," *Washington Post*, October 25, 2004, A3.

97. Lisa Napoli, "Recovery Room," *New York Times*, October 21, 2004, G3.

98. Betsy Spethmann, "Playing Politics," *Promo Magazine*, July 1, 2004, 8.

99. Personal interview with Chandler Spaulding, Communications Director; Lynne Lyman, Program Director; Sarah Rosenberg, Development Director, Rock the Vote headquarters, Los Angeles, California, June 4, 2002.

100. "Rock the Vote Rocks the Web with Strategic Alliance," *Business Wire*, July 15, 2004. To increase its Web visibility and traffic, Rock the Vote relied on the expertise of top consultants in the Web advocacy business, integrating online registration tools into its database systems. Spethmann, "Playing Politics," 8.

101. Rock the Vote, "Join the Street Team Today," http://www.rockthevote.com/cst/index.php (accessed November 26, 2005).

102. Rock the Vote, "Official Merchandise" http://www.giantmerchandising.com/retail/web/rockthevote/index.htm (accessed November 26, 2005).

103. Rock the Vote, "RTV Gear," http://www.rockthevote.com/rtv_gear.php.

104. Rock the Vote, "Blog," http://blog.rockthevote.com (accessed November 26, 2005).

105. Rock the Vote, "Partners," http://www.rockthevote.com/partners_rock.php (accessed November 26, 2005).

106. According to online encyclopedia *Wikipedia,* skins are "custom graphical appearances (GUIs) that can be applied to certain software and websites in order to suit the different tastes of different users." http://en.wikipedia.org/wiki/Skin_%28computing%29.

107. Kinley Levack, "Voting Rocks IM," *EContent,* November 2004.

108. "Rock the Vote and MySpace.com Join Forces to Mobilize Young Voters," *PR Newswire,* April 26, 2004.

109. "About Rock the Vote Mobile," http://www.rtvmo.com/about.htm (accessed January 22, 2005).

110. Todd Wasserman, "Moto Rocks Vote Wirelessly," *Adweek,* April 14, 2004.

111. "Rock the Vote and Motorola Team to Mobilize Electorate; Youth Market's the Target in Drive to Excite, Educate and Engage for 2004 Election; Mobile Handsets Provide Easy, Anytime-Anywhere Platform for Political participation," *PR Newswire,* March 2, 2004.

112. Spethmann, "Playing Politics," 8.

113. Ibid.

114. Rock the Vote, "Rock the Vote 2004 Election Campaign."

115. Rock the Vote Mobile, http://www.rtvmo.com/index1.htm (accessed January 22, 2005).

116. Rock the Vote, "Rock the Vote 2004 Election Campaign."

117. A spokesman for the group told the press in July that it already had 400,000 names and was expecting to reach a million by election day. Spethmann, "Playing Politics," 8.

118. "Young Voters Favor Kerry but Find Bush More Likeable," Press Release, September 21, 2004, Center for Information & Research on Civic Learning & Engagement, http://www.civicyouth.org/research/products/data4.htm.

119. "Election Interest from Young Voters Is Up Sharply from 2000," Press Rlease, The Joan Shorenstein Center on the Press, Politics & Public Policy, March 12, 2004, http://www.ksg.harvard.edu/presspol/vanishvoter/Releases/release031104.shtml.

120. Siobhan Mcdonough, "Not a Breakout for Youth After All," Associated Press, November 3, 2004.

121. Mark Lopez, Presentation to American University class. Spring, 2005. Personal communication from Mark Lopez to author (November 28, 2005).

122. C. W. Nevius, "18 to 24s—Unnoticeable at the Polls," *San Francisco Chronicle,* November 6, 2004, B1.

123. "Census Data Shows Youth Voter Turnout Surged More Than Any Other Age Group," Center for Information & Research on Civil Learning & Engagement, Press Release, May 26, 2005. A subsequent report published by the center noted that "increase in youth voter turnout cut across all demographic groups, contributing to the highest youth voter turnout since 1992." Mark Hugo Lopez, Emily Kirby, and Jared Sagoff, *The 2004 Youth Vote,* Center for Information & Research on Civic Learning & Engagement, 2005, http://www.civicyouth.org/quick/youth_voting.htm.

124. Peter Levine, Deputy Director, CIRCLE and Ivan Frishberg, New Voters Project/State PIRGs, "The Re-Engaged Generation 2004 and Beyond," PowerPoint presentation, (ND), http://www.google.com/search?hl=en&lr=&ie=ISO-8859-1&q=Re-engaged+generation&btnG=Search.

125. Lopez, Kirby, and Sagoff, *The 2004 Youth Vote,* 1.

126. Lee Rainie, Michael Cornfield, and John Horrigan, "The Internet and Campaign 2004," Pew Internet & American Life Project and the Pew Research Center for the People and the Press, March 6, 2005, http://www.pewinternet.org/pdfs/PIP_2004_Campaign.pdf.

127. Fair use is a provision of copyright law that allows individuals to copy, for example, a chapter of a book for research or educational purposes. Lawrence Lessig, *Free Culture: How Big Media Uses Technology and the Law to Lock Down Culture and Control Creativity* (New York: Penguin Press, 2004), 141–143.

128. Robin D. Gross, "DMCA Takes Full Effect—Millions of Americans Become Criminals," EFFector Online Newsletter, a publication of the Electronic Frontier Foundation, December 13, 2000, http://www.eff.org/effector/HTML/effect13.11html#1 (accessed June 11, 2005).

129. Even as recently as 2005, a survey by the Pew Internet & American Life Project found that 75 percent of teenagers continued to download music illegally, believing that it was "unrealistic to expect people not to do it." Amanda Lenhart and Mary Madden, "Teen Content Creators and Consumers," Pew Internet & American Life Project, November 2, 2005, http://www.pewinternet.org/PPF/r/166/report_display.asp.

130. Personal interview with Holmes Wilson, Tiffiniy Cheng, and Nicholas Reville, November 18, 2004.

131. Interview with Wilson, Cheng, and Reville. Martha McCaughey and Michael Ayers use the term "cyberactivism" in their 2003 anthology of the same name to describe a variety of organizations and activities using the Web to promote political causes. They see cyberactivism taking on many varied forms:

"[S]mall and large networks of wired activists have been creating online petitions, developing public awareness websites connected to traditional political organizations (e.g., Amnesty International online), building spoof sites that make political points (such as worldbunk.org), creating online sites that support and propel real-life (RL) protest (e.g., a16.org, which stands for April 16, the date of the World Trade Organization (WTO) protest in Washington, DC.), designing websites to offer citizens information about toxic waste, and creating organizations (e.g., Indymedia.org) that have expanded to do traditional RL activities."

Martha McCaughey and Michael D. Ayers, *Cyberactivism: Online Activism in Theory and Practice* (London: Routledge, 2003): 1.

132. Downhill Battle, "itunes," http://www.downhillbattle.org/itunes (accessed November 27, 2005).

133. Sam Howard-Spink, "Grey Tuesday, Online Cultural Activism and the Mash-Up of Music and Politics," First Monday, http://www.firstmonday.org/issues/issue9_10/howard (accessed June 13, 2005).

134. Interview with Wilson, Cheng, and Reville.

135. What a Crappy Present, "CDs Make Bad Gifts for Kids," http://www.whatacrappypresent.com.

136. Interview with Wilson, Cheng, and Reville.

137. Howard-Spink, "Grey Tuesday," http://www.firstmonday.org/issues/issue9_10/howard.

138. Ibid.; Bill Werde, "Defiant Downloads Rise from Underground," *New York Times*, February 25, 2004, E3.

139. A survey in 2005 by the Pew Internet & American Life Project, for example, found that 57 percent of teenagers could be classified as "content creators." Many were creating blogs or Web pages, posting original artwork, photography, stories, or videos online. They were also remixing online content into their own new creations. Amanda Lenhart and Mary Madden, "Teen Content Creators and Consumers," Pew Internet & American Life Project, November 2, 2005, http://www.pewinternet.org/PPF/r/166/report_display.asp.

140. "Open Source Movement, "Wikipedia," http://en.wikipedia.org/wiki/Open_source_movement.

141. http://participatoryculture.org.

142. "Democratizing Television," Participatory Culture Foundation, http://www.participatoryculture.org/channel.

143. *Free Culture: Phase Two, Next Generation Strategy for Media Democracy and Participatory Culture. Conference Proceedings,* American University (March 2006), http://www.centerforsocialmedia.org/files/pdf/Free_Culture_Conference_Report.pdf.

144. Downhill Battle, "Music Activism," http://www.downhillbattle.org/?seenIEPage=1 (accessed November 29, 2005).

145. Future of Music Coalition, "Future of Music Manifesto," http://www.futureofmusic.org/manifesto (accessed November 29, 2005).

146. Public Knowledge, "About Public Knowledge," http://www.publicknowledge.org/about (accessed November 29, 2005). Other organizations involved in the open-access movement include: Consumer Federation of America, Consumers Union, Media Access Project, Free Press, and the Center for Digital Democracy.

147. Electronic Frontier Foundation, "Mission," http://www.eff.org/mission.php.

148. Lawrence Lessig, *Code and Other Laws of Cyberspace* (New York: Basic Books, 1999); Lawrence Lessig, *The Future of Ideas: The Fate of the Commons in a Connected World* (New York: Random House 2001); Lawrence Lessig, *Free Culture.*

149. Lessig, *Free Culture,* 184.

150. Ibid., xiv.

151. Creative Commons, http://creativecommons.org.

152. Lessig, *Free Culture,* 303.

153. For example, author J. D. Lasica echoed the call for reform of digital copyright law, offering further illustrations of how current regulations could curtail many common uses of content that many people took for granted. "Today, as digital media begin to stream through our homes, we want to hold on to that tangible relationship," Lasica explains in his book *Darknet: Hollywood's War against the Digital Generation.* "The songs on our iPods, the television shows we capture on TiVo, the music videos in our new portable video players, the movies we watch in our DVD collections—we believe that these digital slices of media also belong to us in a real sense." J. D. Lasica, *Darknet: Hollywood's War against the Digital Generation* (Hoboken, NJ: John Wiley and Sons, 2005), 16.

154. Personal interview with Nelson Pavlovsky, November 18, 2004; Sabrina Rubin Erdely, "The Paperless Chase," *Mother Jones,* May–June, 2004, http://www.motherjones.com/news/hellraiser/2004/05/04_403.html.

155. Interview with Pavlovsky.

156. Erdely, "The Paperless Chase."

157. FreeCulture.org was officially founded in April 2004. In May 2005, the group incorporated and began developing into a more formal nonprofit organization. FreeCulture.org, "About," http://freeculture.org/about.php.

158. Lawrence Lessig, *Free Culture,* entire book available for free download at http://www.free-culture.cc.

159. These online campaigns combined humor, pop culture, and righteous indignation. For example, on the Barbieinablender Web site, the activists celebrated the

victorious court case won by the ACLU on behalf of the artists sued by the Mattel corporation for photographing the iconic doll in a blender. http://barbieinablender.org (accessed November 30, 2005).

160. The first freeculture gathering was held in early 2005. In May of that year, American University hosted a FreeCulture2 conference, with many of the same participants, along with Washington media-policy advocacy groups. I organized and hosted this second meeting. (See *Free Culture: Phase Two, Next Generation Strategy for Media Democracy and Participatory Culture, Conference Proceedings.*) In April 2006, freeculture.org and other groups convened a student summit at Swarthmore College. http://freeculture.org/blog/2006/03/31/student-summit/.

161. "Free Culture Manifesto," http://www.freeculture.org/manifesto.php (accessed June 13, 2005).

Chapter 8

1. Lana Castleman, "2016: A 10-year-old's Odyssey," *Kidscreen,* January 1, 2006, 38.

2. As the Pew Internet & American Life Project found, some teenagers have chosen not to go online because of negative experiences, while many remain disconnected owing to economic constraints. Amanda Lenhart, Mary Martin, and Paul Hitlin, *Teens and Technology* (Washington, DC: Pew Internet & American Life Project, July 27, 2005), http://www.pewinternet.org/pdfs/PIP_Teens_Tech_July2005web.pdf.

3. Several studies have tracked the inequities in access to the Internet. For example, according to a study by the Leadership Conference on Civil Rights: "Slightly more than half of all black and Latino children have access to a home computer and approximately 40 percent have access to the Internet at home (compared to 85.5 and 77.4 percent of white, non-Latino children). Ethnic and racial disparities in home computer and Internet access rates are larger for children than for adults." Robert W. Fairlie, "Are We Really a Nation Online? Ethnic and Racial Disparities in Access to Technology and Their Consequences," Leadership Conference on Civil Rights, September 20, 2005, http://www.civilrights.org/issues/communication/details.cfm?id=36098. A 2005 survey by the Children's Partnership revealed a significant "digital opportunity gap" affecting minority and low-income children. "While 77% of children in school ages 7–17 from higher-income households (earning more than $75,000 per year) use a home computer to complete school assignments, only 29% of children from households earning less than $15,000 annually do so." The Children's Partnership, "Measuring Digital Opportunity for American's Children: Where We Stand and Where We Go From Here," 2005, http://www.childrenspartnership.org/AM/Template.cfm?Section=Home&Template=/CM/ContentDisplay.cfm&ContentFileID=1089. See also Kaiser Family Foundation, "Issue Brief: Children, the Digital Divide, and Federal Policy," September 2004, http://www.kff.org/entmedia/upload/Children-The-Digital-Divide-And-Federal-Policy-Issue-Brief.pdf.

4. Federal Communications Commission, "Universal Service," http://www.fcc.gov/wcb/universal_service/welcome.html.

5. Ellen Seiter, *The Internet Playground: Children's Access, Entertainment, and Mis-Education* (New York: Peter Lang Publishing, 2005), 1. See also Tony Wilhelm, Delia Carmen, and Megan Reynolds, "Kids Count Snapshot: Connecting Kids and Technology: Challenges and Opportunities," Annie E. Casey Foundation, June 2002, http://www.aecf.org/publications/data/snapshot_june2002.pdf.

6. As media scholar Marsha Kinder argued: "With the aid of journalists, and industry and bipartisan support from Congress, Clinton and Gore could also help fetishize computers and the Internet as the new miraculous media that would somehow enable our children to not merely catch-up with educational achievements of other nations, but leapfrog ahead on what Gore dubbed the world's information superhighway." Marsha Kinder, ed., *Kids' Media Culture* (Durham, NC: Duke University Press, 1999), 14–15.

7. Kinder, *Kids Media Culture,* 13. Henry Giroux points out that the Clinton Administration, by positioning itself as an advocate for children's welfare in the media-culture area, was able to portray itself as pro-child, while at the same time abandoning the policies that had protected poor children through passage of the welfare reform legislation. Henry Giroux, *Stealing Innocence: Corporate Culture's War on Children* (New York: Palgrave, 2000), 41–42.

8. Scholars have found that the category of childhood itself is socially constructed, emerging in the eighteenth century as a product of Western economic and social forces of the era, and evolving over time since then. This idea has been the subject of long-standing academic debate and discussion, with social theorists and historians arguing that prior to the eighteenth century children were considered miniature adults. For a review of some of the key works in this literature, see Juliet Schor, *Born to Buy: The Commercialized Child and the New Consumer Culture* (New York: Scriber, 2004), 199–203. See also Jyotsna Kapur, "Television and the Transformation of Childhood," in Marsha Kinder, ed., *Kids' Media Culture,* 122–136; and David Buckingham, *After, the Death of Childhood: Growing Up in the Age of Electronic Media* (Cambridge, Polity Press, 2000).

9. A search on the library database Proquest, for example, turned up more than 1,200 references to "tween" in a variety of mainstream and specialized publications over a six-month period. Proquest search conducted at American University's library database, March 3, 2006. The concept also has been successfully promulgated in other countries. See Martin Lindstrom and others, *Brandchild* (London, UK: Kogan Page, 2003).

10. Neil Postman, *The Disappearance of Childhood* (New York: Vintage Books, 1994).

11. AOL's Parental Controls section divides control categories into "kids only (12 and under), young teen (13–15) and mature teen (16–17)." All Info about Internet

for Beginners, "AOL's Parental Controls," http://internetbeginners.allinfoabout.com/articles/parentalcontrols.html. For a discussion of the contemporary marketplace of screening software, blocking technologies, and safe zones, see Julie Frechette, "Cyber-Democracy or Cyber-Hegemony? Exploring the Political and Economic Structures of the Internet as an Alternative Source of Information," *Library Trends* 53, no. 4 (2005): 555. "Youth, Pornography and the Internet," Computer Science and Telecommunications Board, The National Academies of Science, 2002, http://books.nap.edu/catalog/10261.html.

12. As Julie Frechette observed, "advertising is overlooked as 'inappropriate content' because it is part of everyday consumer culture, unlike pornographic and hate sites, which exist beyond the boundaries of what is deemed 'good' for children and teenagers." Frechette, "Cyber-Democracy or Cyber-Hegemony?" 555.

13. Amanda Lenhart, "Protecting Teens Online," Pew Internet and American Life Project, March 17, 2005, http://www.pewinternet.org/pdfs/PIP_Filters_Report.pdf.

14. Walter Minkel, "Bush Signs Dot-Kids into Law," *School Library Journal*, January 2003, 18; Amy Lisewski Lavell, "In the Name of In(ternet)decency: Laws Attempting to Regulate Content Deemed Harmful to Children," *Public Libraries*, November/December 2004, 353.

15. William Triplett, "Broadcast Indecency: Should Sexually Provocative Material Be More Restricted?" *CQ Researcher*, April 16, 2004, 323–330.

16. Center for Democracy and Technology, "The Court Challenge to the Child Online Protection Act," http://www.cdt.org/speech/copa/litigation.shtml (accessed March 24, 2006).

17. Gretchen Ruethling, "27 Charged in International Online Child Pornography Ring," *New York Times*, March 16, 2006, 18; Tom Jackman, "Sex Abusers of Children Are Facing Deportation; Agents Target Foreign-Born Criminals," *Washington Post*, March 23, 2004.; Lance Pugmire, "49 Arrested in Internet Child-Molesting Sting; Riverside County Prosecutor Says the Results of the Operation are 'Alarming.'" *Los Angeles Times*, January 14, 2006, B3.

18. As one mother wrote in an advice column: "I didn't realize the harmful nature of the Web site until I received an anonymous copy of my 16-year-old daughter's MySpace page. . . . I will be monitoring my daughter's activity if not deleting it altogether, but what about all of the parents who are unaware of this Web site? What has our culture degenerated to when photos of girls drinking from tequila bottles, imitating sexual acts, and wearing bras and garter belts in the midst of other boys are posted online for anyone to see?" Letter to the "Dear Amy" column in the *Washington Post*, March 17, 2006.

19. While other companies complied, at least in part, Google refused. Though a federal judge supported Google, it did require the search engine to supply some information from its database.

20. Chris Gaither, "U.S. Is Denied Google Queries; Privacy Activists Hail a Federal Judge's Ruling. But He Orders the Search Engine to Reveal Some Information about Web Sites in Its Database," *Los Angeles Times,* March 18, 2006. A1.

21. Jennifer Barrett Ozols, "We Need to Do More," *Newsweek,* September 25, 2004.

22. Michele Greppi and Melissa Grego, "Nets Turn to V-Chip as Savior," *Television Week*, April 2, 2004, 1.

23. Triplett, "Broadcast Indecency," 323–330.

24. John Eggerton, "NBC Adopts Content Ratings," *Broadcasting & Cable,* April 28, 2005.

25. Ibid.

26. Jube Shiver Jr., "Broadcast Violence Gets New Scrutiny," latimes.com, May 30, 2005.

27. Ken Belson, "FCC Sees Cable Savings in a la Carte," *New York Times,* February 10, 2006, C1.

28. The Parents Television Council was established in the mid-1990s, initially as the "Hollywood project of the Media Research Center," garnering support from well-known entertainment-industry celebrities, including comedian Steve Allen. L. Brent Bozell III founded both organizations and serves as president for the PTC and the MRC. Parents Television Council, "Parents Television Council 2005 Annual Report," http://www.parentstv.org/PTC/aboutus/main.asp. Media Research Center, "Media Research Center 2004 Annual Report," http://www.mediaresearch.org/about/annualreport/MRC_AnnualReport2004.pdf

29. Parents Television Council, "About Us," http://www.parentstv.org/PTC/aboutus/main.asp (accessed March 24, 2006).

30. The report, called "MTV Smut Peddlers: Targeting Kids with Sex, Drugs and Alcohol," looked at 171 hours of MTV's 2004 Spring Break programming, concluding that MTV's reality shows averaged 13 sexual scenes per hour, and its music videos averaged 32 instances of foul language per hour. The music network fired back: "It's unfortunate that Mr. Bozell has yet again attempted to unfairly and inaccurately paint MTV with a brush of irresponsibility around sexual and violent content," the network said in a formal statement the press. "He gravely underestimates young peoples' intellect and level of sophistication. To even imply that young people are likely to base their opinions on important issues after watching one hour of television is nothing short of ludicrous." Pat Nason, "Analysis: Raunchy Sex on MTV?" *UPI,* February 1, 2005.

31. Mike Shields, "Forecast 2006: Interactive Media," *Mediaweek,* January 2, 2006, 12.

32. Matt Richtel, "Hungry Media Companies Find a Meager Menu of Web Sites to Buy," *New York Times,* March 16, 2006, C1.

33. Responding to concerns over online safety on the Web site, Newscorp officials promised to appoint a "safety czar" to help assuage parents' fears. The company said nothing about marketing safeguards, although it does discuss advertisers' worries about the safety issues affecting their own effectiveness. Julia Angwin and Brian Steinberg, "News Corp. Goal: Make MySpace Safer for Teens," *Wall Street Journal,* February 17, 2006, B1.

34. Richtel, "Hungry Media," C1. James Detar, "MTV, Intel Will Market in Harmony; Music Channel in 'Overdrive'; Pair Singing the Praises of Intel's Viiv Technology for Helping Multimedia Content," *Investor's Business Daily,* March 14, 2006, A4.

35. The Federal Trade Commission review of the COPPA rules in 2005 left them in place without any changes. Federal Trade Commission, Press Release, "FTC Retains Children's Online Privacy Protection (COPPA) Rule Without Changes," March 8, 2006, http://www.ftc.gov/opa/2006/03/coppa_frn.htm.

36. Larry Dobrow, "Privacy Issues Loom for Marketers," *Advertising Age,* March 13, 2006, S6.

37. After the Center for Media Education closed its doors in 2003, the Electronic Privacy Information Center continued to monitor compliance with COPPA.

38. Commericial Alert, "Our Mission," http://www.commercialalert.org/about.php (accessed March 25, 2006). At the height of the dot-com boom, Commercial Alert led a successful effort against ZapMe!, a Silicon Valley company that planned to offer schools free computers and Internet service in exchange for the opportunity to conduct database marketing and stream online advertising to students. Commercial Alert, "ZapMe," http://www.commercialalert.org/issues/education/zapme.

39. Claire Atkinson, "Commercial Alert Seeks FTC Buzz Marketing Investigation; Singles Out P&G's Tremor for 'Targeting of Minors,'" *AdAge.com,* October 18, 2005. A summary of the complaint can be found at http://www.commercialalert.org/issues/culture/buzz-marketing. According to Gary Ruskin, Executive Director of Commercial Alert, as of April 2006 the FTC had not acted on the complaint. E-mail from Gary Ruskin to author, April 11, 2006.

40. Susan Linn, *Consuming Kids: The Hostile Takeover of Childhood* (New York: The New Press, 2004), 23.

41. The report's position on psychologists' involvement in market research was worded carefully:

> In sum, while we recognize that line-drawing may well prove to be difficult, we nonetheless believe that some research efforts are capable of crossing the line of appropriate sensitivity to the unique vulnerabilities of young people in this realm. Given that judgment, we believe it is important for the field of psychology to help sensitize its members to the potential ethical

challenges involved in pursuing efforts to more effectively advertise to children, particularly those who are too young to comprehend the persuasive intent of television commercials.

"Report of the APA Task Force on Advertising and Children, Section: Psychological Issues in the Increasing Commercialization of Childhood," February 20, 2004, http://www.apa.org/releases/childrenads.pdf.

42. As communications professor Dale Kunkel, one of the advocates who participated in the FCC rulemaking, explained, among the new rules was an extension of the commission's long-standing "host-selling" prohibition. In its new digital-television ruling, the FCC agreed to prohibit programs from promoting Web sites in their programming, if the site "uses characters from the program to sell products or services." As Kunkel explains, "if the Nickelodeon Web site features *Spongebob Squarepants* in a product ad for Kraft Macaroni & Cheese, the Web site address could not be displayed during the *Spongebob Squarepants* television program nor could an ad for the Nickelodeon Web site appear during the same program." Dale Kunkel, "Kids Media Policy Goes Digital: Current Developments in Children's Television Regulation," in J. Alison Bryant and Jennings Bryant, eds., *The Children's Television Community: Institutional, Critical, Social Systems, and Network Analyses* (Mahwah, NJ: Lawrence Erlbaum Associates, 2006). The commission's actions in issuing the rules, however, triggered a backlash from the television industry, which filed a suit against the FCC, arguing the rules were too strict, and the advocates filed their own suit, charging the rules were not strong enough. Ultimately, the children's advocates conducted a series of negotiations with representatives from the major broadcast and cable TV networks, agreeing in December 2005 to a compromise that upheld the commercial limits as described above. In March 2006, the FCC issued a Notice of Proposed Rulemaking, asking for comment on the agreement. With a buy-in from major parties in interest, an approval was expected. John Eggerton, "FCC Takes Up Kids Compromise," *Broadcasting & Cable,* March 10, 2006.

43. Campaign for a Commercial-Free Childhood, http://www.commercialfree childhood.org (accessed March 25, 2006). For a list of some of the groups involved in anticommercialism efforts on behalf of children, see Linn, *Consuming Kids,* 221–232.

44. The lawsuit planned to ask the Massachusetts court to "enjoin the companies from marketing junk foods to audiences where 15 percent or more of the audience is under age eight, and to cease marketing junk foods through Web sites, toy giveaways, contests, and other techniques aimed at that age group." Campaign for a Commercial-Free Childhood, Press Release, "Parents and Advocates Will Sue Viacom and Kellogg: Lawsuit Aimed at Stopping Junk-Food Marketing to Children by Kellogg and Viacom's Nickelodeon," January 18, 2005, http://www .commercialfreechildhood.org/pressreleases/nickkellogglawsuit.htm.

45. See Kaiser Family Foundation, "Issue Brief: The Role of Media in Childhood Obesity," February 2004, http://www.kff.org/entmedia/upload/The-Role-Of-Media -in-Childhood-Obesity.pdf.

46. See Institute of Medicine of the National Academies, *Preventing Childhood Obesity: Health in the Balance.* September 30, 2004. http://www.iom.edu/?id=25048.

47. Press Release, "Food Marketing Aimed at Kids Influences Poor Nutritional Choices, IOM Study Finds; Broad Effort Needed to Promote Healthier Products and Diets," Press Release, The National Academies, December 6, 2005. Institute of Medicine of the National Academies, *Food Marketing to Children and Youth,* National Academy of Sciences, 2006, http://www.iom.edu/?id=31330&redirect=0.

48. "Executive Summary," *Food Marketing to Children and Youth,* 12.

49. "Food Marketing Aimed at Kids Influences Poor Nutritional Choices."

50. Burger King, "Subservient Chicken," http://www.subservientchicken.com (accessed March 25, 2006). Dan Sewell, "Companies Use Online Magazines, Entertainment to Lure Customers," *Associated Press,* January 2, 2006.

51. *Communications Daily,* March 24, 2006. Julie Bosman, "Chevy Tries a Write-Your -Own-Ad Approach, and the Potshots Fly," *New York Times,* April 4, 2006, C1.

52. The Web site also uses Google Analytics to track user behavior on the site. "CKE Restaurants, Inc. Drives Brand Lift with Google Analytics," http://www.google.com/analytics/case_study_cke.html (accessed March 25, 2006).

53. T. L. Stanley, "P&G, Kraft Brands Adopt 'OddParents'; Nickelodeon Ready to Cross-Promote Latest hit Show With Food, Toy and Internet Tie-Ins," *Advertising Age,* February 16, 2004, 18.

54. Alice Z. Cuneo, "Marketers Dial In to Messaging; With Success Overseas, Chance to Snag Hip U.S. Consumers More Enticing," *Advertising Age,* November 1, 2004, 18.

55. Advertisement in *Advertising Age,* February 27, 2006.

56. Deborah Roedder John, "Consumer Socialization of Children: A Retrospective Look at Twenty-Five Years of Research," *Journal of Consumer Research* 26 (December 1999): 183–213.

57. See Schor, *Born to Buy;* Linn, *Consuming Kids;* Alisa Quart, *Branded: The Buying and Selling of Teenagers* (New York: Basic Books, 2003).

58. At the end of her book, Susan Linn observes that "no one is recommending that it be banned until the age of 16. But, *in the interests of children,* that's not unreasonable. The frontal cortex, which controls higher cognitive processes—including those that affect judgment—is not fully developed until the late teens." Susan Linn, *Consuming Kids,* 218.

59. See Angela J. Campbell, "Restricting the Marketing of Junk Food to Children by Product Placement and Character Selling," *Loyola of Los Angeles Law Review* 39:1523–1580.

60. Both Susan Linn and Juliet Schor, for example, have called for regulations on children's market research. Schor urges requirements for "full disclosure in children's marketing. Congress should pass a federal act mandating disclosure for all sponsored product placements in television, movies, videos, books, radio, and the Internet." She also calls for "disclosure of who carried out the market research, writing and production for any ad directed at children under twelve years of age." Schor, *Born to Buy,* 195.

61. For an overview of major privacy-policy issues in the electronic media, as well as proposed actions, see the Electronic Privacy Information Center, http://www.epic.org.

62. One of the groups working to promote more media literacy in U.S. schools is Action Coalition for Media Education (ACME), http://www.acmecoalition.org. (I serve on the advisory board of this organization.)

63. I attended this event, March 21, 2006, at Kaiser headquarters in Washington, D.C. See Joseph Jaffe, *Life after the 30-Second Spot: Energize Your Brand with a Bold Mix of Alternatives to Traditional Marketing* (Hoboken, NJ: John Wiley & Sons, 2005.)

64. *New Media and the Future of Public Service Advertising, Case Studies,* the Henry J. Kaiser Family Foundation. March 2006. http://www.kff.org/entmedia/7469.cfm.

65. For an overview of some of the research in the United States on children and media, see, for example, Dorothy G. Singer and Jerome L. Singer, eds., *Handbook of Children and the Media* (Thousand Oaks, CA: Sage Publications, 2001); and Victor C. Strasburger and Barbara J. Wilson, *Children, Adolescents, and the Media* (Thousand Oaks, CA: Sage Publications, 2002). Though cultural-studies scholars in the United States have conducted numerous studies of children and television with a much more expanded perspective, their work has not received the official sanction or financial support from government agencies and policymakers that traditional social-science research has received. See Willard Rowland, *The Politics of TV Violence* (Beverly Hills, CA: Sage Publications, 1983). For examples of qualitative research on children and media, see Ellen Seiter, *Sold Separately: Children and Parents in Consumer Culture* (New Brunswick, NJ: Rutgers University Press, 1993); Marsha Kinder, ed., *Kids' Media Culture;* Henry Jenkins, ed., *The Children's Culture Reader* (New York: New York University Press, 1998).

66. Kaiser Family Foundation, "Study of Entertainment Media and Health," http://www.kff.org/entmedia/index.cfm. Pew Internet and American Life Project, http://www.pewinternet.org.

67. Sonia Livingstone, *Young People and New Media* (Thousand Oaks, CA: Sage Publications, 2002), 24–25.

68. From a speech by Henry Jenkins summarized in *E-Marketer,* March 21, 2006, http://www.emarketer.com/Article.aspx?1003881. See also Henry Jenkins, *Conver-*

gence Culture: Where Old and New Media Collide (New York: New York University Press, 2006).

69. Sharon R. Mazzarella, ed., *Girl Wide Web: Girls, the Internet, and the Negotiation of Identity* (New York: Peter Lang, 2005). See also David Buckingham and Rebekah Willett, eds, *Digital Generations: Children, Young People, and New Media* (Mahwah, NJ: Lawrence Erlbaum Associates, 2006), 131–147.

70. Children's Digital Media Center, http://www.digital-kids.net (accessed March 26, 2006). The Children's Digital Media Center is a consortium at four U.S. universities that brings together "scholars, researchers, educators, policy makers, and industry professionals in a community whose goal is to improve the digital media environment in which children live and learn." The funds for the consortium came from a five year, $2.45 million grant from the National Science Foundation, with additional grants from U.S. foundations. See Ellen Wartella, June H. Lee, and Allison G. Capolovitz, *Children and Interactive Media Compendium,* November 2002. http://www.digital-kids.net/modules.php?op=modload&name=Downloads&file=index&req=viewdownload&cid=2. See also Sandra Calvert, *Children's Journeys through the Information Age* (New York: McGraw-Hill, 1999).

71. John D. and Catherine T. MacArthur Foundation, "In Focus: Building the Field of Digital Media & Learning," http://www.macfound.org/site/c.lkLXJ8MQKrH/b.1074781/k.D7EC/In_Focus.htm?tr=y&auid=2082264 (accessed November 7, 2006).

72. Several bills have been introduced that would authorize federal funding for long-term studies of the impact of electronic media on children. For example, a bill was passed in the Senate Committee on Health, Education, Labor, and Pensions in March 2006 to create a program under the auspices of the CDC to study the impact of television, films, DVDs, video games, the Internet, and cell phones. The legislation was introduced by long-time video-game critic Senator Joseph Lieberman(D=CT). John Eggerton, "CDC Media Study Bill Passes in Committee," *Broadcasting & Cable,* March 10, 2006.

73. David Buckingham, *After the Death of Childhood: Growing up in the Age of Electronic Media* (Cambridge, UK: Polity Press, 2000), 199.

74. An analysis of more than 300 youth civic Web sites by scholars at American University found an abundance of online efforts already underway that are designed to harness these tools in engaging young people more fully with their communities and their government. See also Kathryn C. Montgomery and Barbara Robles Gottlieb, "Youth as e-Citizens: The Internet's Contribution to Civic Engagement," in David Buckingham and Rebekah Willett, eds., *Digital Generations,* 131–147.

75. For example, the MacArthur Foundation, as part of its project on Digital Media and Learning, funds collaborative work to examine the role of digital media in youth

civic engagement. http://www.digitallearning.macfound.org/site/c.enJLKQNlFiG/b.2029199/k.BFC9/Home.htm. See also University of Washington's Center for Communication and Civic Engagement, http://depts.washington.edu/ccce/Home.htm.

76. Constance A. Flanagan and Nakesha Faison, "Youth Civic Development: Implications of Research for Social Policy and Programs," *Social Policy Report* 15, no. 1 (2001): 3–14.

77. As Constance A. Flanagan and Nakesha Faison point out, preoccupation with the self and materialism are tendencies that can undermine youth civic engagement. Constance A. Flanagan and Nakesha Faison, "Youth Civic Development: Implications of Research for Social Policy and Programs," Social Policy Report 15, no. 1 (2001): 3–14.

78. A number of scholars and nonprofits are focused on a variety of policy goals for establishing an "electronic commons," that will sustain and support civic and nonprofit efforts. For example, see "The 'Dot Commons' Concept: Making the Internet Safe for Democracy," Center for Digital Democracy, http://www.democraticmedia.org/issues/digitalcommons. See also Peter Levine, "Building the e-Commons," *The Responsive Community* 13, no. 4 (Fall 2003): 28–39. For a discussion of the network-neutrality issue, see Jeff Chester, "The End of the Internet," *The Nation.com,* February 1, 2006, http://www.thenation.com/doc/20060213/chester. For digital-rights management and other intellectual-property issues, see Public Knowledge, http://www.publicknowledge.org. See also Free Press, http://www.freepress.net.

79. Among the private efforts to bridge the digital divide is the Center for Media and Community, a project of the nonprofit Education Development Center in Newton, Massachusetts. Digital Divide Network, http://www.digitaldivide.net.

Index